Bird Families of the World

The Megapodes
Megapodiidae

DARRYL N. JONES
Griffith University, Brisbane

RENÉ W. R. J. DEKKER
National Museum of Natural History, Leiden

CEES S. ROSELAAR
Zoological Museum, Amsterdam

Illustrated by
BER VAN PERLO

Oxford   New York   Tokyo
OXFORD UNIVERSITY PRESS
1995

Oxford University Press, Walton Street, Oxford OX2 6DP

*Oxford   New York
Athens   Auckland   Bangkok   Bombay
Calcutta   Cape Town   Dar es Salaam   Delhi
Florence   Hong Kong   Istanbul   Karachi
Kuala Lumpur   Madras   Madrid   Melbourne
Mexico City   Nairobi   Paris   Singapore
Taipei   Tokyo   Toronto
and associated companies in
Berlin   Ibadan*

Oxford is a trade mark of Oxford University Press

*Published in the United States
by Oxford University Press Inc., New York*

© Text: Darryl N. Jones, René W. R. J. Dekker, and Cees S. Roselaar, 1995;
© Plates, maps, and drawings, Oxford University Press, 1995

*All rights reserved. No part of this publication may be
reproduced, stored in a retrieval system, or transmitted, in any
form or by any means, without the prior permission in writing of Oxford
University Press. Within the UK, exceptions are allowed in respect of any
fair dealing for the purpose of research or private study, or criticism or
review, as permitted under the Copyright, Designs and Patents Act, 1988, or
in the case of reprographic reproduction in accordance with the terms of
licences issued by the Copyright Licensing Agency. Enquiries concerning
reproduction outside those terms and in other countries should be sent to
the Rights Department, Oxford University Press, at the address above.*

*This book is sold subject to the condition that it shall not,
by way of trade or otherwise, be lent, re-sold, hired out, or otherwise
circulated without the publisher's prior consent in any form of binding
or cover other than that in which it is published and without a similar
condition including this condition being imposed
on the subsequent purchaser.*

A catalogue record for this book is available from the British Library

*Library of Congress Cataloging in Publication Data
Jones, Darryl N.
The megapodes: Megapodiidae/Darryl N. Jones, René W. R. J.
Dekker, Cees S. Roselaar; illustrated by Ber van Perlo.
(Bird families of the world; 3)
Includes bibliographical references and index.
1. Megapodiidae.   I. Dekker, René W. R. J.   II. Roselaar, Cees S.
III. Title.   IV. Series.
QL696.G25J66   1995   598.6\12–dc20   94-24732
ISBN 0 19 854651 3*

Typeset by EXPO Holdings Malaysia

Printed in Hong Kong

Bird Families of the World

A series of authoritative, illustrated handbooks, of which this is the third volume to be published

Series editors
C. M. PERRINS Chief editor
W. J. BOCK
J. KIKKAWA

THE AUTHORS Darryl Jones is a Lecturer at Griffith University, Queensland. In addition to his long-term research on megapodes, he has an active interest in the behaviour and ecology of birds in rainforests and in wildlife–human conflicts. He has studied birds in East Africa, New Guinea, the Canadian Arctic, and Australia. René Dekker is Curator of Birds at the National Museum of Natural History, Leiden, and Chairman of the Megapode Specialist Group. He has conducted field studies of megapodes in Sulawesi, the Moluccas, and the Nicobar Islands. Cees Roselaar is in the bird department of the Zoological Museum, University of Amsterdam. He is one of the co-authors of the nine-volume *Birds of the Western Palearctic* (Oxford University Press), and has published a number of articles on systematics and geographical variation.

THE ARTIST Ber van Perlo graduated from the Agricultural University of Wageningen, and worked as a geographer and physical planner for the Dutch National Forest Service until 1991. He has had a lifelong interest in birds and hopes soon to publish a guide to the birds of eastern Africa.

Bird Families of the World

1. The Hornbills
   ALAN KEMP

2. The Penguins
   TONY D. WILLIAMS

3. The Megapodes
   DARRYL N. JONES, RENÉ W. R. J. DEKKER, and CEES S. ROSELAAR

The information contained in this book is founded on specimens and data collected by a great number of nineteenth-century naturalists, many of whom never returned from the tropics. Much of the research conducted today would not have been possible without the work of these pioneers. We dedicate this book to their memory.

Preface

The first monograph to be published on the birds then known as thermometer or incubator birds was Oustalet's *Monographie des oiseaux de la famille des Mégapodiidés*, which appeared between 1879 and 1881. More than a century was to pass before the next monograph, the present book, on these fascinating birds appeared. Although the unique incubation activities of these birds (which utilize the heat generated by rotting vegetation, volcanic heat, and the sun) have attracted scientific attention throughout the last 100 years, much of the published work has been fragmentary and difficult to locate. The most influential studies and reviews were those of R. W. Shufeldt on morphology, osteology, and oology, Ernst Mayr on the taxonomy of *Megapodius*, George A. Clark Jr on ontogeny and evolution, and last, but not least, Harry J. Frith, who unravelled many mysteries about the breeding biology of megapodes in general, and of the Malleefowl *Leipoa ocellata* in particular. (See Bibliography for references.) Frith's book *The Mallee-fowl* (1962a) is still fascinating reading.

In the 1980s, megapode research experienced a resurgence of interest through the studies of an increasing number of scientists. This resulted in, among other things, the formation of the Megapode Specialist Group by René Dekker and Darryl Jones in 1986, which established links with the World Pheasant Association (WPA), the World Conservation Union (International Union for the Conservation of Nature), and, more recently, BirdLife International (formerly International Council for Bird Preservation). New information on a wide variety of topics and on some of the rarer species, as well as the increasing contacts between megapode workers, have combined to provide a valuable opportunity for the writing of this new monograph. As well as giving a comprehensive coverage of the vast amount of published data, this work also includes much recent and unpublished material. Without the assistance of our colleagues in the Megapode Specialist Group, the preparation of this monograph would never have been possible.

It is the intention of Oxford University Press that this new series on bird families of the world should be accessible to amateurs as well as professionals. We have tried on the one hand to be as detailed and complete as possible within the scope of this book, and on the other hand to keep the text of the general chapters understandable and readable. The descriptive paragraphs of the Species accounts may sometimes look rather technical, but in our opinion this was the only way to cover the data comprehensively, and we feel that, with the help of the list of abbreviations (p. xiv), even the text of these paragraphs should be fully understandable.

We hope that our efforts have been successful and that this monograph will be an inexhaustible source of data as well as an 'interesting book to read' for all those, professionals and amateurs, who are or will certainly become as fascinated by megapodes as we are.

June 1994

Leiden R. W. R. J. D.
Brisbane D. N. J.
Amsterdam C. S. R.

Acknowledgements

This book originated in 1988 when Darryl Jones and Jiro Kikkawa were discussing possible Australian subjects for the forthcoming Oxford University Press bird families series. The possibility of a complete monograph on the megapodes seemed premature at the time, but subsequent research soon indicated an unexpectedly large literature, much of it in obscure sources. Moreover, the dramatic increase in scientific interest was yielding a vast amount of new information. This body of both little-known and recent data provided a sound basis for the project. The enthusiastic support of our colleagues in the Megapode Specialist Group gave invaluable stimulus as well.

The production of this book has truly been an international cooperative effort. So many people have assisted us in so many ways that it is inevitable that some are forgotten here; please accept our apologies for any omissions. To all we are extremely grateful.

Thanks are due to Peter Colston (BMNH, Tring), Siegfried Eck (SMTD, Dresden), Renate van den Elzen (ZFMK, Bonn), H. Hoerschelmann (Zoologisches Institut und Zoologisches Museum, Universität Hamburg), G. Mauersberger (ZMB, Berlin), Gerlof Mees (NNM, Leiden), Dewi Prawiradilaga and S. Somadikarta (MZB, Bogor), Jan Wattel (ZMA, Amsterdam), and David E. Willard (FMNH, Chicago) for allowing us to study megapode skins under their care.

We are also grateful to Marc Argeloo, Herman Berkhoudt, Jared Diamond, David Gibbs, Jeroen Goud, Chris Healey, Mohamad Indrawan, Ron Johnstone, Catherine King, Robert Lucking and other members of the University of East Anglia Taliabu Expedition, Kate Monk, Carsten Niemitz, Shane Parker, Dr F. Robiller, Günther Schleussner, Ronald Sluijs, Matthias Starck, Jaap Vermeulen, and D. P. Vernon for supplying additional information.

Many people provided tapes, slides, and photographs for the preparation of the colour plates and sonograms. These include Marc Argeloo, Han Assink, Max van Balgooij, Arnoud van den Berg, Sharon Birks, David Bishop, Jörg Böhner, Marc Craig, Ken Davis, Mohamad Indrawan, Ben King, Carola Kloska, Robert Lucking and other members of the University of East Anglia Taliabu Expedition, Nigel Redman, Dieter Rinke, Jowi de Roever, Jelle Scharringa, Derek Stinson, Robert Stuebing, Werner Suter, David Todd, United States Fish and Wildlife Department–Mariana Islands, and Dennis Yong. Sonograms were prepared at the Expert-center for Taxonomic Identifications (ETI), Amsterdam, by Gideon Gijswijt.

Marc Argeloo, Joe Benshemesh, Sharon Birks, David Bishop, Jörg Böhner, David Booth, John Brickhill, Brian Coates, Leo Joseph, Cécile Mourer-Chauviré, David Priddel, Dieter Rinke, Roger Seymour, Derek Stinson, David Todd, Charlotte Vermeulen, and Gary Wiles, specialists on a certain topic or megapode species, each reviewed one or more of the chapters and species accounts, often adding new and unpublished data. We thank Wendy Arthur for the production of the additional illustrations in Chapter 7 at short notice.

Finally, thanks are due to Margaret Barber, Peter van Bree, Peter van Dam, Eugène van Esch, Ingrid Henneke, Jiro Kikkawa, B. Majumdar, J. P. Misra, Jelle Paul, Caroline Pepermans, Florence Pieters, Tineke Prins, Marianne van der Wal, Jan Wattel, and colleagues at the Griffith University, National Museum of Natural History, and Institute of Taxonomic Zoology, who have all contributed in one way or another to the realization of the book over the years. We also thank our families for their support, especially during the later hectic months of writing.

Contents

List of colour plates		xiii
List of abbreviations		xiv
Plan of the book		xv
Topographical diagrams		xviii

PART I *General chapters*

1	Introduction to the megapodes	3
2	Taxonomy and relationships	11
3	Distribution, biogeography, and speciation	21
4	General biology and behaviour	33
5	Megapode incubation sites	44
6	Ecophysiology and adaptations	52
7	Reproductive behaviour and mating systems	65
8	Evolution of megapode incubation strategies	73
9	Conservation	78

PART II *Species accounts*

Genus *Alectura*		89
Australian Brush-turkey	*Alectura lathami*	89
Genus *Aepypodius*		96
Wattled Brush-turkey	*Aepypodius arfakianus*	97
Bruijn's Brush-turkey	*Aepypodius bruijnii*	103
Genus *Talegalla*		105
Red-billed Talegalla	*Talegalla cuvieri*	106
Black-billed Talegalla	*Talegalla fuscirostris*	110
Brown-collared Talegalla	*Talegalla jobiensis*	116
Genus *Leipoa*		121
Malleefowl	*Leipoa ocellata*	122
Genus *Macrocephalon*		130
Maleo	*Macrocephalon maleo*	130

Genus *Eulipoa*		139
Moluccan Megapode	*Eulipoa wallacei*	140
Genus *Megapodius*		145
Polynesian Megapode	*Megapodius pritchardii*	146
Micronesian Megapode	*Megapodius laperouse*	152
Nicobar Megapode	*Megapodius nicobariensis*	159
Philippine Megapode	*Megapodius cumingii*	165
Sula Megapode	*Megapodius bernsteinii*	175
Tanimbar Megapode	*Megapodius tenimberensis*	179
Dusky Megapode	*Megapodius freycinet*	181
Biak Megapode	*Megapodius geelvinkianus*	189
Forsten's Megapode	*Megapodius forstenii*	193
Melanesian Megapode	*Megapodius eremita*	198
Vanuatu Megapode	*Megapodius layardi*	205
New Guinea Megapode	*Megapodius decollatus*	208
Orange-footed Megapode	*Megapodius reinwardt*	213
References		228
Index		255

Colour plates

Colour plates fall between pages 122 and 123.

Plate 1 *Alectura* and *Aepypodius*
Plate 2 *Talegalla*
Plate 3 *Leipoa*
Plate 4 *Macrocephalon*
Plate 5 *Eulipoa* and *Megapodius*
Plate 6 *Megapodius*
Plate 7 *Megapodius*
Plate 8 *Megapodius*

Abbreviations

♂/♂♂	male(s)	p1	innermost primary
♀/♀♀	female(s)	p10	outermost primary
ad	adult(s)	R.	River
AMNH	American Museum of Natural History, New York	RMNH	Rijksmuseum van Natuurlijke Historie, Leiden, The Netherlands (now National Museum of Natural History, NNM)
BMNH	British Museum of Natural History, Tring, United Kingdom (now The Natural History Museum)		
		s	secondary (feather)
c.	circa or about	s.d.	standard deviation
cm	centimetres	sec	second(s)
FMNH	Field Museum of Natural History, Chicago, USA	SMTD	Staatliches Museum für Tierkunde, Dresden, Germany
g	gram(s)	sp.	species (singular)
h	hours(s)	spp.	species (plural)
ha	hectare(s)	ss	secondaries
imm	immature(s)	ssp.	subspecies (singular)
in litt.	unpublished information received in writing	sspp.	subspecies (plural)
		t	tail feather
I./Is	Island(s)	t1	central tail feather
juv	juvenile(s)	W	watts(s)
kg	kilogram(s)	ZFMK	Zoologisches Forschungsinstitut und Museum Alexander Koenig, Bonn, Germany
km	kilometre(s)		
m	metre(s)		
mm	millimetre(s)		
Mt./Mts	Mount/Mountain(s)	ZMA	Zoologisch Museum, Universiteit van Amsterdam, The Netherlands
MZB	Museum Zoologicum Bogoriense, Bogor, Indonesia		
n	number in sample	ZMB	Museum für Naturkunde der Humboldt-Universität zu Berlin (Zoologisches Museum), Germany
NNM	see RMNH		
p	primary (feather)		
pp	primaries		

Plan of the book

This monograph is divided into two parts.

PART I consists of a first chapter introducing the megapodes, followed by eight chapters on different topics including taxonomy, biogeography, breeding biology, ecophysiology, and conservation. It summarizes the literature as well as much recent unpublished information.

PART II, the Species accounts, describes all seven genera and 22 species, including subspecies, introduced in Chapter 2. It will probably not surprise those acquainted with megapode taxonomy that after extensive re-examination of specimens we have felt compelled to alter the taxonomy and sequence of the genera and species slightly. Explanations for these decisions are given in the text under each species. In addition, new English names have been introduced for *Megapodius* and *Talegalla* for reasons of uniformity and clarity, as explained in Chapter 2.

Information on each species is presented under a standard set of headings and subheadings, outlined below. Some of these may seem to give extensive lists of almost similar data. However, we did not want to be arbitrary in using or omitting data from labels of museum skins and the literature. Unevenness of information between the various headings for the (sub)species throughout the species accounts is due to gaps in our knowledge or lack of materials.

Nomenclature

Each species account starts with the English and Latin name for the species. In a second line the scientific name in its original spelling is followed by the name of the author who made the description and the year in which it was published for the first time. Then the source of this original description is given. Next, alternative English names are given when available. Local names are in a separate paragraph at the end of each species account.

Next, for polytypic species, the Latin name, author, year of description, and distribution of subspecies are given. Synonyms, if any, with author and year of description are noted after each subspecies. When no subspecies have been recognized, the species is simply described as 'monotypic'.

Species names are quoted exactly as originally published, which leads to occasional variations in spelling.

Description

This section includes plumages, bare parts, moults, measurements, weights, and geographical variation. Plumage and moult descriptions (also given for megapodes in general in Chapter 4), measurements, and geographical variation are based on skins from museums mentioned separately under each species, while body weights and coloration of bare parts are from the literature and skin labels. The *Naturalist's color guide* (Smithe 1975, 1981), a standard reference work published by the American Museum of Natural History in New York, has been used for the description of plumages and (occasionally) bare parts, as its samples give names to colours rather than numbers, and detailed cross-references to colour codes used in other standard colour books, such as Ridgways (1912) *Color standards and color nomenclature*, now hard to obtain, and the *Munsell book of color* (Munsell Color Co., Baltimore, Maryland).

Plumage and bare parts are described for the following categories and in the following sequence: ADULT MALE, ADULT FEMALE (only when different from the adult male), CHICK, and IMMATURE. Any subspecific differences are highlighted here as well.

Measurements of wing, tail, tarsus, middle toe, middle claw, bill to skull, bill to nostril, and bill depth, taken by us from museum skins of adult megapodes during the preparation of this monograph, are given in tables (in millimetres) for each subspecies, but with both sexes pooled. However, when sample size was sufficient, these measurements are given for male and female separately or even for different populations of a single subspecies. Where samples of species or subspecies were small, and data from the literature added substantially to our data, these have been included directly below the tables. For newly hatched chicks, only wing and tarsus length are given.

Measurements were taken following methods described in Cramp (1988, pp. 34–8). Bill depth was measured at the middle of the nostril, perpendicular on cutting edges and only when the bill was closed and in a natural position. The middle toe was measured along the upper surface, from the base of the tarsus to the insertion of the claw, but only when the toe was fully straight. The middle claw was measured in a straight line from the insertion on the top to the tip.

Under 'Geographical variation', the differences between the subspecies as described under 'Plumages' and 'Measurements' are summarized, compared, and emphasized, while justification is given for including certain

NOTE: when measuring megapodes, it is most important to know the age of the bird one is examining. It may take over a year before wing, tail, leg, foot, and bill are fully grown. However, the plumage of the head and body is similar to the adult from a few weeks after hatching. Adding such birds to the morphometric data can be an important source of error, making subspecific identifications impossible.

The presence of any first generation primary is crucially important for ageing. Characteristically, the fully grown first generation p9, the second outermost primary, is the longest of the ten primaries until the innermost feathers are replaced by a second generation feather (in adults, p6 is usually the longest and p9 is distinctly shorter than p8). The outermost primary p10 is short and narrow with a long and tapering tip ending in a sharp point.

When the immature megapode reaches approximately one-third of the average adult weight of its population, the second generation primary p2 and the growing p3 are often longer then the wing tip formed by p9 of the first generation. When taking measurements, wings with this configuration should be dismissed, also because the bones of the hand-part of the wing are not yet fully grown either. A classic mistake among students is to measure the wing from the wing bend to the longest tertial, rather than to the longest primary.

When the immature bird reaches approximately three-quarters of the average adult weight of its population, the outermost primaries are of the second generation, the central ones of the third, and the innermost of the third or fourth generation. Correct ageing is crucial now. The easily identifiable first generation p10 on which smaller immatures could be identified has been lost and the bird is easily mistaken for a smaller full adult, although its wing length may be up to 15 per cent below average adult wing length. Ageing is possible only with the help of the outermost primary p10 of the second generation. This p10 is intermediate in shape and length between the p10 of adults and that of juveniles. Its tip is rather narrow and pointed, neither as broad and rounded as in adults nor as sharply pointed and tapering as in juveniles. When the entire wing tip is of the second generation, thus p6 or p7 to p10 all belong to the same moult series, wing, tail, bill, leg, and foot are still 10–15 per cent smaller than in adults. Birds with a second generation wing tip have therefore not been used in the tables of measurements.

When the wing tip is of the third generation, thus p10 (or p9 and p10) are of the second generation and p6/p7 to p8/p9 of the third generation, bill, leg, foot, and usually tail have attained average adult length. Such birds have been used in the tables, even though the wing length is still $c.$ 5 per cent below average adult length. Full adult wing length is not attained until the wing tip (usually p6) is of the fourth or fifth generation.

Thus, megapodes should not be measured without thorough knowledge of moult (see section on moult in Chapter 4). Birds with at least a slightly pointed p10 and a wing tip formed by feathers of the same generation as this p10, as well as immatures with a juvenile p10 and adults missing the wing tip due to moult, should be dismissed from the sample.

samples (from different islands) under a particular subspecies.

Topographical diagrams are presented on pp. xviii–xix.

Range, status, and maps
The distributional ranges of the (sub)species are shown on a series of specially prepared maps, which have been compiled by plotting all locations from museum data and the literature. This means that in some cases the range shown must be regarded as historical, where no recent data were available. Some species may have disappeared from some areas and islands due to human interference (see Chapter 9). The maps have no special captions other than the keys, and should be used in conjunction with the detailed information given in the text, under both 'Range and status' and 'Geographical variation'. The size of small islands has been exaggerated in order to allow shading to be shown to indicate a species' presence. Since altitude may be an important determinant of the distribution of certain megapodes in New Guinea, regions over 1000 m in altitude are shown as diagonally striped areas. Names of countries, islands, etc. conform to those in *The Times atlas of the world*, 1977. The species distribution maps show only the relevant regions. Their locations and boundaries are shown on p. xx.

Field characters
This section gives a brief description of the general appearance of a particular species, and summarizes how to distinguish it from relatives, and from species similar in appearance, such as rails (Rallidae). Approximate body length, from the tip of the bill to the tip of the tail, is given in centimetres.

Voice
Sonograms of 13 species have been specially prepared for this book from tape recordings made by various colleagues. Except for the sonogram of the Malleefowl *Leipoa ocellata*, all are published here for the first time. They are presented in a standard format along a time axis of 5.85 sec (the minimum duration in which the longest call could be illustrated) to allow direct comparison, even though this is at the cost of some detail in the sonograms of species with much shorter calls. The vertical axis shows the frequency in kilohertz (kH). The oscillogram directly below each sonogram shows the intensity of the sound. For general information on sonograms, see Cramp and Simmons (1977, pp. 19–26).

For those species for which sonograms are included the descriptions of the vocalizations are based on these sonograms and original recordings, augmented with data from the literature and unpublished data from field researchers. In all other cases, vocalizations are based entirely on the literature.

Remaining sections
All remaining sections (Habitat and general habits; Food; Displays and breeding behaviour; Breeding; and Local names) are based on the literature, museum-skin labels, personal observations, and unpublished data from the authors as well as from others. These sections are self-explanatory and need no further comment.

References
The references (author and year only) at the end of each species account give a quick overview of all literature consulted for that particular species. Sources consulted but not quoted in the text are included, so that this section summarizes the most important literature as a starting point for additional information on each species (since this information cannot easily be extracted from the main Bibliography).

Colour plates
The colour plates show, in general, an adult and a chick (if available for illustration) of each species. Adult females have only been illustrated in cases of sexual dimorphism. Similarly, subspecies have been illustrated only when differences in coloration could clearly be visualized. Lack of reference materials for some species certainly made this work problematic.

Bibliography
The Bibliography lists books, articles, and reports which relate to megapodes, either cited or uncited in this monograph. Some more general (non-megapode) references which are mentioned in the text have also been included.

Topographical diagrams

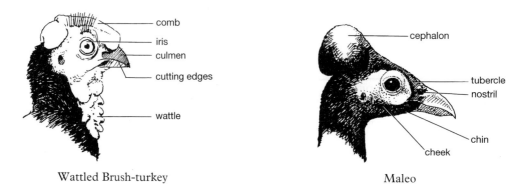

Wattled Brush-turkey

Maleo

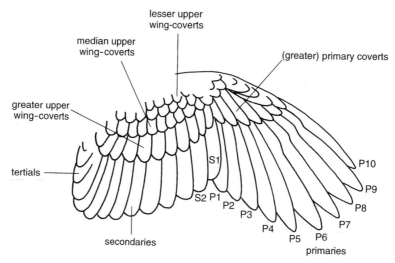

Eulipoa (Moluccan Megapode) wing from above

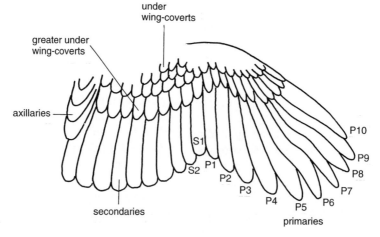

Eulipoa (Moluccan Megapode) wing from below

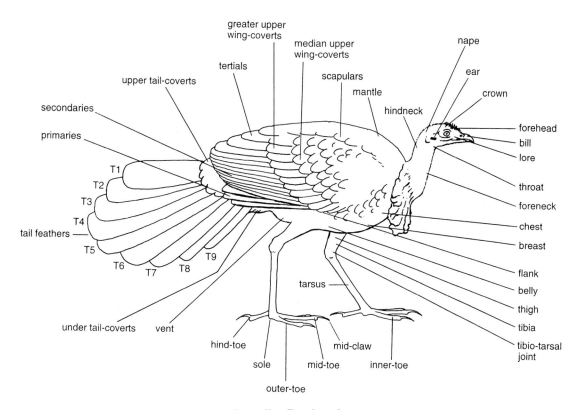

Australian Brush-turkey

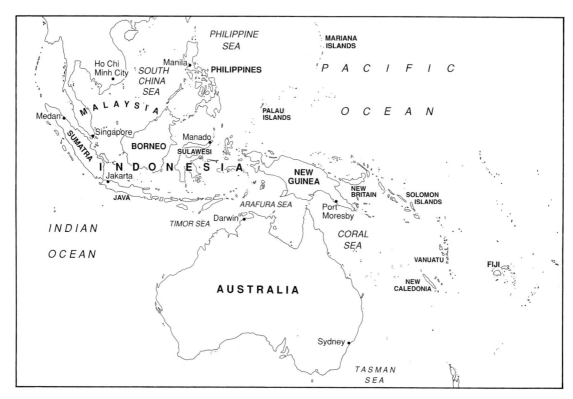

The region within which megapodes occur.

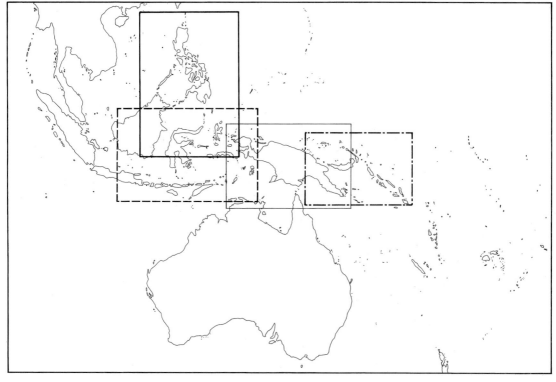

Boundaries of the regional maps used in the species accounts.

PART I

General chapters

1

Introduction to the megapodes

This chapter aims to provide a brief overview of some of the more important features of megapodes. It is not intended as a synopsis of the entire book, but rather serves as a simple introduction to the family.

In 1521, when Antonio Pigafetta, a member of Magellan's ill-fated voyage to the Far East, recorded his observations of a strange bird he had seen in the Philippines, he was impressed not so much by such features as plumage or voice, as by the bird's unusual nesting behaviour. The species he had seen (very probably the Philippine Megapode) was a rather drab, small, fowl-like bird, unremarkable except for the fact that it appeared to bury its eggs in a large hillock of soil. An inspection of this nesting site would have revealed that the soil was warm, the heat required for incubation of the eggs coming from decomposing leaves within the mound. Although the incubation of eggs by using sources of heat found in the environment is common among reptiles, it was not known in any birds. It is primarily this form of incubation, unique to megapodes among bird species, that naturalists and scientists have long found and still find puzzling and fascinating.

The megapode family

Throughout the centuries of exploration in the Far East and the southern Pacific, stories abounded of birds that 'laid their eggs in volcanoes' or 'buried their chicks in the ground'. In the museums of Europe such tales were largely dismissed as sailors' myths, although the many specimens collected were examined with care. Later naturalists, however, began to gather more detailed information about these birds. Often aided by the local people, they found that eggs were laid either in holes in the ground or in mounds of leaf litter. The latter were of particular interest because they were actually constructed by the birds themselves, and some of them contained many tonnes of material. It soon became obvious that these mounds provided the heat required for the incubation, in place of the body heat of the brooding adult.

Despite their strange nesting activities, these birds were found to be similar in most morphological features to the gamebirds, the galliforms. One distinctive feature of the first megapodes to be closely examined (mainly species belonging to the genus *Megapodius*) was their long toes. This characteristic provided the scientific family name for the group, Megapodiidae, meaning simply 'large feet'. Collectively, however, they were more commonly known as 'thermometer birds', 'incubator birds' or 'mound builders', in reference to their incubation habits. Nowadays the name 'megapode' is preferred, as a more useful label for all of the species within the family.

In general, most megapodes resemble other galliforms in body shape and plumage. Most are fairly drab brown, blackish, or grey, and relatively few show detailed feather patterns, the most notable exception being the Malleefowl *Leipoa ocellata*. Many have virtually naked areas on their face or neck, and this exposed skin may be coloured yellow, blue, or dull red. However, one group of species (the genera *Alectura* and *Aepypodius*, known collectively as the brush-turkeys) have vividly coloured heads and necks and possess a variety of wattles and combs. This group includes the only species in which sexual dimorphism is evident, with the males being slightly larger and more colourfully ornamented than the females; in most species the sexes are virtually indistinguishable.

Megapodes also resemble galliforms in being heavy-bodied birds of the forest floor. All are opportunistic ground foragers, eating a wide variety of foods such as insects, seeds, and fallen fruits. Although all are able to fly, and some make considerable flights on a daily basis, most species move primarily by walking.

The number of species recognized as belonging to the Megapodiidae has been a matter of debate for centuries (see Chapter 2 for details). The major source of controversy has been the genus *Megapodius*. This genus contains the smallest species within the family; it is also the most widespread, being distributed from the Nicobar Islands (India) in the west to Tonga (western Pacific) in the east. Within this range they are (or, in some cases, until recently were) found on virtually every island large enough to support some form of forest (except Sumatra, Java, Borneo, and Bali). This distribution indicates a remarkable ability to disperse; indeed, several observations exist of birds landing on ships at a considerable distance from the nearest island. As a result, speciation within this genus has been extensive, although the various species remain quite similar in general appearance, as the colour plates show. In this book, 13 species are recognized as belonging to the genus *Megapodius*; together with the single and closely related species belonging to the genus *Eulipoa*, these species are here given the common name, megapode, in preference to the former name, scrubfowl.

As for the rest of the family, there has generally been agreement on taxonomy. Apart from the 14 megapode species (*Megapodius* and *Eulipoa*) mentioned above, the family consists of the following genera: *Alectura* (one species) and *Aepypodius* (two species), the brush-turkeys, *Talegalla* (three species), here given the name talegallas, *Leipoa* (one species), the malleefowl, and *Macrocephalon* (one species), the maleo. Thus there are seven genera in all, encompassing a total of 22 species. It is also known that other species, primarily belonging to the genus *Megapodius*, have become extinct in relatively recent times, mainly in various small islands in the southern Pacific.

Incubation sites: mounds and burrows

Undoubtedly, the most notable feature shared by all megapodes is their technique of incubation. This involves the utilization of some form of naturally occurring heat to incubate their eggs, rather than the body heat of the adult. Basically, three main sources of heat are used: microbial decomposition of organic matter, geothermal activity, and solar radiation. These are utilized in a variety of ways, and the amount of effort required for their utilization varies greatly. Although most species use only one of these heat sources, some may use several; such species may, for example, construct mounds in some areas, and lay their eggs in burrows in warm soil in other areas. Birds that gather piles of organic matter to be used as an incubation site are called 'mound builders'; those that tunnel into pre-existing locations in volcanic areas or hot sands are called 'burrow nesters'. Features of these two main incubation techniques are described below.

Mound building

The decomposition of damp organic matter provides the most common source of heat utilized by megapodes. This heat source usually takes the form of a large mound of leaf litter and soil, gathered from the forest floor. The vast and diverse community of organisms such as fungi, bacteria, and minute invertebrates found naturally within the humus increase their activities and populations enormously within the favourable environment of the mound. By maintaining a supply of fresh material and ensuring optimal moisture levels within the mounds, the birds constructing them promote microbial decomposition, and use the heat resulting from the metabolic activities of the organisms involved.

This form of heat production often necessitates a considerable amount of work, although the effort varies depending on the environment and climate. The mounds most laborious to construct and maintain are those of the only megapode found in arid areas, the Malleefowl. In this species, the male spends months of often daily work gathering leaf litter (the female does not participate), opening the mound and mixing the material to maintain the appropriate levels of microbial activity; remarkably, it is able to detect slight changes in temperature and to manipulate heat production by removing or adding material. In this extremely dry environment, conserving moisture within the mound is critical; microbial activity is limited directly by the availability of moisture. This is achieved by the raking of a deep insulative layer of sand and soil over the region of heat production. Of course, this must be removed and replaced whenever access to the warm core is required, a particularly time- and energy-consuming task.

Most of the other mound-building species are found in much wetter areas and are normally confined to forested country where suitable mound material is abundant and moisture is not usually limiting. The building of mounds involves substantial effort for most of these species during the initial construction period, when moist leaf litter and soil are raked together into the pile; this construction feat may take several weeks and involve the relocation of several tonnes of material. Microbial activity usually commences quickly and internal temperatures may rise to very high levels before stabilizing at a level suitable for incubation, typically at 32–35 °C. Once thermal stability is reached, maintenance of the mound as a potential incubator usually requires considerably less effort, involving only the addition and mixing of fresh, damp material at regular intervals (see Fig. 1.1).

Two main types of mound are constructed, the difference relating to the number of consecutive breeding seasons in which the mound

1.1 A mound of the Australian Brush-turkey from subtropical rainforest in southern Queensland, Australia. This mound was composed of rainforest leaf litter, twigs and a large proportion of soil. Many larger sticks and tree limbs were also incorporated into the surface material. The scale indicates a mound height of about 1.5 m.

is used. Most species construct a new mound each year, although these are commonly placed on the site of an old mound, which has been reduced by decomposition to a low pile of compressed soil. These 'annual mounds' rarely exceed 1.5 m in height though they may be several metres in circumference (see Fig. 1.1).

In contrast, some species maintain what may be termed 'perennial mounds'. These species remain near their mounds for many years, adding and mixing fresh material more or less regularly throughout the year. Such mounds appear to be used continuously by successions of pairs and may grow into enormous structures many metres high; some are known to have been attended regularly since early this century (see Fig. 1.2).

Megapode mounds are found in a variety of wooded environments, though well shaded sites with good supplies of suitable leaf litter are preferred. However, mounds may be successfully tended in locations offering minimal amounts of organic matter and moisture; compared to typical nests, they are remarkably resilient in the face of disturbance and are able to withstand considerable climatic change.

The bulk of the mound itself provides a very effective buffer between the eggs within and the vagaries of the environment outside. None the less, as the largest 'nests' of any bird, they do require a great amount of time and energy in construction and maintenance (see Fig. 1.2).

Burrow nesting

Some species have evolved ways of utilizing naturally occurring heat so that incubation requires comparatively little work. There are two such heat sources: geothermal activity and solar radiation. Birds using these sources burrow into the warm substrate and lay their eggs at a depth providing the appropriate incubation temperature. Sites heated by these sources and suitable for burrow nesting tend to be very localized, and often attract large numbers of birds. The largest concentrations of any megapode species occur in a few locations on the island of New Britain near New Guinea, where volcanically heated streams provide nesting grounds for tens of thousands of birds; as far as is known, these birds are

1.2 An enormous mound of the Orange-footed Megapode from coastal rainforest in southern Papua New Guinea. This mound consisted of compacted soil and leaf litter but only the very top of the mound was being used as an incubator. The mound had apparently been used annually for over 50 years, probably by a succession of pairs. The scale is 4.5 m in height.

normally distributed widely in the mountains around the area, but all make brief visits to the site in order to lay.

Solar-heated incubation sites tend to be areas of exposed sand or soil in tropical areas, such as clearings in the forest or beaches along rivers or the seashore. Species using these sites must also travel from their normal home ranges in the forests. Eggs are laid in shallow pits or in much deeper burrows which are often used communally and which have been excavated on earlier occasions by other members of the same species. Being dependent on the sun for heat means that these sites are particularly vulnerable to the effects of unsuitable weather, even extended periods of heavy cloud cover (see Fig. 1.3).

Because the processes that provide the heat for incubation occur naturally and are entirely external to the birds themselves, incubation sites tend to be available for extended periods during the year. Indeed, for most species the 'breeding season' appears to be limited mainly by adverse climatic conditions rather than by their own physiology. For instance, mound building typically commences following good rains, while many burrow-nesting species lay in the drier periods following the wet season. Thus, megapodes may continue laying for many months. No incubation sites appear to be used during sustained wet periods such as monsoons.

Adaptations to underground incubation

Megapodes exhibit a surprising and interrelated array of physiological, ecological, and behavioural adaptations related directly to the special conditions of their unique incubation technique. Perhaps the most remarkable of these adaptations are those associated with the incubation process itself. This is because the egg, the developing embryo within, and the newly hatched chick must cope with conditions dramatically different from those associated with any normal nest. Most critically, the gaseous environment surrounding the egg is extremely humid and is very low in oxygen and very high in carbon dioxide. Megapodes have solved these serious problems by means of several specializations of their eggs and their physiology.

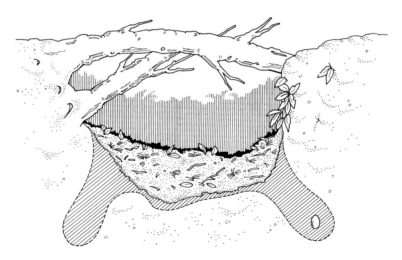

1.3 One of many nesting burrows of the Maleo, at a large communal nesting ground at Tambun in North Sulawesi, Indonesia. The soil is heated by geothermal hot water streams. The depth and complexity of these burrows is variable, due to the presence of hard soil areas, tree roots and large rocks and depending on ground temperatures. After egg-laying, the burrows are filled again with loose sand and soil.

Megapode eggshells are very thin compared to those of birds of similar size. In addition, the pores in the shell through which the gases pass actually change shape during incubation, becoming wider as the growing embryo assimilates the calcium for bone formation. This progressive change in pore shape is a very unusual phenomenon in birds and greatly facilitates gas exchange between the embryo and the atmosphere immediately surrounding the egg. Therefore, despite the extreme humidity and high gas pressures within the incubation site, the embryo's tissues develop at rates similar to those of birds incubated in a normal nest.

Although the rates of growth of the megapode embryo are similar to those of other birds, the chick that finally hatches is relatively large. This is associated with the very long period of incubation which may be between 50 and 70 days. During this time the chick continues to develop within the egg, finally hatching by cracking through the eggshell with its feet rather than by using an egg-tooth as in other birds. The hatchling may spend many hours or even several days battling upwards through the material of the incubation site. The chick that eventually emerges is, in terms of its behavioural and physiological capabilities, the most advanced of any bird; it is able to run, feed, regulate its body temperature, find food, and even fly on the day of hatching. This extreme precociousness is both a necessity and a consequence of the specialized incubation technique; unlike any other birds, young megapodes receive absolutely no assistance from their parents and must live independently from the moment of emergence.

Social behaviour

One of the many implications of megapodes leaving their eggs in warm material, rather than incubating a clutch in a nest, is the loss of direct contact between the adults and their eggs, and consequently a lack of contact with the hatchlings. The eggs begin to develop as soon as they are laid in the mound or burrow and will hatch when the embryo has reached a certain stage of development. This means that the adults have no control over the time of emergence, and the eggs hatch completely independently of each other. Megapode eggs are relatively very large (in the range 75–230 g and representing about 20 per cent of the female's body weight) and females lay their eggs at intervals of several days over a breeding season of many months. A female typically lays from about 12 to perhaps 30 eggs each season. These eggs hatch at more or less regular intervals, and hatchlings may emerge from the incubation site at any time. This makes it impossible for adults to keep the chicks together. Under these circumstances, natural selection has favoured young that are able to survive alone from the start.

Following emergence, hatchlings remain solitary for some time. Very little is known about these young birds; they are well camouflaged and secretive. Certainly, a large proportion do not survive, due to predation or starvation. By the time they are a few months old, however, most survivors will have joined other birds in loose groups.

Although the social organization of many species of megapode is not well known, there are apparently two main patterns of breeding association, which relate to the extent to which breeding adults form pair bonds. Most species appear to be monogamous, with males and females forming close pair bonds that may last for many years. The sexes in monogamous species are almost always indistinguishable, and few have any form of bright colouring. In these species the pair is the primary unit and paired birds seem to stay together permanently. None the less, many of these species are highly social and large numbers of individuals may congregate regularly, especially for roosting. Species within this group may be mound builders or burrow nesters and a few are both, with either mounds or burrows being used in different parts of their range. The pair tend to visit the incubation site together, and in many species both members of the pair may be involved in constructing and maintaining it.

A few species, however, do not show any form of pair bonding and the sexes are visually different, especially during the breeding season when the males develop colourful wattles. In these species (the brush-turkeys *Alectura* and *Aepypodius*), social groups are also common, but during the breeding season the males tend to remain near their mounds. Interestingly, in all of these species, single males construct mounds and these are defended vigorously. Females requiring a site in which to lay their eggs must visit these mounds, and the resident male usually only allows females that have mated with him to lay.

Conservation problems

Pigafetta's journal entry, dated April 1521, is the first known sighting by a European of the family of birds now known as the megapodes. However, these birds have always been well known to indigenous people for much the same reasons that have attracted the attention of foreigners: their strange nesting behaviour and large eggs. Megapodes lay very large eggs indeed and each female may lay a considerable number in the course of the breeding season. Sites that are used for laying (and some are extensive communal nesting grounds visited by many birds) represent a wonderfully rich food source for humans. The laying sites of virtually every species of megapode have been harvested for eggs for millennia. For many peoples, megapode eggs have provided an important food and sometimes the principal source of protein, as they are extremely rich in yolk. Over-exploitation of this resource has commonly been prevented by traditional laws governing ownership and rates of harvesting. It is noteworthy that many of these indigenous restrictions have broken down only in the last few decades. Today, human interference, in the form of uncontrolled egg harvesting or the destruction of habitat, threatens many species.

Throughout their distribution, megapodes are now adversely affected by habitat clearance and fragmentation, egg predation, predation on chicks (especially by introduced mammals), and possibly by declining fertility in the case of isolated populations. Currently, at least half of the species are regarded as being vulnerable or seriously endangered. Detailed conservation recovery programmes are under way for particular species such as the Maleo *Macrocephalon maleo* and the Malleefowl, but reliable assessment of the conservation status of many species is difficult because so little is known about them. There is a great need for further field studies of these birds.

Observing megapodes

The distribution of the megapodes makes observation of most species in the wild rather difficult. Not only are they confined chiefly to remote and often inaccessible locations, but in many cases hunting and disturbance by man have driven the birds far from human settlements. There are, however, still places where they remain common and access is relatively straightforward.

Probably the easiest species to observe are the three occurring in Australia; these are all mound builders. Most notably, the Australian Brush-turkey *Alectura lathami* is found in many locations in southern Queensland, where it is common in National Park picnic grounds and even in the suburbs of large cities. In tropical Australia, Orange-footed Megapodes *Megapodius reinwardt* are conspicuously present around many resorts and islands used by visitors. In inland southern Australia, the now endangered Malleefowl may be readily observed in Wyperfeld and Little Desert National Parks.

Elsewhere, opportunities to encounter megapodes in the wild will depend on the extent to which interested persons are willing to travel to remote areas. Papua New Guinea is certainly the richest area in terms of species, but most of the sites remain difficult of access. Locations such as the volcanic communal egg grounds at Pokili and Garu in western New Britain provide a spectacular example of high-density burrow nesting, but require considerable effort to reach.

Megapodes are kept in only a few zoological collections around the world, notably at Frankfurt Zoological Gardens and Vogelpark Walsrode in Germany, Audubon Park Zoo in New Orleans, USA, and Adelaide and Perth Zoological Gardens in Australia.

Although comprehensive collections of live megapodes are fairly rare, skins and mounted specimens are surprisingly abundant in museums throughout the world; some particularly fine collections are housed at The Natural History Museum, Tring, UK, the American Museum of Natural History, New York, and the National Museum of Natural History, Leiden, The Netherlands.

2

Taxonomy and relationships

Introduction

This chapter gives a chronological review of megapode classification as proposed by many ornithologists during the past nearly 200 years, concluding with our most recent classification based on studies prior to and during the preparation of this monograph. It provides a general overview of megapode taxonomy for readers with a general interest, and reference to the most important classifications, authors, and the kind of research they did, for the more advanced scientist. For that reason, characters used in these classifications are merely noted, not evaluated; for evaluations, the reader is referred to the literature cited. Justification for the division of the megapode family into seven genera and 22 species in the present work is given in Chapter 3, as well as in the species accounts.

Interfamilial relationships

The inter- and intrafamilial relationships of megapodes are, after nearly two centuries, still the subject of much debate. The position of the family within the Galliformes, their relationship with the Neotropical guans and curassows (Cracidae), and the number of megapode genera and species have received much attention from taxonomists. Early in the nineteenth century, megapodes were believed to have affinities with waders (Charadriiformes), pigeons (Columbiformes), lyrebirds (Menuridae, Passeriformes), or even vultures (Falconiformes). It was at that time that the Australian Brush-turkey *Alectura lathami* was called the 'New Holland Vulture' because of its bare head. The famous Dutch ornithologist Coenraad Jacob Temminck was reminded of the South American tinamous (Tinamidae,

Orange-footed Megapode.

Tinamiformes) when he saw his first *Megapodius* species. In about 1840, however, the view gained general favour that megapodes belonged to the Galliformes, and some believed that within the galliforms they were most closely related to the guans and curassows (Cracidae).

Based on comparative studies of anatomy, chromosomes (karyology), egg white proteins, and DNA–DNA hybridization, megapodes are commonly regarded as a monophyletic group, that is, a group that has developed from a single ancestral type. Traditionally, they have been given family rank in classifications.

Historically, the megapodes have been placed between the cracids and phasianids (pheasants, etc.) by Gray (1840), together with cracids, pigeons, and doves in the order 'Pullastrae' by Lilljeborg (1866), with cracids, buttonquails, phasianids, and sandgrouse in the 'Alectoromorphae' by Huxley (1867), and with the cracids in the suborder 'Peristeropodes' by Huxley (1868) and Sclater (1880). Huxley could not discover any important osteological differences between the skeletons of megapodes and cracids, although he did consider that the pneumatics of the bones differed considerably. He was of the opinion, however, that this character did not have any 'systematic' value. Sundevall (1872) divided the megapodes into two groups, the Catheturinae and Megapodiinae, under the cohort Macronyches, order Gallinae. This order included not only the galliforms as we know them now, but also, among other birds, the tinamous (Tinamidae), seedsnipes (Thinocoridae), and sheathbills (Chionididea). The French zoologist Oustalet (1881) published his monograph on megapodes around this time, following Huxley's ideas. Shortly after, Reichenow (1882) placed the megapodes with the cracids, hoatzin, phasianids, partridges, and grouse in the order 'Rasores', while Seebohm (1888) was of a different opinion and placed them with the cracids and phasianids in the suborder Gallinae. In that same year, Fürbringer (1888) placed the megapodes together with the Cracidae, Gallidae, and Opisthocomidae in his suborder 'Galliformes'. In the *Catalogue of the birds in the British Museum*, Vol. 22, Ogilvie-Grant (1893) followed Huxley (1868) and Sclater (1880), and regarded the Megapodiidae as a separate family. At the turn of the century, after nearly 80 years of debate, Sharpe (1899) divided the Galliformes into several suborders, the first one being formed by the 'Megapodii'.

During the twentieth century ornithologists have remained interested in the question of the systematic position of the megapodes, partly stimulated by the lack of hard evidence obtained by their predecessors. In his classification of the birds of the world, Wetmore (1930) placed the Megapodiidae with the Cracidae and the extinct Gallinuloididae in the superfamily Cracides, suborder Galli. Stresemann (1927–34) recognized three families within the Galliformes, the Megapodiidae, the Cracidae, and the phasianidae, and was supported many years later by Verheyen (1956). The latter author also pointed out that megapodes and cracids were more similar to one another than to the phasianids. Peters (1934) followed Wetmore, classifying the Megapodiidae with the Cracidae in the Cracoidea beside the other galliforms, the Phasianoidea. Study of the gross anatomy of wing muscles (Hudson and Lanzillotti 1964), biochemical analysis of ovomucoids (glycoproteins in avian egg whites) (Laskowski and Fitch 1989) and DNA–DNA hybridization experiments by Sibley and Ahlquist (1985) and Sibley *et al.* (1988) led to the same division. Most recently, Sibley and Ahlquist (1990) united the Cracidae and Megapodiidae in the order Craciformes; together with the Galliformes, which in their definition included the Phasianidae, Numididae, and Odontophoridae, the Craciformes form the superorder Gallomorphae.

Not everyone has agreed with the above ideas. Osteological characters led Cracraft (1972a, 1973, 1980) to the conclusion that the Megapodiidae are the sister group of all remaining galliforms. He stated that, although the 'study of the skeletal anatomy of galliforms does indeed reveal that the Megapodiidae and Cracidae share a large number of striking features', characters used to unite megapodes and cracids 'consist of primitive characters which are devoid of phylogenetic information since they were inherited from their common ancestor' (Cracraft 1972a, pp. 383–4). Morphological study of megapodes made Clark (1964a) conclude that 'megapodes, cracids, and phasianids had evolved as three separate lines from an unknown gallinaceous ancestor'. Ectoparasites (von Kéler 1958), the chemical composition of uropygial (oil gland) secretion, which in the Malleefowl *Leipoa ocellata* is 'the simplest diester wax yet examined' (Edkins and Hansen 1971), biochemical analysis of egg white proteins (Sibley and Ahlquist 1972; Sibley 1976), the structure of the eggshell (Board *et al.* 1982), and karyology (the study of chromosomes) (Sasaki *et al.* 1982; Belterman and de Boer 1984) also pointed to the megapodes being the sister group of all remaining galliforms. Recently, this idea has been further supported by investigations into feather structure; in megapodes, the downy barbules at the base of contour feathers lack the so-called detached or multiple nodes (Brom 1991a). These nodes are characteristic of all other galliform birds, including the cracids.

Based on the above evidence (explained in detail in Brom and Dekker 1992), the megapodes are treated in this monograph and in particular in the discussion on the evolution of incubation strategies (Chapter 8) as the sister group of all other galliforms rather than as the sister group of the cracids.

Intrafamilial relationships

Gray (1840) recognized the following megapode genera: *Talegallus* Lesson, *Leipoa* Gould, *Megapodius* Quoy and Gaimard, *Mesites* Geoffrey, and *Alecthelia* Lesson. This last genus in fact involved a chick of *Megapodius*. Later (Gray 1849), he divided the megapodes into two subfamilies based on the shape of the bill: Talegallinae (including *Talegallus* and *Megacephalon*) and Megapodinae (including *Megapodius*, *Alecthelia*, and *Leipoa*). Soon after, Charles Lucien Bonaparte recognized Gray's two subfamilies, but proposed a different division: Talegallinae (including *Leipoa*, *Catheturus* (= *Alectura*), and *Talegalla*) and Megapodinae (including *Megapodius*, (*Alecthelia*), and *Macrocephalon*). Oustalet (1881) recognized four genera: *Leipoa*, *Megacephalon*, *Talegallus*, and *Megapodius*, but preferred to consider them as four parallel types in contrast to the division into two subfamilies proposed by Gray and Bonaparte. Schlegel (1880a), in his review of megapodes in the Leiden collection, also recognized four genera: *Megapodius*, *Mégacephalon*, *Tallegallus*, and *Leipoa*.

At the present day, the megapodes are divided into either six or (as in this monograph) seven genera. The brush-turkeys, *Aepypodius* and *Alectura*, originally included in the genus *Talegallus* (= *Talegalla* as explained in the species account under *Talegalla*), are considered to form two separate genera. The number of genera further depends on whether *Eulipoa* Ogilvie-Grant, 1893, is treated separately or included under *Megapodius* Gaimard, 1823. Ogilvie-Grant was the first to regard this species, originally described by Gray (1860) as *Megapodius wallacei*, as truly distinct from all other *Megapodius* species and erected a separate, monotypic genus for it. Apart from its aberrant plumage, the main structural difference between *Megapodius* and *Eulipoa* lies in the relative length of the secondaries, which are much shorter than the primary quills in *Eulipoa*, but equal in length in *Megapodius*. Also, the length of the primaries relative to one another is different in *wallacei*, and its legs are relatively shorter but its toes and nails longer than in any *Megapodius* species. Feather-mite studies have further revealed that '*wallacei* was separated from the ancestral *Megapodius* stock quite early' (Atyeo 1992). Siebers (1930), Peters (1934), and Mayr (1938) were among those who followed Ogilvie-Grant and treated *wallacei* as representative of a separate genus, *Eulipoa*. Ripley (1964) merged *Eulipoa* again with *Megapodius* because 'in identical size and proportion, this species resembles *Megapodius* so closely that only the brighter banding on the wings and back, and white undertail coverts separate it' from the other *Megapodius* species. He judged these characters of

14 The Megapodes

insufficient taxonomic importance to require generic separation, which, he noted, 'continued to date largely because no taxonomist has reviewed the status of this rare species'. Clark (1964a), and more recently White and Bruce (1986) and Sibley and Monroe (1990), have followed Ripley. However, re-examination of characters such as the structure and shape of the wing, the relative length of tarsus, toes, and nails, coloration, and the fact that it is sympatric with (that is, occurs in the same geographical area as) two *Megapodius* species, made the present writers decide to treat *wallacei* as the representative of a distinct genus *Eulipoa* as proposed by Ogilvie-Grant (1893). For further details, see the Species account.

The following seven genera are recognized in this monograph: *Alectura* Latham, 1824, *Aepypodius* Oustalet, 1880, *Talegalla* Lesson, 1828, *Leipoa* Gould, 1840, *Macrocephalon* S. Müller, 1846, *Eulipoa* Ogilvie-Grant, 1893, and *Megapodius* Gaimard, 1823. Within these genera 22 species are recognized; these are listed at the end of the chapter.

The phylogeny of megapodes has received very little attention. An initial attempt was made by Clark (1964a), who separated them into two subgroups, with *Macrocephalon* and *Megapodius* (including *Eulipoa*) in one subgroup and *Aepypodius*, *Talegalla*, *Leipoa*, and *Alectura* in the other (Fig. 2.1). Clark's classification was based chiefly on overall similarity between the species, using characters

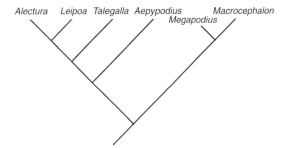

2.1 Phylogeny of the family Megapodiidae according to Clark (1964a).

such as proportions (relative length) at hatching, tarsal scutellation (presence of scales), foot webbing, the presence or absence of the fifth secondary (eutaxy versus diastataxy), and oil gland feathering, but without comparing them with the condition in related families such as cracids and phasianids. Recently, Brom (1991b) and Brom and Dekker (1992) reviewed these and other variable morphological characters, selected those which have phylogenetic significance (Table 2.1) and compared these selected characters with the condition in other galliforms. The result is illustrated in Fig. 2.2. The megapodes are not separated into two subgroups as in Clark's phylogeny (Fig. 2.1), and *Aepypodius* and *Alectura* are considered as closely related genera. Owing to lack of suitable characters the exact position of *Talegalla* and *Leipoa* is not resolved. On the basis of pterolichoid feather mites, Atyeo (1992) found some congruence between his

Table 2.1 Variable characters in megapodes. (Adapted from Brom and Dekker 1992.)

Genus	Wing	Nostril	Wattles	Uropygial gland	Penis	Egg	Yolk content (%)
Aepypodius	eutaxic	round	present	naked	yes	white	c. 55
Alectura	eutaxic	round	present	naked	yes	white	48–52
Talegalla	eutaxic	oval	absent	naked	?	red	?
Leipoa	eutaxic	oval	absent	naked?	yes	red	51–54
Macrocephalon	diastataxic	oval	absent	tufted	?	red	61–64
Eulipoa	diastataxic	oval	absent	tufted	?	red	65–67
Megapodius	diastataxic	oval	absent	tufted	no	red	63–69

Note: diastataxis is the structural absence of the fifth secondary flight feather in the wing; wings are eutaxic when the fifth secondary flight feather is present.

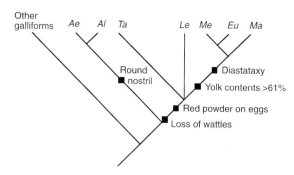

2.2 Phylogeny of the family Megapodiidae. (Adapted from Brom and Dekker 1992.)

data and Clark's proposed phylogeny; of particular interest are the links between *Talegalla*, *Aepypodius*, and *Alectura* with three sets of shared sibling species of mites, and between *Talegalla* and *Aepypodius* with three pairs of shared sibling species.

Species and subspecies

Except for *Megapodius*, there has generally been agreement about the number of species and subspecies within all genera. Among non-*Megapodius* genera, three new subspecies have been described during the preparation of this book: *Talegalla cuvieri granti* Roselaar, 1994, *T. fuscirostris aruensis* Roselaar, 1994, and *T. fuscirostris meyeri* Roselaar, 1994.

The number of species and subspecies within the genus *Megapodius* is still the subject of much debate and has puzzled ornithologists from Oustalet, Ogilvie-Grant, and Hartert, to Stresemann, Siebers, Mayr, White and Bruce, and the present writers. The number of species recognized has varied over the years from a minimum of four to a maximum of 19.

It was Oustalet (1879–1881) who recognized the largest number of species: *M. freycineti* Gaimard, 1823, *M. geelvinkianus* Meyer, 1874, *M. duperreyi* Lesson, 1826, *M. affinis* Meyer, 1874, *M. forsteni* Temminck, 1847, *M. eremita* Hartlaub, 1867, *M. brenchleyi* G. R. Gray, 1870, *M. gilberti* G. R. Gray, 1861, *M. sanghirensis* Schlegel, 1880, *M. cumingii* Dillwyn, 1853, *M. dillwyni* Tweeddale, 1877, *M. nicobariensis* Blyth, 1846, *M. bernsteini* Schlegel, 1866, *M. laperousii* Gaimard, 1823, *M. senex* Hartlaub, 1867, *M. stairi* G. R. Gray, 1861, *M. layardi* Tristram, 1879, *M. brazieri* Sclater, 1869, and *M. wallacei* G. R. Gray, 1860.

Schlegel (1880a) divided *Megapodius* into two groups, based on leg coloration: one group with bright (yellow, orange, or reddish) legs: *duperreyi*, *geelvinkianus*, *bernsteini*, *pritchardi*, and *senex*, and the other with dark legs: *freycineti*, *forstenii*, *lowei*, *gilbertii*, *sanghirensis*, *nicobariensis*, and *wallacei*. Salvadori (1882) separated *Megapodius* into three groups, also based on leg coloration: one group with bright (reddish) legs: *duperreyi* and *macgillivrayi*, another with legs which are dark in front and reddish on the back: *geelvinkianus*, and a third with dark legs: *freycineti*, *forstenii*, *affinis*, *eremita*, *brenchleyi*, and *wallacei*. Compared with Oustalet's division, Ogilvie-Grant (1893) united *affinis* with *forsteni*, omitted *brenchleyi*, *gilberti*, *dillwyni*, *brazieri*, and *stairi*, did recognize *tenimberensis*, *brunneiventris*, and *macgillivrayi*, and considered *wallacei* to represent a separate genus, *Eulipoa*. A few years later, Sharpe (1899) largely followed Ogilvie-Grant (1893), though recognizing *affinis* as a distinct species, thus separating it from *forsteni* again.

Hartert (in Rothschild and Hartert 1901) did not attach as much importance to the coloration of the legs as had Schlegel and Salvadori. He divided *Megapodius* for the Papuan region into two species and a number of subspecies based on general coloration, feathering of the head, and sympatry. His division was as follows:

(1) *duperreyi*, including *duperreyi*, *tumulus*, *macgillivrayi*, *forsteni*, *affinis*, *brunneiventris*, and *eremita*, with brownish feathering and a normally-feathered head;

(2) *freycineti*, including *freycineti* and *geelvinkianus*, with blackish feathering and 'scantily feathered forehead and throat'.

Siebers (1930) did not agree with Hartert, who in his opinion was not consistent in using the characters. He rejected all evidence of sympatry, a conclusion later supported by Mayr (1938), and united the *Megapodius* species recognized by Hartert into a single species. For reasons of convenience, he defined the following groups, based on coloration of the upperparts and legs: (1) upperparts reddish to olive-brown, (a) legs dark: *cumingi* group, (b) legs red: *bernsteini* group; (2) upperparts greyish-brown to grey, (a) legs dark: *forsteni* group, (b) legs red: *duperreyi* group; (3) upperparts greyish-black, (a) legs dark: *freycineti* group, (b) legs red: *laperousii* group.

Stresemann and Paludan (in Rothschild *et al.* 1932) recognized three species for the Papuan region: (1) *M. f. freycinet* and *M. f. geelvinkianus*; (2) *M. a. affinis*, *M. a. decollatus*, and *M. a. huonensis*; (3) *M. reinwardt* subspecies. Peters (1934) followed the division proposed by Stresemann and Paludan, and worked this out for the entire genus. He recognized nine species and various subspecies.

1. *M. nicobariensis nicobariensis* Blyth, 1846; *M. n. abbotti* Oberholser, 1919; *M. n. pusillus* Tweeddale, 1877; *M. n. tabon* Hachisuka, 1931; *M. n. cumingii* Dillwyn, 1853; *M. n. balukensis* Oberholser, 1924; *M. n. sanghirensis* Schlegel, 1880; *M. n. gilbertii* G. R. Gray, 1861, and *M. n. bernsteinii* Schlegel, 1866.
2. *M. tenimberensis* P. L. Sclater, 1883.
3. *M. reinwardt reinwardt* Dumont, 1823; *M. r. buruensis* Stresemann, 1914; *M. r. forstenii* G. R. Gray, 1847; *M. r. macgillivrayi* G. R. Gray, 1861; *M. r. tumulus* Gould, 1842, and *M. r. yorki* Mathews, 1929.
4. *M. affinis affinis* A. B. Meyer, 1874; *M. a. jobiensis* Oustalet, 1881; *M. a. decollatus* Oustalet, 1878, and *M. a. huonensis* Stresemann, 1922.
5. *M. eremita eremita* Hartlaub, 1867, and *M. e. brenchleyi* G. R. Gray, 1870.
6. *M. freycinet freycinet* Gaimard, 1823, and *M. f. geelvinkianus* A. B. Meyer, 1874.
7. *M. laperouse laperouse* Gaimard, 1823, and *M. l. senex* Hartlaub, 1867.
8. *M. layardi* Tristram, 1879.
9. *M. pritchardii* G. R. Gray, 1864.

Stimulated by Peters (1934, p. 3), who indicated that 'the genus *Megapodius* is badly in need of revision in order to determine the validity of many of the species and subspecies, their relationships and distribution', and by the fact that his own attempt to arrange the various forms from his New Guinea checklist caused confusion, Mayr (1938) made a revision of the genus *Megapodius*. He reviewed most forms from the Lesser Sundas to the Solomons and New Hebrides (Vanuatu), but did not discuss the westernmost taxa. He was of the opinion that 'the greatest difficulty presented by this genus is the tremendous individual variation of colour and particularly of size' (Mayr 1938, pp. 1–2), while 'differences between adults and immatures are not always well developed and small size of the tarsus is sometimes the only character by which an immature can be told'. At that time Mayr was not yet aware of the possibility of ageing the birds by the length and shape of the tip of the outermost primaries p9 and p10, which are short and pointed in immatures but longer and rounded in adults (see under Moult in Chapter 4). Further difficulties, according to Mayr, related to the fact that *Megapodius* is a social species, and that subspecies are not uniform in characters, 'but rather conglomerates of slightly different populations'. He concluded that all forms were allopatric and conspecific. Of the species recognized by Peters, he accepted only *pritchardii* G. R. Gray, 1864, and *laperouse* Gaimard, 1823 (with *senex* Hartlaub, 1867) as full species, and treated the other taxa as conspecific with *freycinet*. He considered *M. wallacei* to form a separate genus, *Eulipoa*.

Mayr was convinced that most modern authors would not hesitate to regard all forms as subspecies of one species, if not for one serious difficulty: in at least six localities, two different forms have been reported to live sympatrically (see Chapter 3 for a detailed discussion on this). The classification in Rothschild *et al.* (1932) and Peters (1934) created more difficulties than it resolved since 'forms which are exactly intermediate between two "species" are arbitrarily placed in one of the two', and there are no morphological characters that can be used to split *Megapodius* from the Papuan region into several subspecies (Mayr 1938, pp. 2–3). To further support his conclusion, he cast suspicion on the six localities of overlap. He rejected all of them, including the presumed sympatry on Japen Island between *affinis* and *geelvinkianus*, because of lack of reliable evidence to prove that two forms of *Megapodius* occur in exactly the same locality without mixing. Furthermore, the most divergent forms are connected by intermediates, which makes it necessary to consider them all as subspecies of one species. The best support for his views came, he felt, from the population on Dampier Island, where the two forms *affinis* and *eremita* meet and mix completely to form a hybrid population.

Mayr's view was usually adopted until the 1970s (for example, by Wolters 1976), although Stresemann (1941) recognized *M. nicobariensis*, *M. cumingi*, and *M. bernsteinii* as distinct species, but without giving specific reasons. Schodde (1977) preferred the arrangements in Rothschild *et al.* (1932) and Peters (1934) to the revision of Mayr (1938) and recognized four species for Papuasia: *M. freycinet*, *M. reinwardt*, *M. affinis*, and *M. eremita*. He did not agree with Mayr's rejection of sympatry between *reinwardt* and *affinis* in West and East New Guinea where *reinwardt* 'holds the low ground, while *affinis* is apparently pushed upwards to altitudes above which it normally occurs': the Papuasian populations of *Megapodius* thus behave as good species towards each other in some parts of their range, and not in others. Schodde therefore preferred to treat these forms as members of a superspecies.

White and Bruce (1986) followed Schodde and expanded his arrangement, treating as separate species the eight forms previously united under *M. freycinet* by Mayr. They were of the opinion that 'whilst there is no doubt that all the taxa are allopatric and form a superspecies, some strongly differentiated forms replace each other abruptly and it is convenient to treat them as species'. This resulted in the following arrangement: (1) *M. nicobariensis* Blyth, 1846; (2) *M. cumingii* Dillwyn, 1853; (3) *M. bernsteinii* Schlegel, 1866; (4) *M. reinwardt* Dumont, 1823; (5) *M. freycinet* Gaimard, 1823; (6) *M. affinis* Meyer, 1874; (7) *M. eremita* Hartlaub, 1867; (8) *M. layardi* Tristram, 1879; (9) *M. laperouse* Gaimard, 1823; (10) *M. pritchardii* G. R. Gray, 1864; (11) *M. wallacei* G. R. Gray, 1860. Most recently, Sibley and Monroe (1990) have followed this division.

The division of the genus *Megapodius* in this monograph is to some extent congruent with White and Bruce, although three subspecies have been given species rank as in some previous arrangements; these are *forstenii*, *geelvinkianus*, and *tenimberensis*. The species *wallacei* is regarded here as representative of a separate genus, *Eulipoa*, as proposed earlier by Ogilvie-Grant (1893) (see above). For justification of these decisions, see Chapter 3 and also the Species Account.

The seven genera, 22 species, and 44 subspecies treated in this monograph are as follows. (For an explanation of the English names, see below.)

Alectura Latham, 1824
 Alectura lathami, Australian Brush-turkey, polytypic, two subspecies: *A. l. lathami* J. E. Gray, 1831 and *A. l. purpureicollis* (Le Souëf, 1898).

Aepypodius Oustalet, 1880
 Aepypodius arfakianus, Wattled Brush-turkey, polytypic, two subspecies: *A. a. arfakianus* (Salvadori, 1877) and *A. a. misoliensis* Ripley, 1957.
 Aepypodius bruijnii, Bruijn's Brush-turkey, monotypic: *A. bruijnii* (Oustalet, 1880).

Talegalla Lesson, 1828

Talegalla cuvieri, Red-billed Talegalla, polytypic, two subspecies: *T. c. cuvieri* Lesson, 1828 and *T. c. granti* Roselaar, 1994.

Talegalla fuscirostris, Black-billed Talegalla, polytypic, four subspecies: *T. f. fuscirostris* Salvadori, 1877, *T. f. occidentis* White, 1938, *T. f. aruensis* Roselaar, 1994, and *T. f. meyeri* Roselaar, 1994.

Talegalla jobiensis, Brown-collared Talegalla, polytypic, two subspecies: *T. j. jobiensis* A. B. Meyer, 1874 and *T. j. longicauda* A. B. Meyer, 1891.

Leipoa Gould, 1840

Leipoa ocellata, Malleefowl, monotypic: *L. ocellata* Gould, 1840.

Macrocephalon S. Müller, 1846

Macrocephalon maleo, Maleo, monotypic: *M. maleo* S. Müller, 1846.

Eulipoa Ogilvie-Grant, 1893

Eulipoa wallacei, Moluccan Megapode, monotypic: *E. wallacei* (G. R. Gray, 1860).

Megapodius Gaimard, 1823

Megapodius pritchardii, Polynesian Megapode, monotypic: *M. pritchardii* G. R. Gray, 1864.

Megapodius laperouse, Micronesian Megapode, polytypic, two subspecies: *M. l. laperouse* Gaimard, 1823 and *M. l. senex* Hartlaub, 1867.

Megapodius nicobariensis, Nicobar Megapode, polytypic, two subspecies: *M. n. nicobariensis* Blyth, 1846 and *M. n. abbotti* Oberholser, 1919.

Megapodius cumingii, Philippine Megapode, polytypic, seven subspecies: *M. c. gilbertii* G. R. Gray, 1861; *M. c. cumingii* Dillwyn, 1853; *M. c. dillwyni* Tweeddale, 1877; *M. c. pusillus* Tweeddale, 1877; *M. c. tabon* Hachisuka, 1931; *M. c. talautensis* Roselaar, 1994, and *M. c. sanghirensis* Schlegel, 1880.

Megapodius bernsteinii, Sula Megapode, monotypic: *M. bernsteinii* Schlegel, 1866.

Megapodius tenimberensis, Tanimbar Megapode, monotypic: *M. tenimberensis* Sclater, 1883.

Megapodius freycinet, Dusky Megapode, polytypic, three subspecies: *M. f. freycinet* Gaimard, 1823; *M. f. oustaleti* Roselaar, 1994, and *M. f. quoyii* G. R. Gray, 1861.

Megapodius geelvinkianus, Biak Megapode, monotypic: *M. geelvinkianus* A. B. Meyer, 1874.

Megapodius forstenii, Forsten's Megapode, polytypic, two subspecies: *M. f. forstenii* G. R. Gray, 1847 and *M. f. buruensis* Stresemann, 1914.

Megapodius eremita, Melanesian Megapode, monotypic: *M. eremita* Hartlaub, 1867.

Megapodius layardi, Vanuatu Megapode, monotypic: *M. layardi* Tristam, 1879.

Megapodius decollatus, New Guinea Megapode, monotypic: *M. decollatus* Oustalet, 1878.

Megapodius reinwardt, Orange-footed Megapode, polytypic, five subspecies: *M. r. reinwardt* Dumont, 1823; *M. r. tumulus* Gould, 1842; *M. r. yorki* Mathews, 1929; *M. r. castanonotus* Mayr, 1938, and *M. r. macgillivrayi* G. R. Gray, 1861.

English names

The English names of two genera have been changed here; they are in accordance with some of the older literature (e.g. Ogilvie-Grant 1893, 1915), but not with recent publications. 'Scrubfowl', as the name for most of the *Megapodius* species, has been changed to 'Megapode', to conserve this well-known and meaningful name in the literature for all representatives of this genus as well as for *Eulipoa wallacei*, the Moluccan Megapode.

Also, to emphasize the separation between the brush-turkeys *Aepypodius* and *Alectura* on the one hand, and *Talegalla* on the other (see Fig. 2.2), the English name of the *Talegalla* genus has been changed to 'Talegalla', a name which is in common use in, for example, the German language for the three representatives of this genus.

Megapode etymology

In Table 2.2, the meaning and origin of all 51 generic, specific, and subspecific scientific names is given in alphabetical order. The descriptions are based chiefly on Jobling (1991).

Table 2.2 Megapode etymology

abbotti	After William L. Abbott (1860–1936), US explorer, naturalist, and collector of, in particular, this subspecies of *Megapodius nicobariensis*.
Aepypodius	From Greek *aipus* (steep or lofty) and *pous* or *podos* (foot).
Alectura	From Greek *alektor* (domestic cock) and *oura* (tail).
arfakianus	After the Arfak Mountains, Irian Jaya, Indonesia.
aruensis	After the Aru Islands, Indonesia.
bernsteinii	After Heinrich Agathon Bernstein (1828–1865), German physician, zoologist, and collector in Indonesia.
bruijnii	After Anton August Bruijn (died 1885). Dutch merchant of Ternate (Moluccas), engaged in the New Guinea plume trade. His collectors shot all except one of the 15 skins of *Aepypodius bruijnii* known to science.
buruensis	After the island of Buru, Indonesia.
castanonotus	From Latin *castaneus* (chestnut-coloured) and Greek *-notos* (backed).
cumingii	After Hugh Cuming (1791–1865), English naturalist and collector in Polynesia, the East Indies, and the Philippines.
cuvieri	After Frédéric Georges Cuvier (1775–1838), French zoologist.
decollatus	From French 'decolleté', low-necked (dress), because the species has an extensive bare, red foreneck.
dillwyni	After L. Llewellyn Dillwyn (1814–1892), British ornithologist who described *Megapodius cumingii* from specimens collected on Labuan by J. Motley, with whom he wrote the *Natural history of Labuan* in 1855.
eremita	From Latin *eremita* (hermit).
Eulipoa	From Greek *eu-*(good), *leipo* (abandon), and *oon* (egg). (Compare *Leipoa* below.)
forstenii	After Eltio Alegondas Forsten (1811–1843), Dutch medical doctor and collector in Sulawesi, Indonesia, between 1838 and 1843.
freycinet	After Louis Claude Desaulses de Freycinet (1779–1842), French navigator and explorer in the Pacific between 1817 and 1820. Captain on the expedition during which Jean René Constant Quoy and Paul Gaimard, French surgeon-naturalists and explorers, collected the first megapode specimen in December 1818 (on Waigeu Island, Irian Jaya, Indonesia).
fuscirostris	From Latin *fuscus* (dark or dusky) and *-rostris* (billed).
geelvinkianus	After Geelvink Bay, Irian Jaya, Indonesia.
gilbertii	Probably refers to John Gilbert (died 1845), John Gould's great assistant who collected the majority of Gould's new species of Australian birds. He was killed by aboriginals in Queensland on 28 June 1845.
granti	After Claude Henry Baxter Grant (1878–1958), ardent collector in South and East Africa and South America, member of the British Ornithologists' Union Expedition 1909–1911 to New Guinea, collector of the type specimen of *Talegalla cuvieri granti*.
jobiensis	After Jobi (Japen) Island, Geelvink Bay, Irian Jaya, Indonesia.
laperouse	After Captain Jean François de Galaup Comte de la Pérouse (1741–1788), French explorer in the Pacific between 1785 and 1788.
lathami	After John Latham (1740–1837), English ornithologist.
layardi	After Edgar Leopold Layard (1824–1900), English diplomat and naturalist.
Leipoa	From Greek *leipo* (abandon) and *oon* (egg).
longicauda	From Latin *longus* (long) and *cauda* (tail).
macgillivrayi	After John MacGillivray (1821–1867), Scottish naturalist and collector.
Macrocephalon	From Greek *makrocephalos* (long- or great-headed).
maleo	From the Malayan names *maleo* or *moleo*, the original, local name for megapodes in Indonesia.
Megapodius	From Greek *megas* (great or large) and *pous* or *podos* (foot).
meyeri	After Adolf Bernhard Meyer (1840–1911), Director of the Zoological–Anthropological Museum of Dresden, traveller to the Moluccas and western New Guinea, 1870–1873; the specialist of birds of Indonesia and New Guinea of his time, author of *The birds of Celebes* (1898, with L. W. Wiglesworth).
misoliensis	After Misol Island, Irian Jaya, Indonesia.

Table 2.2 *cont.*

nicobariensis	After the Nicobar Islands, India.
ocellata	From Latin *ocellatus* (ocellated).
occidentis	From Latin *occidens* (west).
oustaleti	After Emile Oustalet (1844–1905), French zoologist who wrote the first monograph on megapodes in 1879–1881.
pritchardii	After William Thomas Pritchard (1829–1909), British consul to Fiji between 1857 and 1862, who collected the first specimen of *Megapodius pritchardii* on Niuafo'ou (near Fiji) from which Gray described the species.
purpureicollis	From Latin *purpureus* (purple-coloured) and *-collis* (necked).
pusillus	From Latin *pusillus* (tiny or very small).
quoyii	After Jean René Constant Quoy (1790–1869), French naturalist, explorer in the Pacific 1826–1829. See also *freycinet*.
reinwardt	After Caspar Georg Carl Reinwardt (1773–1854), Dutch ornithologist and collector in Indonesia between 1817 and 1822.
sanghirensis	After Sangihe Island, Indonesia.
senex	From Latin *senex* (old person), which refers to the white or grey crown of the species.
tabon	From the Philippine word *tavon* or *tabon*, the local name of this subspecies in the Philippines.
talautensis	After Talaut Island, Indonesia.
Talegalla	From *talève*, French name for *Porphyrio* genus of rails Rallidae (compare Malagasy *talavana*).
tenimberensis	After Tanimbar Islands, Indonesia
tumulus	From Latin *tumulus*, meaning (a) burial mound or (b) small hill, possibly due to close resemblance of *Megapodius reinwardt*'s nesting mound to stone-age graves (see Stone 1991).
wallacei	After Alfred Russel Wallace (1823–1913), the famous British zoologist who discovered *Eulipoa wallacei* during his voyage to the Malay Archipelago.
yorki	After Cape York District, Australia.

3

Distribution, biogeography, and speciation

Introduction

The distribution of megapodes in Indo-Australia and on islands in the Pacific is still not fully understood. Over the years, some ornithologists have tried to explain irregularities in the distribution and to find areas of origin and routes of dispersal. Others have focused on areas of contact or exclusion to find arguments for lumping or splitting species. However, palaeontological research has recently shown that the distribution of megapodes in the Pacific was one much more extensive than it is now, and that human activ-

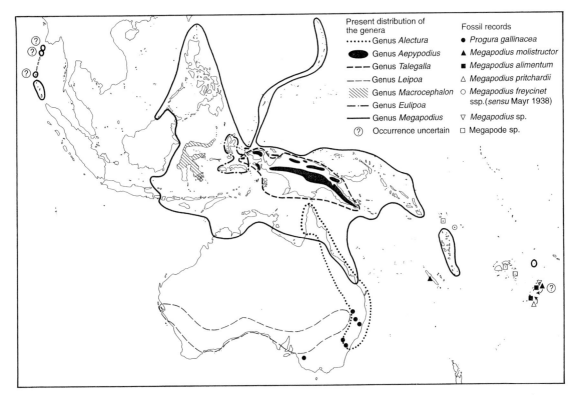

3.1 Distribution of megapode genera and locations of fossil records of extinct and some extant species. '*Megapodius* sp.' indicates remains that have been identified as a megapode of the genus *Megapodius*. 'Megapode sp.' indicates megapode remains of an unknown genus.

ity has resulted in their extermination on many southwest Pacific islands. Although this discovery has given answers to questions about the regional presence or absence of megapodes, controversy has continued about the origin of the family. This chapter will discuss the present and historic distribution of megapodes (shown in Fig. 3.1), and will summarize old and new theories, some of which are still speculative, about their place of origin, routes of dispersal, and patterns of speciation.

Present distribution

Megapodes are mainly confined to Indo-Australia east of Wallace's Line (the imaginary boundary between Borneo and Sulawesi separating the Oriental and Australasian flora and fauna), ranging from Australia, New Guinea and its surrounding islands, through eastern Indonesia to the Philippines. No megapodes have ever been recorded from Bali, Java, Sumatra, the greater part of Borneo, or the mainland of Southeast Asia. Three *Megapodius* species are well isolated from the rest and are found on the remote island of Niuafo'ou (Kingdom of Tonga, southwest Pacific), the Palau and Mariana Islands (US Trust Territory), and the Nicobar and Andaman Islands in the Gulf of Bengal (India). New Guinea and Australia show the widest diversity of genera and species, with *Aepypodius*, *Talegalla*, and *Megapodius* in New Guinea and *Alectura*, *Leipoa*, and *Megapodius* in Australia.

Apart from *Megapodius*, all genera have a restricted range. *Aepypodius* and *Talegalla* are endemic to New Guinea and some of its western islands. *Alectura* and *Leipoa* are Australian endemics, with the former found mainly in closed forests along the east coast and the latter restricted to dry Mallee habitat in the south. *Macrocephalon* is restricted to Sulawesi and *Eulipoa* to the Moluccan islands. *Megapodius* is found throughout the entire range of the family, from the Nicobars in the west to Niuafo'ou in the east. *Megapodius* species are strong flyers and adults are regularly observed flying from one island to another and covering distances of several kilometres. Even newly hatched chicks have been observed landing aboard ships well offshore. The Orange-footed Megapode *Megapodius reinwardt* has been found as a recent invader on remote tiny islands, such as Manuk, Gunung Api, and the Luciparas in the Banda Sea, which are widely separated from any larger landmass.

Historic distribution

Unconfirmed accounts of megapodes outside the boundaries of their present range exist for several islands in the Pacific. In the nineteenth century, mounds which probably belonged to a species of the genus *Megapodius* were reported from Sunday Island in the Kermadec Group between New Zealand and Tonga (Lister 1911*a*); a volcanic eruption on the islands in 1876 is said to have been responsible for the extermination of this species. Reports of young megapodes from Lord Howe Island, east of Australia, have been found to have derived from this locality being mistaken for New Hope Island, which in fact is a synonym for Niuafo'ou where the Polynesian Megapode *Megapodius pritchardii* occurs. Another unconfirmed account is from New Caledonia, where a megapode was described by George Robert Gray (1861) as *Megapodius andersoni* on the basis of a very brief description of a black grouse-like bird with unfeathered legs. This bird, named '*Tetrao australis*' by William Anderson, was seen during Captain Cook's second voyage to the South Pacific, 1772–1775. Since that time, no megapodes other than fossils have been recorded on New Caledonia (see below). Finally, Gray (1861) described two species of *Megapodius* based on two eggs from islands where there has never been any other indication of their presence: *Megapodius stairi* from Samoan or Navigator's Islands and *Megapodius burnabyi* from an island in the Ha'apai group of Tonga. Both names have since then always been regarded as synonyms of *Megapodius pritchardii*. However, these eggs, now in the collection of The Natural History Museum in Tring, have recently been examined by David W. Steadman (1991). He concluded that neither *stairi* nor *burnabyi* is unequivocally synonymous with *pritchardii* and that they should be regarded as *nomina dubia* (names of doubtful application). The eggs might indeed have belonged to extinct species from these islands. However, they might also have been imported from other islands and therefore have originated from either one or two other species still alive today. The dimensions of both eggs *(M. stairi* 78.8 × 47.6 mm, *M. burnabyi* 77.3 × 43.9 mm) fall within the range of eggs of the Polynesian Megapode from Niuafo'ou, the Melanesian Megapode *M. eremita* from the Solomons, and the Vanuatu Megapode *M. layardi* from Vanuatu. The former existence of recently extinct *Megapodius* on Samoan and Tongan islands therefore remains doubtful and cannot be concluded from these two eggs alone.

The fossil evidence

In southeast Australia, remains of *Progura gallinacea* De Vis, 1888, have been found in Pleistocene deposits (van Tets 1974, Rich and van Tets 1985). This megapode, originally believed to be a giant crowned pigeon, a bustard, or even a stork and described under the synonyms of *Chosornis praeteritus* De Vis, 1889 and *Palaeopelargus nobilis* De Vis, 1891, was much larger than any of the present-day megapodes. It must have reached a body weight of approximately 4 kg in females to 7 kg in males, which is two to three times the weight of the Australian Brush-turkey *Alectura lathami* or the Malleefowl *Leipoa ocellata*. *Progura naracoortensis* Van Tets, 1974, initially presumed to be a close relative of *P. gallinacea* of which bones have been discovered in the same area, is now regarded as a (smaller) female *P. gallinacea* (cf. Rich and van Tets 1985), although it has also been suggested that it might represent a different genus (Olson 1985). These examples illustrate the difficulties in the interpretation of skeletal fragments. No fossil remains of megapodes have ever been found in Tasmania or southwestern Australia.

A large and flightless extinct bird, exterminated by man less than 2000 years ago (Balouet and Olson 1989), has been described from New Caledonia as *Sylviornis neocaledoniae* Poplin, 1980, and was originally believed to have ratite affinities (Poplin 1980). Three years later it was considered to be a megapode (Poplin *et al.* 1983). However, its identification has recently been questioned and the remains are no longer regarded as those of either a ratite or a megapode, although it is still admitted that it must have had galliform relationships. C. Mourer-Chauviré (*in litt.*), a French palaeontologist, proposes that it would be better to place *Sylviornis* in a family of its own, though more closely related to the megapodes than to any other galliform family.

Also in New Caledonia the extinct *Megapodius molistructor* Balouet and Olson, 1989, has been described from bones found in association with human remains in Holocene deposits (Balouet and Olson 1989). This species, the largest and most robust representative of its genus known so far, is believed to have been exterminated between 2000 and 3000 years ago by the early Polynesian invaders of the island. It is interesting to compare this record with the observation by William Anderson of a blackish-brown grouse-like bird with bare legs during Captain Cook's second voyage and described as *M. andersoni* by Gray in 1861 (see above). The remains described by Balouet and Olson (1989) as *M. molistructor* might even refer to *M. andersoni*. However, as the description by Anderson is very brief, and as New Caledonia is large enough to harbour more than one species of megapode, Balouet and Olson felt they could not be certain that the bird Anderson saw was in fact the same species as they described as *M. molistructor*. This or a closely related species, as well as another extinct megapode, *Megapodius alimentum* Steadman, 1989, have been found on Lifuka (Ha'apai group, Tonga)(Steadman 1989a). Again, man is presumed to have been responsible for the extinction of *M. alimentum*. Qualitative similarity between the tarsometatarsi of *M. alimentum* and *M. pritchardii* suggests that *M. alimentum*, which has also been claimed for the island of 'Eua (Tonga), may have been more closely related to this sole surviving Polynesian species than to *M. freycinet* subspecies (*sensu* Mayr 1938). The discovery of an ulna and femur of a *Megapodius* on Ofu (Neiufu group, Samoa), provided the easternmost record for the megapodes so far (Steadman 1991, and in press). The bones are larger than those of *M. pritchardii* but smaller than those of *M. alimentum* and *M. molistructor*. Fossil remains from Tikopia, Solomon Islands, resemble those of a small *Megapodius* possibly related to *M. freycinet* subspecies (*sensu* Mayr 1938). Finally, *M. pritchardii* was apparently once more widespread; its bones have been found on 'Eua, Tonga, approximately 700 km south of Niuafo'ou.

Megapode bones excavated on Lakemba and Naingani, Fiji, and the Santa Cruz group of the Solomon Islands have not yet been identified to generic or species level. Also, leg bones of an extinct but as yet unidentified species, presumed to have been a 'giant' megapode, have been found in a Polynesian midden on a tiny island in the Fiji Archipelago. It appears to have been even larger than *P. gallinacea* (Rich and van Tets 1985).

Despite extensive palaeontological research no fossil megapodes have been found in New Zealand. This suggests either that ancestral megapodes were not yet present on the Australian landmass 80 million years ago, when New Zealand separated from it and drifted to its present position, or that they never reached New Zealand after its separation from the Australian landmass.

The discovery of fossil fragments considered to have megapode affinities from Eocene–Oligocene deposits of France was remarkable and against all expectations. However, these fossils, *Quercymegapodius depereti* (Gaillard, 1908), and *Quercymegapodius brodkorbi* Mourer-Chauviré, 1992, have recently been placed in a new, extinct family, Quercymegapodiidae, by Mourer-Chauviré (1992). The humerus, carpometacarpus, coracoid, tibiotarsus, and tarsometatarsus resemble those of the Megapodiidae. Others also questioned the original identification and considered the fragments to represent extinct Gallinuloididae (Crowe and Short 1992), an idea which is not supported by C. Mourer-Chauviré (*in litt.*).

The origin of megapodes

Following earlier ideas of Alfred Russel Wallace, Lister (1911*a*) proposed that megapodes on the Nicobar Islands, Niuafo'ou, Palau, and the Mariana Islands had been introduced by man. His theory was based on the opinion that the Nicobar *Megapodius nicobariensis*, the Polynesian *Megapodius pritchardii*, and the Micronesian Megapode *Megapodius laperouse* could not have reached these islands by themselves, and by evidence of domestication and the transport from one island to another of *Megapodius* species. Recent findings of extinct *Megapodius* species on several Polynesian islands, however, allow his theory to be rejected with regard to the occurrence of the Polynesian Megapode on Niuafo'ou. For the Nicobar and Micronesian Megapode also, Lister's theory gains little support. These two species could probably not have diverged so much if they had only been isolated for 3000 years or less, which would be the case if they were brought to the Nicobar, Palau, and Mariana Islands by man. Ironical as it seems, the opposite of what Lister had in mind has happened: far from introducing megapodes to western Pacific islands, the Polynesians and Melanesians in fact exterminated them, just as happened with so many other species such as rails Rallidae, pigeons Columbidae, and, of course, the New Zealand moas (Dinornithiformes).

In contrast with Lister, Oustalet (1881), Croizat (1958), and Dekker (1989*c*) considered the presence of *Megapodius* as far west as the Nicobars to be a relic of an ancient distribution. This theory assumes that within historical times megapodes had a wider distribution along the western part of their present range, and possibly occurred on Bali, Java, Sumatra, Borneo, and perhaps even on the mainland of Southeast Asia. This conjecture has not yet been supported by fossil evidence from the region, however. The presence of (ancestral) megapodes in this region has also been suggested by Olson (1980), who was of the opinion that megapodes reached Australasia from the north; the fossils of *Quercymegapodius* from France, at that time still considered to have had megapode affinities, must have supported him in this scenario.

The competition theory versus the predation theory

Two theories have recently been presented to explain the western limits of the megapodes' range. Olson (1980) regarded competitive exclusion (for example, competition for food) between phasianids (pheasants and their allies) and megapodes as an explanation for the absence of megapodes from the Greater Sunda Islands (Borneo, Sumatra, and Java) and the mainland of Southeast Asia. Phasianids (excluding *Coturnix* quails and derivatives), who have their centre of taxonomic diversity in Asia (Olson 1980), do not occur east of Wallace's Line. In contrast to *Megapodius* species, phasianids are incapable of dispersing over water, and they are thought to have excluded megapodes from the mainland of Southeast Asia, or prevented them from expanding their range towards it, by means of competition for food. This resulted in an almost perfectly complementary distribution of the two families. There are, however, three areas of apparent overlap where species from both groups co-exist. In the Lesser Sunda Islands, the Green Jungle Fowl *Gallus varius* is sympatric with the Orange-footed Megapode. In Palawan, the Palawan Peacock-pheasant *Polyplectron emphanum* and the Philippine Megapode *Megapodius cumingii* meet. In north Borneo, a number of phasianid species occur on the mainland, while the Philippine Megapode is resticted to small offshore islands; it is only occasionally recorded from the mainland, apparently to avoid interaction with these phasianids. Despite their great differences in structure and habits, phasianids and megapodes are considered as ecological counterparts that cannot co-exist (Olson 1980).

The alternative theory proposes that the expansion of megapodes westward is prevented by carnivores, especially cats (Felidae) and civet-cats (Viverridae) (Dekker 1989c). In the same way as the distribution of phasianids and megapodes is complementary, so the distribution of certain cats and civet-cats which are potential predators of mound-building megapodes is also complementary to that of megapodes. Adult mound-building megapodes are highly susceptible to predation, because of the need for prolonged periods of activity at their mounds. The high predation pressure associated with the wide variety of large predators on the Greater Sunda Islands and on the mainland of Southeast Asia renders these regions unsuitable for mound-building megapodes. The fact that the Nicobar Islands have never had a land connection and are thus devoid of carnivores could explain the survival of the Nicobar Megapode as a relic there. Several generalist carnivores of the genera *Felis*, *Neofelis*, *Panthera*, *Macrogalidia*, and *Viverricula* are considered to be potentially capable of exterminating populations of mound-building megapodes, or of preventing stragglers from founding new populations. As with the competition theory, a few localities are known where potential predators and megapodes do actually overlap or appear to overlap: Sulawesi, three Philippine islands (Palawan, Negros, and Panay), North Borneo and its offshore islands, and the Kangean Islands. However, a closer inspection shows that within the Kangean Archipelago and in North Borneo megapodes and carnivores are in fact separated, with the megapodes confined to the smaller (carnivore-free) islands and the carnivores restricted to the larger islands and the mainland. Thus, within the Kangean Archipelago, the mound-building

Orange-footed Megapode is common on Saubi, Sepanjang, Bangkan, Saebus, and probably other small islands, whereas it is absent from the main island Kangean which holds a population of Leopards *Panthera pardus* and Little Civets *Viverricula indica*. The megapode species which inhabit Sulawesi and the three Philippine islands, (the Maleo *Macrocephalon maleo* and the Philippine Megapode respectively), are burrow nesters rather than mound builders; their breeding behaviour differs from that of mound builders as it is normally communal and less labour intensive, allowing visits to the nesting sites to be brief. Adult burrow-nesting megapodes appear therefore to be less prone to predation, especially when the variety of predator species or their numbers are limited as in Sulawesi and the Philippines.

In 1987, two Little Civets and one Leopard Cat *Felis bengalensis* were collected on Lombok in the Lesser Sunda Islands (Kitchener *et al.* 1990). These observations, which represent new records for the island, are at variance with the predation hypothesis, since the Orange-footed Megapode is presumed to be a mound builder on Lombok. Whether these two carnivores are recent introductions, or whether they are present in such small numbers that they do not pose any danger for the megapode population, remains to be investigated.

Centre of origin: north or south?

As indicated above, part of the debate on the western limits of the megapodes' range results from controversy about their origin. Either megapodes reached Australia from the north via Asia, or they are derived from ancestors which were present in Australasia from the time of the break-up of Gondwana. One of the reasons for this controversy is the disagreement about their inter- and intrafamilial relationships (see Chapter 2).

The first theory places the centre of origin of the galliforms in the northern hemisphere, and proposes that megapodes probably reached Australia from the north by the Miocene (26 to 7 million years ago) or even in the early Tertiary as Mayr (1944) claimed. This idea is supported, among other things, by the presence in the northern hemisphere of fossils from the Tertiary of three extinct galliform families with about 30 genera and 60 species (C. Mourer-Chauviré, *in litt.*), whereas the oldest galliform fossils from Africa (which continent is of Gondwana origin) are more recent and date back to the Miocene. No galliform fossil remains have been found (as yet) in South America and Australia. Mayr (1938, 1944) and Olson (1980, 1985), who support the theory that megapodes reached Australia from the north and that megapodes are most closely related to the cracids (guans and curassows), are led in their opinion by records of fossil cracid remains from the middle Eocene and early Oligocene of North America (for reference, see Cracraft 1973, p. 507). They therefore place the centre of origin of the cracids as most probably in tropical North America. The present distribution of cracids in the Americas and their degree of morphological divergence and isolation made Vuilleumier (1965) (also) suggest that their presence in South America is of more recent origin, probably post-Pliocene or even mid-Pleistocene.

Others, such as Darlington (1957), Cracraft (1973) and Crowe and Short (1992) were of the opinion that these fossils show more resemblance to phasianids and Gallinuloididae than to cracids and concluded that the widely held belief that the Cracidae originated in North America is doubtful. Cracraft (1973) therefore proposed a Gondwana origin for the galliforms and suggested that megapodes were derived from ancestors which were present in Australasia from the time of the break-up of Gondwana during the Cretaceous. In his opinion, this theory accords well with the distribution of various taxa and would most easily explain the present distribution patterns of the families. The two most primitive families, the megapodes and cracids, which Cracraft did not consider as sister groups, are distributed in the southern hemisphere. Furthermore, he was of the opinion that megapodes, being primitive in many features but also very distinct morphologically, must have been isolated for a considerable period of time. Suggestions that megapodes arrived in Australia from Southeast Asia were considered unrealistic since during much of the early to middle Tertiary Australia was situated much further south than at the present day. Moreover, dispersal of megapodes in Indo-Australia appeared to be from south to north, though no evidence for this theory was given.

Both theories are speculative in that neither is based upon convincing evidence of relationships or adequate (old) fossil records (Sibley 1976; Rich *et al.* 1985). However, they both agree that megapodes, wherever they came from, must have been isolated in Australasia for a considerable period, at least since the Miocene. During that period, the Australian landmass was within 10° of latitude of its present position and came under the influence of increasingly warmer climates. In the late Tertiary the climate became cooler and more arid, while during that period (Pliocene) the Australian plate had moved up to its present position, closer to Asia than ever before, but connected with it only indirectly by a series of large and small islands separated by much open ocean. During the Pleistocene, when the sea levels were lowered, Australia had a land connection with New Guinea and its surrounding islands, such as Waigeu, Aru, and Japen. This land connection existed as recently as 20 000 years ago. The subsequent rise of the sea-level created islands and isolated populations. The present distribution of *Macrocephalon*, *Eulipoa*, and *Megapodius*, and speciation in *Megapodius*, suggest that colonization and dispersal in a northwest direction from the Australian landmass has taken place several times since the common ancestor of these three genera split off from the remaining megapodes which are all confined to Australia–New Guinea (see Fig. 2.2).

The most detailed account of dispersal and evolution within *Megapodius* was given by Croizat (1958). He considered the triangle Bismarck/Admiralties–D'Entrecasteaux–South Solomons as one of the centres of evolution for the genus *Megapodius*. Dispersal took place from the Solomons to New Guinea over two distinct tracks, one in a northwesterly direction through New Britain, Manam (Vulcan Islands), and Karkar (Dampier Islands) to northern New Guinea and the other in a westerly direction through the D'Entrecasteaux and Louisiades to southeastern New Guinea. From this centre in the Solomons *Megapodius* must have reached Vanuatu as well. Also Diamond and Marshall (1976) suggested that *Megapodius* must have reached Vanuatu from the Solomon Islands. However, they placed the origin of the Vanuatu Megapode in New Guinea and not in the Solomons.

Croizat considered the distribution of *Talegalla* species unlike that of *Megapodius*, which made him state that 'in a time anterior to modern genus-making, the proximal ancestors of the living megapodes had already generally perfected the dispersal that still belongs today to their descendants between the Nicobar Islands in the Bay of Bengal and Niuafou Island in Central Polynesia'. He estimated the 'time anterior to modern genus-making' conservatively as Late Cretaceous–Earliest Tertiary and thought it unlikely that the genera now recorded in New Guinea formed ranks only after the Eocene. This is of much earlier date than the Miocene mentioned previously in the text, but matches with Mayr's idea of timing. Other authors have speculated only incidentally on speciation in and distribution of *Megapodius*. Only Siebers (1930) is worth mentioning; he presumed that Forsten's Megapode *M. forstenii* on Buru reached the island from the east, thus in a northwesterly direction, across Ceram, because it is related neither to the red-legged Sula Megapode *M. bernsteinii* from Sula nor to the dark-legged Dusky Megapode *M. freycinet* from the northern Moluccas.

The following section gives the ideas of the present authors on speciation in *Megapodius* and *Talegalla*, especially in relation to areas of contact where two or more closely related species co-exist or where their ranges touch. Such areas, which have only been described for these two genera, are of special interest from a biogeographic as well as a taxonomic point of view.

Speciation in Talegalla and Megapodius: areas of contact

The Talegalla case

The ranges of the Black-billed Talegalla *Talegalla fuscirostris* and Red-billed Talegalla *T. cuvieri* in New Guinea overlap in the foothills of the Snow Mountains (Iwaka River area and nearby Utakwa River area), where they appear to be segregated altitudinally, with the Red-billed restricted to higher elevations. Specialization and differentiation in the area of overlap is not only expressed in an altitudinal segregation, but also through character displacement which has resulted in the Red-billed Talegalla being larger and heavier here than elsewhere (see Species account). In eastern New Guinea, the range of the Black-billed Talegalla overlaps with the Brown-collared Talegalla *T. jobiensis* for at least 90 km around 147°E above Port Moresby. Here also, the Black-billed is the lowland species while the Brown-collared is confined to higher ground. Further west, a Talegalla, probably the Brown-collared, is recorded locally on the south slopes of the Central Range in Papua New Guinea (Lake Kutubu and Mount Sisa in Southern Highland Province; Karimui area in Chimbu Province) near or just within the range of the Black-billed Talegalla; probably, the Brown-collared Talegalla occurs here over a distance of *c.* 700 km, between *c.* 142° and 148°E, replacing the Black-billed Talegalla on higher levels. Thus, both the northwestern Red-billed Talegalla and the northeastern Brown-collared Talegalla extend in range to the southern slope of the Central Range, meeting the Black-billed Talegalla without apparent intergradation, avoiding competition by showing an increased difference in size and by inhabiting different altitudes. However, the southern lowland species Black-billed Talegalla has extended its range also, reaching northern New Guinea along the southern shore of the Geelvink Bay at 135–136°E. No other species of Talegalla is known to inhabit this area, but the Red-billed Talegalla occurs a little to the west and Brown-collared a little to the east, the non-overlap perhaps due to competitive exclusion. Both northern species differentiated in size when they invaded the range of the southern one, but in contrast to this the southern Black-billed Talegalla evolved in size and structure (but not plumage characteristics) towards the neighbouring northern species when crossing the Central Range: *T. f. meyeri* of the southern Geelvink Bay is larger and has a deeper bill than *T. f. occidentis* occurring just south of the Central Range, but these measurements of *meyeri* are very close to its neighbours *T. c. cuvieri* in the west and *T. j. jobiensis* in the east.

With Diamond (1972), the present authors conclude that *Talegalla* once consisted of a superspecies ring of three allopatric forms round the Central Range, of which the southern Black-billed species has invaded northern New Guinea, while the range of both northern species has extended towards the south, where the forms segregate altitudinally in the overlap zones. There is no evidence of any interbreeding in the area of overlap in the south or where the species touch in the north.

The Megapodius case

As mentioned in Chapter 2, Mayr (1938) listed a number of areas where two *Megapodius* forms were said to co-exist. However, he cast suspicion on most of these records, doubting whether there was any real evidence of overlap at all. The evidence came from only one or a few old specimen records of species said to have been collected within the range of another more common species, but obtained as trade skins without reliable data. As he regarded these reports of overlap as unconvincing, Mayr considered all forms of *Megapodius* to be allopatric (mutually exclusive geographically), forming subspecies of a single large polytypic species, with the exception of the Micronesian Megapode from the Palau and

Mariana Islands and the Polynesian Megapode from Niuafo'ou. Apparently, these species were too isolated to be included.

However, a reconsideration of the situation in areas where forms may perhaps overlap, or where they touch, has led the present writers to believe that many if not all forms of *Megapodius* are probably good species. Not only do some forms overlap without any or only limited signs of hybridization, but several forms which differ greatly in size, structure, and colour replace each other abruptly without mixing. These parapatric forms which do not interbreed point to the occurrence of reproductive isolation and, hence, each has attained species rank. They probably cannot overlap because their ecological requirements are similar in every respect, and thus they show competitive exclusion (Mayr 1963). Some forms hybridize, but the hybridization zone is narrow and the birds within it are individually highly variable, pointing to a secondary contact between two originally separated forms. As all forms of *Megapodius* recognized in this book have their own individual characteristics in colour, size, and (especially) relative length and structure of extremities, all are considered to be separate species, together forming a single superspecies, of which some members show marginal overlap.

Within some of the species recognized some minor variation occurs, often in body size (as expressed in weight, wing length, or bill length) or in colour, but not in relative length of tail, tarsus, or toes (which tend to differ between species). These minor variations are in some cases reason to recognize subspecies, for example, within the Dusky Megapode where populations differ in size, within the Nicobar Megapode and the Australian forms of the Orange-footed Megapode where populations differ in colour, or within the Micronesian Megapode and Philippine Megapode where they differ in both colour and size. In other species, the variations in colour or size are too slight for recognition of a separate subspecies, or the characters are too patchily distributed to deserve a name (for example, within the Melanesian Megapode or within the Indonesian populations of the Orange-footed Megapode); in some cases the various populations are still insufficiently known.

Based on distribution, Mayr (1938) united all *Megapodius* forms within a single polytypic species, *M. freycinet*, except for the Micronesian and Polynesian Megapodes. Based on structural characters measured for the purpose of this book, the most distinctly defined species is the Orange-footed Megapode, which differs greatly from the others in having a relatively long tail and tarsus but has short toes. The Vanuatu Megapode also stands apart; in proportions it does not seem to be close to the Melanesian Megapode occurring nearby, although they both have a relatively large extent of bare skin on the forehead. All remaining species agree fairly closely in structure and relative proportions, although some subgroups can be distinguished between which proportions gradually change. *Megapodius nicobariensis* and *M. pritchardii* seem to form a group, agreeing closely in relative proportions, head pattern, and crest shape. The *cumingii–bernsteinii–tenimberensis* group stands nearby, as does the *laperouse–freycinet* group to these, while the *forstenii–geelvinkianus–eremita* group is not as distant as the *decollatus* group. It is noteworthy that the Nicobar and Polynesian Megapodes seem to be structurally related. The close similarity in relative proportions and structure may also have evolved due to similar local requirements. On the other hand, they may be related descendants of an old wave of early *Megapodius* stock which colonized the Indo-Pacific region from a centre of origin in northern Australia and/or southern New Guinea, with the other groups forming subsequent waves (Roselaar 1994).

If this theory is correct, the Orange-footed Megapode, which lives in the centre of the supposed origin of the group, may be the next successful species starting to invade the range of earlier ones. This is more or less supported by its present-day large geographical range with only a limited amount of pronounced geographical variation. This limited variation may be due to a more intense gene flow even between populations at the extreme ends of its distribution. However, it is more likely due to a geologically recent spread westwards and northwards from a centre in southern New Guinea (towards Vogelkop peninsula and the Lesser Sunda Islands) or perhaps from northern Australia (especially as birds from Sumba are much nearer to *tumulus* from northwest Australia in colour and size than to nominate *reinwardt* from southern New Guinea). The Orange-footed Megapode is still colonizing islands, with the recent occurrence of vagrants on Lucipara and Penju Islands as well as (apparently) recent colonization of Gunung Api and Manuk, all isolated islands in the Banda Sea (G. F. Mees and K. J. de Korte, personal communication). Other species show similar behaviour. The Micronesian Megapode, for instance, is in the process of recolonizing some islands over distances up to 100 km in the Mariana Islands. During this supposed recent spread, the Orange-footed Megapode may have displaced other forms when invading their range, unless these latter were larger and/or better adapted to the local environment. This may have been the case with the large Tanimbar Megapode *M. tenimberensis*, which is now completely surrounded by the Orange-footed Megapode and no longer in touch with its supposed relatives. The disjunct distribution of the presumably related Forsten's Megapode (on the southern Moluccan islands) and the Biak Megapode *M. geelvinkianus* (on Biak and surrounding islands in the Geelvink Bay) may be due to extinction of an intermediate form on the intervening Vogelkop peninsula, which was ousted by the expanding Orange-footed Megapode.

The following areas of contact between various *Megapodius* species occur:

1. Areas of marginal sympatry, with or without a limited amount of hybridization.

(a) The New Guinea Megapode *M. decollatus* occurs locally on the southern slopes of the Central Range of New Guinea, within the range of the Orange-footed Megapode, both in western New Guinea (Mimika–Setakwa area) and in Papua New Guinea. The New Guinea Megapode seems to inhabit higher areas here than the Orange-footed Megapode, parallel to a similar overlap in these areas between various species of Talegalla. Both may also occur on the southern shore of the Geelvink Bay, but more information is needed.

(b) The Dusky Megapode, subspecies *oustaleti*, occurs on small islands and on the low-lying coast of the northwest Vogelkop peninsula of New Guinea. The Vogelkop is also inhabited by the Orange-footed Megapode. From Sorong, both *M. f. oustaleti*, *M. r. reinwardt*, and some hybrids between them are known, but it is uncertain whether this 'Sorong' refers to Sorong town on the mainland or to Sorong islet just offshore. This hybridization seems to be secondary, the hybrids being highly variable in appearance. It may be noteworthy that these hybrids strongly resemble the New Guinea Megapode in plumage, but not in measurements. Apparent stragglers of the Orange-footed Megapode are known from Batanta and Salawati, within the range of the Dusky Megapode, but no hybrids have been reported from here. If the large Orange-footed Megapode is expanding, it may oust the small Dusky Megapode, subspecies *oustaleti*, to marginal habitats or it may eventually swamp it completely.

(c) A single specimen of the Biak Megapode is known from Dorei (now Manokwari) on the northeast Vogelkop peninsula, within the range of the Orange-footed Megapode. This specimen is similar to birds from nearby Numfoor Island. Perhaps the Biak Megapode occurs in the coastal belt of the western Vogelkop as well as on small islands just offshore, as does the Dusky Megapode in the west, but the specimen may have been a straggler or was perhaps shipped by local people.

2. Areas where the ranges of two species are contiguous, but due to competitive exclusion they do not mix, or one of them shows a very slight introgression (inclusion) of characters of the other species at most. It should be noted that most species occur on islands and, thus, ranges are not really contiguous. However, in view of good colonization capabilities (for example, colonization by the Orange-footed Megapode of some Banda Sea islands and recolonization of islands by the Micronesian Megapode mentioned above) one should expect mutual influence on islands lying near to each other.

(a) The Tanimbar Megapode on the Tanimbar group of islands is surrounded by the Orange-footed Megapode. No Tanimbar Megapodes examined showed any influence of the Orange-footed Megapode, although the former species is still poorly known.

(b) Forsten's Megapode on Buru, Ceram, and Gorong occurs as close as 20 km from Orange-footed Megapodes on Kasiui in the Watubela group of islands, without evidence of influence.

(c) The Sula Megapode on Peleng (in the Banggai group of islands) is only 10–15 km from the mainland of Sulawesi, where the Philippine Megapode, subspecies *gilbertii*, occurs. However, there is no proof yet of the occurrence of *M. c. gilbertii* on the eastern peninsula of Sulawesi opposite Peleng. This area is ornithologically poorly known and, thus, information about the possible influence of Peleng birds on those of Sulawesi is still lacking.

3. Secondary hybridization between two species.

(a) Populations of Karkar and Bagabag Island, off northeast New Guinea, are a variable mixture of the New Guinea Megapode of northern New Guinea and the Melanesian Megapode of the Bismarck Archipelago and the Solomon Islands (Mayr 1938; Diamond and LeCroy 1979). Apparently, the islands have been colonized from two directions.

(b) Birds of Mios Num in the Geelvink Bay are apparently intermediate between the Biak Megapode and the New Guinea Megapode. Those of Numfoor are variable intermediates also, but nearer to the Biak Megapode. Birds of Japen are New Guinea Megapodes, perhaps with some introgression of Biak Megapodes.

(c) Birds of the Trobiand, D'Entrecasteaux, and Louisiade islands, off the eastern tip of New Guinea, are intermediate between the Melanesian and the Orange-footed Megapode to a varying extent. In general, birds near the northern and eastern fringe of the archipelago are nearer to the Melanesian Megapode in colour, those nearest to mainland New Guinea nearer to the Orange-footed Megapode. Despite the study of Mayr (1938), the variation in size is not yet elucidated. Originally, the Melanesian Megapode was probably the only inhabitant of the islands, and this species gradually mixed with the Orange-footed Megapode when the latter expanded its range. The orange-yellow leg colour (a character of the Orange-footed Megapode) apparently spread more rapidly than the body colour, as birds examined from Goodenough and Duchateau Islands (not far from the mainland) are nearer to the Melanesian Megapode in colour, while yellow-orange legs occur over all the islands.

(d) The situation in the mountains of southeast New Guinea is not clear. Here, birds occur which show yellow-orange legs (like the Orange-footed Megapode), but which are otherwise very dark and have far heavier feet than any Orange-footed Megapode. One may suppose that they are hybrids of the Orange-footed Megapode with another species. However, the birds are darker than one would expect for a hybrid with the New Guinea Megapode, the only species occurring in eastern New Guinea. Perhaps the dark colour is a remnant character of an unknown species formerly occurring in eastern New Guinea, but now swamped by the expanding Orange-footed Megapode.

(e) The Dusky Megapode from the Obi group of islands in the northern Moluccas differs slightly in colour and size from the neighbouring populations of the same species on Bacan and Halmahera. The difference is probably due to some slight gene flow of Forsten's Megapode of the southern Moluccas, even though the nearest point of Ceram is some 90 km away.

Taken as a whole, the instances of sympatry and parapatry (occupying different but contiguous geographical areas), as well as the secondary character of hybridization between the various forms of *Megapodius*, have convinced the present writers that all these forms are separate species, and not subspecies of a single polytypic species.

4

General biology and behaviour

Introduction

This chapter provides an overview of biological, ecological, and behavioural characteristics of the family. These are discussed in four sections: annual cycles, feeding ecology, behaviour, and communication. Behaviour specifically related to reproduction will be dealt with in Chapter 7. Mounds and other incubation sites are the subject of Chapter 5.

Annual cycles

Eggs and clutch size

Megapode eggs are large and heavy compared with eggs of birds of equivalent size. The average weight ranges from about 75 g for eggs of the Polynesian Megapode *Megapodius pritchardii* and the Micronesian Megapode *M. laperouse*, to about 230 g for eggs of the Maleo *Macrocephalon maleo* (Table 4.1). The eggshells are mainly dull white or cream. However, the shells of all species except the Brush-turkeys (*Aepypodius* and *Alectura*) are covered with a pinkish-brown powder. This powder flakes off during incubation, revealing a white shell beneath. Older eggs, which tend to become dirty during incubation, can therefore easily be distinguished from the normally pristine pink of freshly laid eggs. The eggs are mainly elliptical in shape, but tend to be more elongated in some species. Length to width ratios of eggs range from 1.53 to 1.72 (Table 4.1). Variation in size and shape of eggs is considerable, however, both within species and within an individual female's clutch.

Table 4.1 Relative shape, and weight as a percentage of body weight of some megapode eggs

Species	Average relative length/width	Average weight (g)	Relative egg weight (%)
Australian Brush-turkey	1.59	180	10.3
Wattled Brush-turkey	1.53	192	13.3
Malleefowl	1.56	173	9.5–10.2
Maleo	1.72	231	13.8–17.6
Moluccan Megapode	1.63	106	18.4–21.0
Micronesian Megapode	1.58	77	22.0
Polynesian Megapode	1.67	75	18.0
Melanesian Megapode	1.65	102	15.8–18.0
Orange-footed Megapode	1.62	126	±20.0

Megapode eggs contain very large amounts of yolk, ranging from 48 to 69 per cent of the weight of egg contents, as shown by Dekker and Brom (1990). These are among the highest yolk proportions of any bird, and relate to the extreme precocity of megapode hatchlings (see Chapter 6). Using the data on yolk content, Dekker and Brom calculate that species using incubation mounds have significantly lower yolk proportions (48–55 per cent for the Brush-turkeys and Malleefowl *Leipoa ocellata*) than the burrow-nesting species (greater than 61 per cent for the Maleo, the Moluccan Megapode *Eulipoa wallacei*, and various *Megapodius* species). Not only are megapode eggs large in an absolute sense, they also represent a major investment in egg mass by females relative to their body weight; single eggs represent 9.5–22 per cent of female body weight (Table 4.1).

Discussion about the clutch size of megapodes raises a number of difficulties. First, the term 'clutch' normally refers to a discrete set of eggs that are incubated together and result in a similarly discrete brood of hatchlings. The nature of the megapode incubation, however, is such that hatchlings emerge quite independently of each other, and develop no obvious social relationship to their parents or to each other.

In addition, obtaining useful data on eggs can be problematic. Not only are the eggs deposited deep within the substrate and therefore difficult to locate and count, they are also laid singly at varying intervals and over a breeding period that may be many weeks or months in duration. Moreover, in some species, females may use the same incubation site communally or lay in several sites during the same season. Furthermore, the number of eggs produced by an individual is probably determined seasonally through balances of energy, protein and calcium, and the length of the effective breeding season (see below), rather than being determined by selection for some optimal brood size as in many other birds. Thus, the number of eggs produced per female may vary considerably between years; at times of severe environmental stress egg laying may cease completely. These features render most estimates of 'clutch size' in megapodes unreliable.

The few sound data on egg numbers indicate that females commonly produce in excess of their body weight in egg mass over a typical breeding season. With an annual production of 15–24 eggs, female Malleefowl produce 150–250 per cent of their adult body weight each year (Vleck *et al.* 1984). Maleos produce an estimated 8–12 eggs per season (Dekker 1990*a*) or approximately 120–180 per cent of female body weight, and Australian Brush-turkeys *Alectura lathami* up to three times female weight (Baltin 1969). However, the larger of these figures relate to years of maximum productivity of certain individuals; during any season the number of eggs produced by females may vary widely. Captive megapodes supplied with a high quality diet often lay significantly more eggs than wild birds: the largest number recorded for an Australian Brush-turkey was 56 eggs during a single breeding season (Coles 1937).

Laying is relatively continuous throughout the breeding season. Each egg is laid individually at intervals ranging from 2 to 9 days in Australian Brush-turkeys (Baltin 1969) and averaged about 13 days in Orange-footed Megapodes *Megapodius reinwardt* (Crome and Brown 1979). Although very poorly documented, the interval appears to become longer towards the end of the breeding season. These intervals are consistent with the rate of tissue formation of such large eggs and similar to those of other galliforms. There is also some evidence that the size of eggs declines late in the season, which, along with the greater interval between eggs, is probably related to depletion of reserves.

Breeding seasons

There are several problems associated with the definition of the beginning, end, and duration of the breeding season among megapodes. Even for the mound-building species, where there is conspicuous reproductive activity in the form of mound building, there may be great differences in the duration of mound attention and the actual period of egg laying. Many species appear to tend their mounds to varying degrees throughout the year, yet may lay eggs over a limited period. Frith (1959a) defined the breeding season as the period during which eggs are being incubated. This is often difficult to determine without seeing eggs being laid or *in situ*. Most of the references in the literature are less specific, often simply mentioning sightings of birds at the incubation site. In this chapter the 'breeding season' refers to the general period of attendance of birds at the incubation site. The period of egg laying, if known, will be referred to specifically.

The age of sexual maturity is not known for most species. Australian Brush-turkey females are known to have bred within their first year, but maturity is later in other species (see under 'Captive breeding' in Chapter 9).

The breeding seasons of many megapodes are determined by environmental features that are mostly independent of the usual seasonal determinants of reproductive phases, such as day length. In general, the period of reproductive activity is limited by climatic conditions, especially rainfall. The importance of these conditions varies with the source of environmental heat utilized for incubation. There are two major ways in which these sources of heat are used: by the construction of mounds of decomposing material and by burrowing into existing warm substrates (see Chapter 5).

Mound builders

The breeding seasons of species that utilize the heat generated by decomposing organic matter are determined to a large extent by the local rainfall pattern. The rate of microbial activity is initially influenced by ambient temperatures and the moisture levels of the material being decomposed (see Chapter 6). Once widespread decomposition has commenced within the mound and assuming there to be no shortage of suitable organic material, the generation of heat will be governed by the amount of moisture in the substrate.

For both Malleefowl and Australian Brush-turkeys the period of mound attendance extends throughout the winter, spring, and summer (May to February), that is, 8–10 months, provided that the leaf litter added to the mound is sufficiently moist. This is most critical at the start of the season when moist material is gathered into the mound. Thereafter, moisture can be conserved fairly effectively through the siting of the mound and the use of outer insulation layers of dry material. Jones (1988a) proposed that mound building in Australian Brush-turkeys followed the first falls of rain in excess of 100 mm occurring after May (late autumn). Similarly, Malleefowl commence the excavation of the depression over which their mounds are formed (see Fig. 5.2) following the normal autumn (March–April) rains. Should these rains fail to occur no further mound construction takes place and reproductive activities cease completely.

In all mound-building species, the first evidence of breeding activities is usually the gathering of moist material for mound construction. This does not, however, normally coincide with the start of sexual activity. Environmental variables used by males as cues indicating suitable conditions for mound construction may not be the same as those used by females to prepare for reproduction. In both Australian Brush-turkeys and Malleefowl, early rains often lead to an abortive start to mound construction without any evidence of female sexual receptivity (D. N. Jones, unpublished data). During a severe drought, Booth and Seymour (1984) added water to several Malleefowl mounds that were being tended but had failed to generate adequate heat. These mounds quickly reached temperatures suitable for incubation, but no females in the population entered reproductive condition.

While it is very likely that some minimal amount of rainfall, at least sufficient to moisten organic matter, is a prerequisite for mound construction, the cues that lead females to begin physiological preparations for egg laying are not known. None the less, under normal conditions, appropriate amounts of timely rains may correlate with increases in insect abundance, probably the most important nutritional requirement of females.

The influence of the rainy season on the breeding season of mound builders varies between species. In northern Australia, Orange-footed Megapodes construct their mounds during the dry season. Most eggs are laid between the first rains and the wet season proper, a period that may last for 2 months (Frith 1959a). In regions with a more equitable climate, some species appear to lay eggs throughout the year, although most show shorter periods of intensive activity. Even among species from the same area there may be differences in the timing of egg laying: in central New Guinea, for example, Dwyer (1981) reported that Wattled Brush-turkey *Aepypodius arfakianus* eggs were collected by local people during most months, while the sympatric Brown-collared Talegalla *Tallegalla jobiensis* laid eggs from November to February only.

The end of the breeding season is usually more difficult to determine. Frith (1956b) proposed that for many mound builders the end of the breeding season is brought about by the saturation of mounds associated with the start of the tropical wet season. This is certainly very likely in equatorial areas; indeed, the timing and duration of the breeding period for many species is confined to the months other than those of prolonged monsoonal rains (see Fig. 4.1). For species occurring in areas without extreme rainfall patterns (such as Malleefowl and Australian Brush-turkeys in southern Australia), however, there is often no distinct end to the season. Rather, mounds appear to be progressively neglected around the end of summer until very few are still in operation by early to mid-autumn. In tropical areas the influence of rainfall patterns on breeding seasons is much less predictable, although sustained monsoonal downpours will normally put an end to mound use.

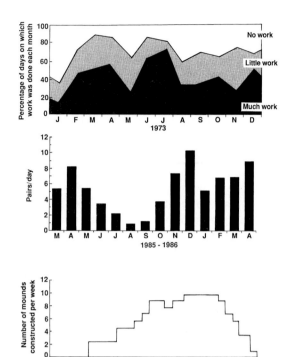

4.1 Seasonal use of incubation sites in three megapode species. *Top*: Orange-footed Megapode pairs tend their mound throughout the year although egg laying is limited to a period prior to the wet season (Sept to Nov in North Queensland) (from Crome and Brown 1979); *centre*: Maleos lay all year round at geothermal incubation sites with the greatest activity occurring in the dry season (from Dekker 1990a); *bottom*: Australian Brush-turkeys begin to construct mounds in winter and eggs are laid from late winter to late summer, the duration of the season being determined by rainfall patterns. (From Jones 1988a.)

Burrow nesters

Megapodes that have developed the ability to exploit the perennial heat produced by the sun or by geothermal activity would appear to have escaped the constraints on reproductive activities experienced by the mound builders. None the less, although the source of incubation heat is independent of the activities of the birds, all incubation sites are influenced to varying amounts by climatic features. Heavy tropical rains may flood burrows and dilute the warm moisture or fill air pockets surrounding the eggs with cool water. Where birds lay their eggs in burrows in sun-warmed beaches, prolonged periods of heavy cloud cover and rain may cool the sands. However, the importance of these events varies greatly with the topography and soil structure of the incubation site.

For burrow-nesting species the most frequent pattern of egg laying appears to be that of starting soon after the end of the wet season when the substrates have warmed. The actual timing of this period will depend largely on the pattern of the local monsoon. In New Britain, Melanesian Megapodes *Megapodius eremita* lay their eggs throughout the prolonged dry season (April–December) with a peak in June–July. Among Maleo populations in Sulawesi, Dekker (1988*d*) found that the more exposed coastal beach-sand sites were used during the dry season, while the inland hot spring sites were used to some extent throughout the year (Fig. 4.1). It is important to note that these prolonged patterns of egg laying do not mean that particular individuals produce eggs all the year round. Rather, these patterns result from the overlap of many birds using the sites for short, irregular periods.

Moult

Megapodes are unusual among galliforms in having feathered rather than downy plumage at hatching. This distinction is based on detailed studies of feather structure by Clark (1964*a*). He concluded that megapode hatching plumage was homologous to that of other galliforms but clearly different in structure. By the time a hatchling has dug its way to the surface of the incubation site, this juvenile plumage (the first generation of true feathering) is fully grown on the head and body. On the wing, the secondaries and the inner eight primaries are still growing. As in most birds, adult megapodes have ten primaries. In hatchling megapodes, however, the outer two primaries are absent, or are present only as tiny pins. The juvenile's tail is fully grown but the feathers are short and rather woolly. This juvenile feathering is the only plumage that is visually different from that of adults, showing a variable amount of barring with rufous or buff, unlike the usual uniformly coloured feathers of older birds. The head, which is more or less bare in adults of most species, is mostly feathered in the hatchlings.

Within a few weeks of hatching, when the inner primaries are fully grown and the outer two are beginning to lengthen, the post-juvenile moult starts. A second generation of feathers appears on the mantle, back, side of the breast, upper belly, and inner primaries. Moult of the primaries is descendant, beginning with the innermost primary (p1) and gradually moving to the outermost (p10). Moult of the secondaries is a combination of ascendant and descendant, with the moult moving inward from the outermost secondary (s1) and outward from the innermost.

When the immature reaches about one-quarter or one-third of adult body weight, the head and body are fully covered with a second generation of feathers, which is similar to adult feathering. Only the face of the immature is more heavily feathered than that of an adult. By the time the immature bird reaches about half of the adult body weight, moult of the second generation of primaries has reached the wing tip, usually about p7 at this age. Primaries p8 and p9 are in various stages of moult, and p10, a characteristically small and pointed primary, is still present. The distance between the tip of the second generation primaries (p1 and p2) is often pronounced at this age because p1 is relatively short compared to its length later in the bird's life. However, usually the innermost primaries of the second generation have already been replaced by those of the third generation. At this time, two primary moult waves may be active in each wing, leading to the simultaneous presence of three generations of feathers. In each of these generations or 'series', the outermost feather is the most recent, the inner the oldest.

This 'serially descendant moult' (or 'Staffelmauser') has been described in storks

(Ciconiidae), eagles (Accipitridae), bustards (Otididae) and many other large birds. In this type of moult, primary p1 is shed at the onset of a first moult season, but the moult wave only reaches p10 several moult seasons later. Smaller species with a more rapid moult and younger (non-breeding) birds have fewer generations of primaries in each wing than larger and older birds.

Moult may also be curtailed during periods of poor food availability and recommence when conditions improve, a phenomenon called 'suspended moult'. In immature megapodes, moult may stop when p3 and p9 are full size, to be resumed later with the shedding of p4 and p10. Simultaneously, a new (fourth) generation may start with the loss of p1. During suspended moult no feathers are actively growing. Each generation of feathers may be difficult to distinguish, especially in older birds where the older feathers of each generation differ little from a neighbouring feather from a different generation. In immature birds the difference in age of feathers between different generations is usually obvious but, unlike among adults, suspended moult is unusual. Suspension of the moult, with most body, wing, and tail feathers in the second generation but with some outer primaries still juvenile (as occurs in many pheasants, Phasianidae), does not occur in megapodes.

By the time an immature megapode is about three-quarters of adult body weight, the outermost primaries are of the second generation, those in the centre are third generation, and the innermost are of the third or fourth generation. This serially descendant moult continues in adults. In some of the larger species, each wing may show up to five generations of primaries.

Moult of the secondaries and tail is also serial in megapodes, showing several generations within each set of feathers. Secondary moult has not yet been studied in detail, mainly because the boundaries between the generations are difficult to detect. This is because moult may start simultaneously on either side of a feather set, and because secondaries are particularly difficult to examine in study specimens. Similar problems have delayed understanding of tail moult.

Feeding ecology

There have been no detailed studies of the diet of any megapode species, but only a few reports giving the stomach contents of small numbers of birds. Although there are numerous accounts of the types of food taken, very few of these are quantitative. Even in the few reports of stomach or crop contents, differential rates of digestion and the disappearance of any trace of various types of food taken can make the data unreliable. Therefore, the following discussion as well as the details on diet provided in the Species accounts must be regarded as preliminary.

Megapodes forage by raking and scratching among the debris on the ground, opportunistically taking food items as they are exposed. A wide variety of food items has been reported, including items of both plant and animal origin; in all species, that diet may be described as general and omnivorous.

The two types of food most frequently mentioned are leaf-litter invertebrates (a wide variety of insects and freshwater and terrestrial snails being the most important) and plant material (primarily fruit and seeds). However, there are also numerous references to unexpected food items such as ants, scorpions, centipedes, phasmids (stick insects), and even small snakes.

There are many accounts of Australian Brush-turkeys feeding on fruit in the tree canopy (for example, North 1914), on one occasion 15 m above the ground (Gill 1970). Certainly, this species appears to be particularly versatile in its feeding methods; birds have been reported flying up to tree-borne fruits to dislodge them with their wings (for example, North 1914) and seizing the fruit of prickly pear *Opuntia* sp. while flying over the plants (Brookes 1919a).

The unique reproductive adaptations of the megapodes impose serious and specific physiological stresses on breeding birds and these are likely to differ significantly between the sexes. For females, the production of many large eggs throughout each prolonged breeding season will be dependent upon an adequate intake of certain nutritional components, especially calcium and protein. The availability of these foods is likely to place important limitations on female fecundity.

In general, the nutritional and energetic requirements of female megapodes appear to necessitate prolonged periods of foraging. In most situations this inevitably means that the female must spend significant amounts of time away from the incubation site. Whether the male also leaves to accompany the female depends on whether the incubation site is defended. It is likely that this feature of female physiology has had a primary influence on the type of mating system exhibited by each species (see Chapter 7). In some monogamous species, it is clear that males assist their mates in locating food items; in the Maleo, Polynesian Megapodes, and Nicobar Megapodes *Megapodius nicobariensis* males have been observed offering food items to females (see Species accounts).

For males, the most obvious nutritional need is for foods providing sufficient energy for the sustained work required for the construction and maintenance (and in some cases the defence) of the incubation site. Of course, this will vary greatly among the species: a male Malleefowl, working his mound for many hours daily for much of the year, is presumably more prone to energetic stresses than, for example, a male Maleo sharing the digging of a nesting hole in a beach perhaps once a week.

The ease with which males are able to meet the energetic requirements associated with utilizing their incubation site also varies with the environmental conditions of the species' range. Again the most extreme example is likely to be that of Malleefowl; its dry, arid habitat must impose particularly serious stresses on both sexes. For the majority of megapode species, however, their tropical habitats probably provide much easier foraging opportunities, at least outside the dry season.

The environmental features of incubation sites may also influence the foraging activities of the birds. Many solar or geothermal areas that are utilized as communal nesting sites are some distance away from suitable foraging areas. For example, Melanesian Megapodes in New Britain and Maleos in Sulawesi only visit their nesting grounds briefly for egg laying before returning to their feeding territories in the adjacent rainforests. On a smaller scale, the process of gathering leaf litter for building and maintaining an incubation mound often leaves the surrounding area entirely bare. In such cases, both sexes must leave the immediate area to feed, although some food may be obtained from the mound itself. During the breeding season, Australian Brush-turkey males usually do not feed during the first few hours of the day but remain at their mounds, apparently awaiting the arrival of females. This prolonging of the overnight fast must increase their nutritional stress; this phenomenon appears to be uncommon among birds where feeding soon after waking is normally vital. It is not known whether this also occurs in other mound-defending species.

Not only are the nutritional requirements of males and females likely to differ markedly, but the capacity of different environments to supply all the birds' needs will tend to differ also. Because reproduction in megapodes seems to be limited primarily by the number of eggs females are able to lay, the best option would appear to be to allow females unrestricted foraging. Comparisons of fecundity in species where females forage alone and those where males accompany them would be of interest, particularly in species where males enhance the female's feeding success by locating and bringing food items. Such feeding behaviour has been described so far in three species (see above) but needs to be investigated in other monogamous species.

As in many other aspects of megapode ecology, virtually nothing is known about the feeding activities of chicks. In captivity, they quickly dispose of a wide variety of live foods and will peck at anything unusual.

Behaviour

The postures adopted by megapodes during most individual activities such as preening, dust bathing, perching, roosting, and feeding appear to be similar to those of other galliforms, but no detailed comparisons have yet been made. There are, however, numerous postures and behavioural characteristics exhibited by megapodes that are worthy of special attention. These relate to the behaviour of birds at incubation sites and during social interactions.

Incubation-site behaviour

The use of the feet in scratching aside leaf litter during foraging has particular significance among mound-building megapodes, having been expanded into the technique used in gathering material for mound building or digging a nest burrow. There appear to be no differences in the digging or scratching behaviour of males and females, although only one sex may be involved in mound work in some species. The pattern of raking described for Australian Brush-turkeys (Baltin 1969), indicating no preference for the right or left leg and four to ten strokes in succession per leg, is similar to that of Maleos and numerous *Megapodius* species.

All megapodes exhibit behavioural characteristics associated with determining the temperature of the incubation site. In all species this involves inserting the head into the warm material at the base of narrow holes excavated into the core of the mound. This 'temperature testing' is always performed by females prior to laying but is also common for males during their regular maintenance of the incubation site.

Despite many assertions that various parts of the anatomy are used in temperature assessment, for example the foot (Coles 1937), the bare skin of the head (Fleay 1937), the underside of the bill (MacDonald 1973), and the neck sac (Aagard 1980), the actual site has yet to be confirmed by histological studies. It is most likely that the palate or tongue is involved, since many species regularly take a bill-full of substrate during this activity.

Maintenance of mounds involves the gathering of fresh material and mixing this into the central section of the structure. This work may be done solely by the male (in the Brush-turkeys), or jointly by members of a pair (as in Orange-footed Megapodes). Non-mound incubation sites usually require little or no work to maintain.

It is usually the female that digs the hole into which the egg will finally be deposited, though the male may also be involved to some extent. However, the assistance of males varies from an approximately equal share in work among Maleos and Orange-footed Megapodes (MacKinnon 1978; Crome and Brown 1979) to no help or even deliberate disruption of the female (see below) in Australian Brush-turkeys. In the Malleefowl, the male removes the vast insulation layer of sand from the mound but leaves the digging of the actual laying hole to the female. In several species utilizing geothermal areas (for example, Polynesian and Micronesian Megapodes), females appear to dig their holes entirely alone, even though their mates are probably nearby.

Aggressive interactions

Megapodes exhibit a complex array of behavioural traits associated with dominance and submission, competition, and defence. Interactions may often be intense, although fighting is not usually a common part of them on a daily basis. The most serious aggressive behaviour occurs during conflicts among males over a defended resource. The most obvious example is a dispute over the ownership of a defended mound, but fights over certain prized food resources are not unusual in some species. Aggressive acts include pecks, bites, and wing blows. Cockfights are the most extreme form of these interactions. They are not limited to males, however; female Australian Brush-turkeys often fight viciously if competing for access to a particular mound. These fights are rarely prolonged and usually rapidly turn into a chase.

Submission is typically signalled by a withdrawn neck and movement away from the dominant bird. In species with coloured neck sacs or collars, withdrawal of the neck hides these prominent colours from the aggressor, often leading to cessation of the interaction.

Aggressive interactions are not limited to intrasexual conflicts. In both Australian and Wattled Brush-turkeys, males commonly peck and buffet females seeking to lay in their mounds. These acts appear not to be intended to drive the female away, but may be an attempt to ensure that only females with eggs stay at the mound. Additional copulations, often unsolicited, may occur while the female is excavating the laying hole. Once the egg has been laid, however, the female is promptly chased from the mound. Among captive Australian and Wattled Brush-turkeys, where females are unable to escape the harassment of the male, prolonged interactions of this kind can lead to the death of the female (Seth-Smith 1934).

Among the few consistent observations made of megapode chicks in general is their antisocial behaviour. At least during their first few weeks of life hatchlings avoid each other, commonly giving threat displays similar to those of adults when meeting other chicks. Interactions between adults and chicks appear to be non-existent.

Communication

Visual signals

Although vocalizations are the most obvious form of communication, several species of megapodes also employ skin colours of head and neck, inflatable neck sacs, and head combs as communication signals. This is most evident among the Brush-turkeys, in which all three species possess conspicuous neck sacs, wattles and/or combs as well as extensive areas of highly coloured skin on their heads and necks. Although other megapode species also have coloured facial skin (see the Plates), only the Brush-turkeys possess such large areas of brightly coloured skin or ornaments. Moreover, in each of the Brush-turkey species, the intensity of the colour of the skin becomes considerably heightened during the breeding season and especially during sexual activities. This occurs in both sexes. Similarly, the neck sacs and combs of males become greatly enlarged and more pronounced during the breeding season. These structures can also be elongated at will by the male, and are extended fully during both sexual encounters with females and aggressive interactions with other males.

Vocalizations

Megapodes produce a variety of vocalizations and these are described in detail in the Species Accounts. Unfortunately, there are few studies of the functions of these calls in the wild and no terminology has been suggested that allows useful comparisons between species. They may be loosely described as either clucking, crowing, or booming. In this section, some of the main vocalizations that have been described are discussed briefly in terms of their possible function.

All megapode species appear to produce typical galliform-like sounds including 'clucks', 'squawks', 'gulps', and 'grunts'. These are often described as 'contact calls' but may be uttered in other contexts as well. They are generally of a low frequency, deep and low in pitch and volume, and mainly given with a closed bill. In most situations these calls are associated with intraspecific communication over a short distance. This may be between paired individuals or among birds aggregated at a feeding site or roost. Low grunts have also been detected among groups of very young birds in captivity. Clucking is common among foraging pairs, and appears to assist in maintaining contact between the individuals (for example, in various *Megapodius* species; Lincoln 1974). Malleefowl pairs grunt quietly while working together at the mound (Frith 1959b). In another context, Australian Brush-turkeys cluck when disturbed or in the presence of a potential predator (Jones 1987b).

A number of the megapodes produce very loud, raucous, and repetitive calls that may be heard over considerable distances. In several species these vocalizations have been termed 'crowing'. When giving these calls, the body is typically strained upwards, the neck outstretched and the bill wide (Fig. 4.2). These calls may be given at any time during the year but are primarily associated with the breeding season. Often a number of birds are involved and the sound produced may be very loud. Maleos crow noisily from the trees surrounding a nesting site during the afternoon and early morning preceding egg laying, but are also heard calling from their feeding territories well away from the incubation sites (R. Dekker, personal observation). Orange-footed Megapodes and others in the genus are particularly vocal, calling both day and night throughout the year. Many mound-building species (including numerous *Megapodius* and *Talegalla* species) also crow when nearing their mounds or when foraging near the boundaries of a territory. These calls are often given repeatedly by the pair, and are answered by neighbouring birds. The function of many of these calls has been interpreted as being territorial, with birds advertising their presence in feeding territories or incubation sites to potential competitors within hearing range (Crome and Brown 1979; Coates 1985). However, it is probable that these loud vocalizations also serve as contact calls between neighbouring birds, allowing pairs to assess the movement and locations of other pairs. Many of the species producing crowing vocalizations deliver these calls as a duet by paired birds; as in many other duetting bird species, it is likely that these megapodes are monogamous.

Another group of vocalizations is given by only two species, the Australian Brush-turkey and the Malleefowl (Frith 1959b; Jones 1987a). These vocalizations, usually described as 'booming', are produced by the forcing of a large volume of air through the vocal chords. The resulting sound is a deep, resonant 'boom'. The call is preceded by the inflation of air sacs in the neck and, in Australian Brush-turkeys, by pumping up the large neck wattle (Fig. 4.3). When delivering the call, both species bow, with the body horizontal; in Malleefowl, the head almost touches the ground beneath the bird's breast (Fig. 4.4).

The function of this call appears to differ markedly between the Australian Brush-turkey and Malleefowl. In Australian Brush-turkeys, booming occurs in two distinct contexts: during male–male interactions, and also when males are attending a mound in the absence of females. This suggests two separate functions for this call, with the former being a form of social signal and the latter being used for ad-

4.2 The posture of a male Wattled Brush-turkey during the crowing vocalization.

4.3 The posture of a male Australian Brush-turkey during the booming vocalization.

4.4 The posture of a male Malleefowl during the booming vocalization.

vertising the male's presence at the mound, possibly as a method of attracting females. Similar functions have also been suggested for the crowing vocalization of Wattled Brushturkeys, a species whose behaviour appears to resemble closely that of Australian Brushturkeys (Kloska and Nicolai 1988). In Malleefowl, the boom is given by the male on arrival at the mound or when foraging nearby and may be a form of contact call between paired birds (Frith 1959*b*). Although Böhner and Immelmann (1987) described Malleefowl delivering booming calls as part of a duet, their observations appear to be atypical for the species (D. Priddel, personal observation).

5
Megapode incubation sites

Introduction

The paramount evolutionary achievement of the megapodes has been their utilization of naturally occurring heat-producing phenomena for the incubation of their eggs. The implications of this development have been profound, influencing virtually every aspect of megapode physiology, ecology, and behaviour. The physiological adaptations that have accompanied the evolution of megapode incubation are detailed in Chapter 6 and ideas on possible origins of this process are dealt with in Chapter 8. Here, the types of heat sources used for incubation within the family are described.

Incubation heat sources

The heat utilized for incubation is generated by three main sources:

(1) microbial respiration, where heat is produced through the decomposition of organic matter by micro-organisms;
(2) geothermal activity, where soil is warmed by proximity to hot springs or hot gases associated with volcanism;
(3) solar radiation, where substrates such as sandy beaches or soil are warmed by the sun.

Generally, those megapode species that construct incubation sites (in order to harness the heat of microbial decomposition) are termed 'mound builders', while those exploiting geothermal or sun-warmed sands are termed 'burrow nesters'. While these terms are retained here, they are somewhat simplistic: all megapodes, including those constructing mounds, must burrow into the incubation site in order to deposit their eggs in substrate at an appropriate incubation temperature. Similarly, it is misleading to imagine that only mound builders actively work on the incubation site. Many burrow-nesting species remove and replace considerable amounts of material during egg laying and some also rake organic materials into their burrows. Burrow nesters do not, however, construct mounds of any size. Table 5.1 gives a summary of the types of incubation sites used by the various species (see Species accounts for further details).

Mound builders

Most species of megapode construct some form of mound of decomposing organic matter as an incubation site. The details of composition, construction, and maintenance, however, vary widely among the species. None the less all incubation mounds are designed to concentrate, enhance, and prolong the normal processes of microbial decomposition occurring within the leaf litter of the species habitat. In all species leaf litter is gathered when damp, piled into large mounds, and usually covered in some form of drier material that assists in the retention of heat as well as the conservation of moisture.

Table 5.1 Incubation methods of megapodes

Species	Incubation method
Australian Brush-turkey *Alectura lathami*	M
Wattled Brush-turkey *Aepypodius arfakianus*	M
Bruijn's Brush-turkey *Aepypodius bruijnii*	M?
Red-billed Talegalla *Talegalla cuvieri*	M
Black-billed Talegalla *Talegalla fuscirostris*	M
Brown-collared Talegalla *Talegalla jobiensis*	M
Malleefowl *Leipoa ocellata*	M
Maleo *Macrocephalon maleo*	Bg, Bb
Moluccan Megapode *Eulipoa wallacei*	Bb, Bg?
Polynesian Megapode *Megapodius pritchardii*	Bg
Micronesian Megapode *Megapodius laperouse*	M, Bg
Nicobar Megapode *Megapodius nicobariensis*	M, Mr
Philippine Megapode *Megapodius cumingii*	M, Mr, Bb?
Sula Megapode *Megapodius bernsteinii*	M
Tanimbar Megapode *Megapodius tenimberensis*	M
Dusky Megapode *Megapodius freycinet*	M
Biak Megapode *Megapodius geelvinkianus*	M?
Forsten's Megapode *Megapodius forstenii*	M
Melanesian Megapode *Megapodius eremita*	M, Mr?, Bg, Bb
Vanuatu Megapode *Megapodius layardi*	M, Mr, Bg?
New Guinea Megapode *Megapodius decollatus*	M, P
Orange-footed Megapode *Megapodius reinwardt*	M

Incubation method: M, incubation mound; Mr, decaying roots of trees; B, burrow nesting in: g, geothermal sites; b, solar-heated beaches; P, known to parasitize other species' mounds; ?, incubation method needs confirmation or is undescribed.

Contrary to previous suggestions that decomposition is due to fermentation, it is now known that the heat produced results entirely from microbial respiration; no part of the mound is oxygen deficient as is necessary for fermentation (Seymour and Ackerman 1980). Although very little is currently known about the composition of these diverse communities of microorganisms, various types of heat-adapted fungi appear to be particularly prolific (Seymour 1985). (See Chapter 6 for further details.)

The composition of mounds is determined directly by the environmental condition of the site. Mounds are built in locations providing suitable material for decomposition; it is the leaf litter immediately surrounding the mound that is gathered for incorporation into it. Even within apparently uniform habitats certain locations may be favoured. Australian Brush-turkeys usually place their mounds on the earthen bases marking old mounds rather than in new locations (Jones 1988a). This leads to a recycling of frequently used sites, though the same site is rarely used by the same male in consecutive years. As well as being very shady, sites with dense canopies of leaves above the mound also provide a good source of future leaf litter. Also, Australian Brush-turkeys avoid areas dominated by decomposition-resistant *Eucalyptus* trees (Jones 1988a). The substrate type may also be important. In one area Malleefowl sited their mounds preferentially on sandy rather than loam or clay soils, apparently because of better drainage features (Frith 1959b); in other areas, however, no preferences were detectable (Booth 1987b).

All mounds are made of an admixture of organic matter, usually leaf litter, and varying amounts of soil, sand, or, in certain species building close to beaches, coral debris. Other materials such as large sticks, stones, or human refuse are often included, presumably inadvertently. The ratio of litter to substrate will depend on the age of the mound, the composition of the available material surrounding the mound, and the individual constructing the mound. Mounds containing large proportions of organic matter may decompose more rapidly than those with a greater ratio of soil. Similarly, mounds incorporating many large sticks are likely to be less suitable as incubators. Although some writers have attributed certain mixtures of these materials to particular species, the variation between and within species is so extreme that this seems of little value.

Mounds can vary greatly in size, shape, and dimensions depending upon the age of the mound, the material used, and other features of the immediate location. Thus there is considerable variation in the size of mounds within species. This may also reflect differences in the construction style or abilities of individuals. For example, several individually marked Australian Brush-turkeys constructed mounds each year that were visually distinctive in shape and composition (S. Birks, personal observation). Other variability results from the common habit among some species of siting mounds on the remains of mounds used during previous seasons. These bases of compacted soil vary greatly in size. Another important feature of many species are the smaller, often incomplete mounds built by younger, inexperienced males.

The largest mounds are those that have been used repeatedly, with new material being added year after year. Orange-footed Megapodes construct the largest mounds of any megapode; new material is added each breeding season leading to truly enormous structures (see Fig. 5.1). Small hillocks 12 m across and 5 m high are not uncommon in northern Australia, and may represent incubation sites used continuously for extremely long periods (Frith 1956b). Reports of continual use for in

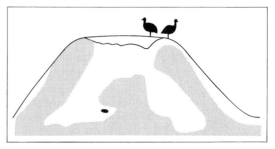

5.1 Structure of the mound of the Orange-footed Megapode on Komodo, showing the compacted material and the burrows leading to the egg chamber. (From Lincoln 1974; used with permission.)

excess of 40 years are not unusual (Banfield 1913). Indeed, Stone (1991) has recently dated 'fossilized' Orange-footed Megapode mounds at more than 1500 years old, remaining as permanent parts of the environment. In general, the mounds of other *Megapodius* species are somewhat smaller, 1–3 m high and 6–10 m in diameter.

Other megapode species construct new mounds each year. These mounds are usually less than 2 m in height and of a domed cone shape. These mounds are typical of species from inland tropical rainforest such as the brush-turkeys and talegallas. Jones (1988a) described two phases of mound building in the Australian Brush-turkey, and similar processes probably occur in the other species that construct new mounds annually. These are the 'construction' phase, during which material is gathered into the growing mound but little work is done on the mound itself, and the 'maintenance' phase, when the internal temperature has stabilized allowing the mound to be used as an incubator. The shape of the mound during these phases is characteristic: construction phase mounds are conical with a definite apex, while during maintenance activities the cone is flattened to a plateau shape. It is possible that the confusingly diverse descriptions of the mounds of brush-turkey and talegalla species may represent mounds in differing phases (see Ogilvie-Grant 1897; Barrett and Crandall 1931; Coates 1985).

Mounds may be constructed entirely by males, or by a pair with varying amounts of assistance from the female. This pattern will be determined by the mating system of the particular species (see Chapter 7). Among the mound-building species that are monogamous, both sexes work on the mound to a similar extent and typically remain together at all times. The exception is the Malleefowl in which the male remains near the mound whereas the female forages away from it rather than staying near her mate. However, early in the breeding season both Malleefowl work together on the mound for long hours (Frith 1962a).

Crome and Brown (1979) studied individually marked Orange-footed Megapodes in northern Australia and confirmed that several pairs may use a single mound simultaneously. This species is essentially territorial, with each pair sharing the same mound throughout the breeding season but avoiding contact with other pairs. Because one of these pairs undertook by far the greatest amount of mound maintenance work, the use by the other pairs may be regarded as parasitic. Similarly, several pairs of Nicobar Megapodes use the same mounds (Dekker 1992). However, in this species there is no evidence of mound maintenance by any pair and therefore none of the pairs sharing the mound could be described as parasitic.

Megapodes do not always need to construct a mound in order to utilize the heat generated through decomposition. Among *Megapodius* species especially, there are many reports of eggs being laid into decomposing organic matter in a variety of situations. From Frith's (1956b) original review of these reports the following seem to be plausible: holes scratched into decaying tree stumps and roots, and eggs placed between tree buttresses and covered with leaf litter in amounts varying from a thin covering to virtually a true mound. The extent to which individual species use any of these methods is only poorly known. Dekker (1992) confirmed that the Nicobar Megapode exhibited a variety of mound constructions both away from and adjoining tree roots and buttresses.

There are also observations of megapodes utilizing decomposing material other than that found naturally on the forest floor. Melanesian Megapodes in New Ireland and Philippine Megapodes on Labuan Island off Borneo are both known to lay their eggs in pits of decomposing garbage (Bishop 1978; A. van den Berg, personal communication). In Brisbane, Queensland, Australian Brush-turkeys commonly take over compost heaps of lawn clippings in suburban house yards (Jones and Everding 1991).

The amount of effort required to activate a suitable pile of damp material relates directly to the environmental conditions, especially the climate, in which the species lives. Malleefowl mounds are undoubtedly the most complex and aberrant mounds produced by a megapode, and the sophistication of the behaviour needed to maintain them as an incubation site is remarkable for any animal (Frith 1956a, 1957, 1962a). Because of the extreme aridity and unpredictable patterns of rainfall in the Malleefowl's range, the importance of moisture conservation within the mound is critical. Males are chiefly responsible for the preparation of the site. Leaf litter is gathered prior to the breeding season and deposited in a large depression excavated by the male (see Fig. 5.2). Immediately

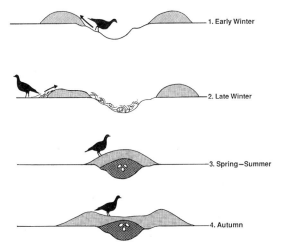

5.2 Stages of construction and structure of a mound of the Malleefowl. (Adapted from Rowley 1974.)

following suitable rainfalls the moist material is promptly covered with a deep layer of sand. While this is an effective means of preventing the escape of heat and moisture, it does necessitate the removal of an enormous amount of material whenever access to the incubation chamber is required. Opening the mound requires the removal of about 850 kg of material (Vleck *et al.* 1984) and males may be occupied with mound work for up to 11 consecutive months (Frith 1959*b*).

Although the heat produced by microorganisms is the major source of incubation heat for all mound builders, solar radiation also assists indirectly in the incubation efficiency of mounds; high ambient air temperatures reduce the differential between the temperatures within the incubation site and the outside air thereby minimizing the loss of heat to the environment (Seymour and Bradford 1992). However, the Malleefowl also utilizes the heat of the sun directly (Frith 1957). Late in the breeding season, most of the organic matter within the mound has been decomposed and is producing little useful heat. In order to take advantage of alternative sources of heat, the male removes the surface layers of sand and soil, thereby exposing the inner egg chamber to the sun throughout the central part of the day. The insulation layers are replaced during the late afternoon so as to minimize heat loss. This species' use of multiple sources of heat, while remaining essentially a mound builder, further emphasizes the complexity of its behaviour.

Burrow nesters

Two main sources of heat are utilized by megapodes that do not construct mounds, the heat associated with geothermal areas and sun-warmed soil or beach sand.

Geothermal heat sources

The distribution of the megapode family encompasses some of the earth's most active seismic locations, owing to the convergence of a number of major lithospheric plates. Zones of subduction, where one plate actively descends beneath another, are areas of sometimes dramatic volcanic activity. There are three major subduction zones running through the region where megapodes are found:

(1) south of the Indonesian archipelago toward New Guinea;
(2) roughly north to south through the Philippine Islands;
(3) east from New Guinea through the Bismarck archipelago and looping around the Fijian Island chain.

Many islands associated with these zones have areas where either recent or continuing volcanic activity provides a continuous source of heat that is utilized by megapodes.

At least four species (Melanesian Megapode, Micronesian Megapode, Polynesian Megapode and Maleo) are known to use geothermal areas as incubation sites. Because such sites tend to be highly localized, the populations using them can become extremely concentrated during egg laying periods. Some of these communal incubation sites (known variously as 'egg grounds', 'breeding grounds' or 'nesting grounds') represent concentrations of breeding birds in spectacular numbers yet often remain relatively unknown outside the local region.

New Britain probably contains the largest of these communal incubation sites, several of which are used by huge numbers of Melanesian Megapodes. At Pokili, the most extensive of the four well-known sites on the island, about 53 000 birds visit the site annually during the June–September laying period (Broome *et al.* 1984). Eggs are deposited in the loose warm soil at the end of tunnels burrowed into the site; apart from being buried into the substrate, they receive no other attention. At Pokili, these tunnels, which occur at an extraordinary density, often run into each other and honeycomb the ground throughout the 2–4 km^2 site (Coates 1985). At the smaller Garu site, studies of the distribution of the burrows indicate that their position and depth in the slopes near the local hot water streams relate to their proximity to the water table (D. N. Jones, unpublished data). As elsewhere, these areas are heavily harvested for eggs by the local people (see Chapter 9).

Both Micronesian and Polynesian Megapodes utilize the warm cinder fields found on the slopes of volcanic islands (Todd 1983; Glass 1988). Areas of intensive laying are concentrated around vents, where eggs are laid in burrows 1–2 m in length. Again, the depth at which eggs are deposited is clearly determined by the temperature of the substrate. The nature of the substrate appears to be of little consequence, apart from areas which are too steep or where the soil is too compacted to dig, although most birds use sites where other birds or human harvesters leave the soil loose and friable. For Polynesian Megapodes, the only available incubation sites are limited entirely to the single volcanic island of Niuafo'ou. In the Mariana Islands, Micronesian Megapodes are burrow nesters on the volcanic islands (Glass 1988), but build typical *Megapodius* mounds elsewhere in the island chain (Yamashina 1932; Frith 1956b.)

Maleos are among the few species using more than one source of heat for incubation (see below), although all involve burrowing into the warm substrate. About 60 incubation sites are known from Sulawesi and about half of these are situated near inland hot streams or wells (Dekker 1990a). In contrast to the relatively narrow burrows and tunnels described above, Maleos excavate broader vertical 'pits', due mainly to the bird's much larger size (see Fig. 5.3). The size and depth of these pits depend on the distance to the water table, the frequency of use, and the age of the pit. Although heavy rains may reduce the temperature of the substrate, geothermal nesting grounds are used throughout the year (Dekker 1988d, 1990a.)

Solar heat sources

The best examples of incubation by solar means are the coastal nesting grounds of the

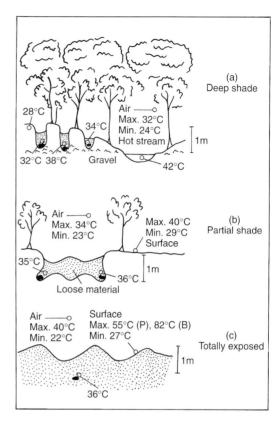

5.3 Various incubation sites used by Maleo in Sulawesi. (From MacKinnon 1978; used with permission.)

Maleo in Sulawesi and of the Moluccan Megapode. These consist of either black sand, derived from a recent volcanic flow or, more commonly, white silaceous sand. Such incubation sites are used only during the dry season, mainly because of the obvious limited effectiveness of solar radiation during the cloudy wet season. Not only is direct sunshine absent for months during this period, but the intense rainfall both cools and saturates the substrate in which the eggs are laid (Dekker 1988d).

Parasitic exploitation of incubation sites

In general, megapodes appear to utilize the heat source that is most readily exploited or else is the only source available locally. In some populations, this could include the mounds of other megapodes. Suggestion of such parasitic use of incubation sites (in most cases mounds), where one species incubates its eggs through the efforts of another, have long been made by indigenous peoples. In the Southern Highlands region of Papua New Guinea, Dwyer (1981) found eggs of the New Guinea Megapode in the mounds of both the Brown-collared Talegalla and the Wattled Brush-turkey. Although this phenomenon has been widely reported (for example, Wallace 1860; Ogilvie-Grant 1897; Rand and Gilliard 1967), unequivocal evidence has not been forthcoming. Just how widespread this is remains unknown.

Species using more than one type of incubation site

Although the majority of megapodes apparently use only one type of heat source, several species have been recorded using more than one. Table 5.1 indicates that 19 of the 22 species construct mounds, with five of these also exploiting other sources at least somewhere in their range. Only three species (Maleo, Moluccan Megapode, and Polynesian Megapode) appear not to construct mounds at all, but rely entirely on geothermal or sun-heated sites.

There is little reliable information available on any association between the various incubation methods used by megapode species and their taxonomic relationships. Although some authorities have attempted to designate subspecies primarily on the basis of type of incubation method used, this is highly problematic. For example, the Melanesian Megapode is monotypic yet within its relatively small distribution exhibits mound building (including laying in rotting tree roots) and burrow nesting in both geothermal areas and sun-warmed beaches. At the current level of knowledge this is the most diverse array of incubation techniques of any megapode. In contrast, the polytypic Philippine Megapode includes some subspecies that appear to use mounds exclusively, while other subspecies use geothermal heat and decomposing tree roots (Dekker 1989c). Whether these contrasting modes of incubation reflect genetic differences within species is not known.

Dekker (1989c) has argued that in the western part of the megapode range the type of site used may be influenced by the relative predation risk associated with each. This assumes that prolonged attendance at an incubation mound increases the risks of predation, as compared to the brief visits necessary for laying at most burrow sites. Thus, several species construct mounds in the absence of certain mammalian predators and use burrow nesting in areas containing predators. Elsewhere, the reasons for one method being used but not another are less obvious.

Detailed descriptions of the specific incubation techniques employed by each species are given in the Species Accounts.

Temperatures of incubation sites

The evolution of the megapode incubation method has been dependent upon the suitability and stability of the temperatures of the incubation sites. In most cases, the temperature range at which eggs are laid is remarkably stable, although the actual temperature may vary considerably between sites. None the less, the physiological and energetic requirements for normal embryo development dictate a fairly narrow range of incubation temperatures. Booth (1987a) found that incubation temperatures in Malleefowl mounds varied between 27 and 38 °C, but hatchability was best at 34 °C. Hatching success dropped dramatically at temperatures below 32 °C and above 38 °C in this species.

Extreme climatic events can, however, seriously affect incubation temperatures. Severe rainfall may cool incubation sites quickly and prevent temperatures from recovering for several days (Jones 1988a). Such sudden reductions in temperature are likely to have serious effects on rates of embryonic development and may be lethal if prolonged. However, megapode embryos are remarkably tolerant of minor changes in incubation temperatures, usually simply showing slower growth and lengthened incubation periods without evidence of lasting effects.

The temperatures of megapode incubation sites have not been recorded consistently and historical data are of unknown reliability. The following data may be regarded as indicative only. Among mound builders, temperatures of 35–39 °C and 33–37 °C have been reported for Orange-footed Megapode mounds in Australia and Komodo, respectively (Frith 1956b; Lincoln 1974). The mean temperature of 50 Australian Brush-turkey mounds was 33.3 °C, (and the average range in temperature of individual mounds was 30.8–35.8 °C (Jones 1988a). Among burrow-nesting species, Polynesian Megapode burrows were 32–38 °C (Todd 1983). Roper (1983) found a temperature of 31 to 33 °C over a vertical range of 1 m in beach-nesting Melanesian Megapodes, while Dekker (1988d) reported temperatures of 31–38 °C between 20 and 50 cm depth in the burrows of Maleos.

In general, then, the incubation temperatures in most megapode incubation sites appear to follow similar trends; most sites are between 32 and 35 °C and are remarkably resistant to daytime temperature fluctuations. This is especially pronounced in mounds, where both the maintenance activities of the tending birds and the physical features of the mound itself provide a distinctly stable incubation environment (see Chapter 6).

6

Ecophysiology and adaptations

Introduction

The evolution of the megapode incubation method has had a major influence on many aspects of the development and structure of eggs and embryos. These features are evidence of the major adaptations that the birds have made to the special problems associated with what is essentially underground incubation. Although this type of incubation is similar to that used by certain reptiles such as crocodiles and turtles, it is now accepted that megapodes have evolved their method from a brood-incubating avian ancestor: it is not a relic of their even more ancient reptilian past as long suspected (Clark 1964a). Rather, it is likely that megapodes have descended from a galliform-like ancestor that probably nested on the ground (see Chapter 8 for further details). The adaptations shown by megapodes to this type of incubation have evolved within the normal constraints of avian physiology and developmental patterns.

The many questions surrounding the adaptations made by megapodes to their incubation method have resulted in the most sustained research interest of any area of megapode study. The bulk of this work has been conducted by Roger Seymour, David Booth, and their colleagues. The aim of this chapter is to provide a summary of this work and to point to some of the more pertinent findings.

Similarities to other underground nesters

No birds other than megapodes are known to utilize environmental heat as the sole source of incubation heat. Although some do bury their eggs, this is usually to conceal them from predators or, as in the Egyptian Plover *Pluvianus aegyptius* (Howell 1979), to prevent overheating and water loss; while covering eggs may also conserve incubation warmth in the nest, all of these species still rely on body heat to provide incubation temperatures. Reptiles, on the other hand, rely entirely on environmental rather than body heat for incubation; most also bury their eggs in sites that vary in construction from those simply covered by leaf litter to the complex mounds and burrows of the larger species (Seymour and Ackerman 1980). This parallels the range of incubation sites utilized by the megapodes (see Chapter 5) and the two groups share many similarities.

Crocodiles (and their relatives) are the only reptiles to construct incubation mounds. These are composed of soil and moist vegetation and often attain temperatures in excess of 35 °C, well above that of the outside air (Neill 1971). The temperatures within these mounds are determined mainly by the existing temperatures of the nesting medium, although some heating via the decomposition of mound material is also likely (Seymour and Ackerman

1980). Crocodiles do not appear to manipulate their mounds as do megapodes, so the prevention of overheating may be limited to the careful selection of the site. The eggs of these reptiles are left to develop without any direct attention, although some species do guard their mounds and may assist in releasing the hatching young (Booth and Thompson 1991).

Incubation temperatures also have important implications for the rate of embryo development and, in some reptiles, even for the sex determination of the young (Bull 1980). In general, reptilian eggs are incubated at temperatures 5–8 °C lower than most birds, and these temperatures are much less stable than for normally-brooded avian eggs (Booth and Thompson 1991). The eggs of both reptiles and megapodes are viable over a much greater range of temperatures than are other birds' eggs (see below) though obviously there is an optimum. In many reptiles this is around 30 °C (Booth and Thompson 1991) while in Malleefowl *Leipoa ocellata*, for example, hatching is most successful at 34 °C (Booth 1987b). The average incubation temperature for birds has been determined as 35.7 °C (Drent 1975).

Reptile eggs have much longer incubation periods than eggs of birds of equivalent size, due in part to their lower incubation temperatures. Megapode eggs, however, are incubated for very long periods compared to those of other birds. This may vary from 44 days to as long as 99 days depending on the incubation temperature, though 49–65 days is the normal range (Bellchambers 1917; Nice 1962). This period is similar to some of the sea turtles, whose eggs are incubated at significantly lower temperatures (Ackerman 1981). The long period of incubation in megapodes, despite higher incubation temperatures, is related to the distinctly slower developmental rates of embryos (Booth 1987b).

The proportion of yolk in reptile and megapode eggs is very high, respectively 32–99 and 48–69 per cent of total egg contents weight (Dekker and Brom 1990; Booth and Thompson 1991), compared to only 14–35 per cent for altricial birds (Sotherland and Rahn 1987). This large component of high energy yolk correlates with the evolution of highly developed hatchlings. Like most reptiles and unlike all other birds, megapode young are completely independent of their parents from the time of emergence.

Incubation processes

As outlined in the previous chapter, megapodes can be divided into two groups based on the way they gain access to an incubation site: 'burrow nesters', species that burrow into naturally occurring warm substrates (such as sun-warmed beaches and volcanically heated soil), and 'mound builders', that construct mounds of decomposing leaf litter. These two main types of nesting technique involve different processes of heat generation. While much is now known about the processes in incubation mounds, very little is known about the mechanisms of heat transfer and management among burrow nesters. Thus, burrow nesting will be discussed briefly here, while most of this chapter will relate to mound building.

Burrow nesting

Burrow nesting would appear to be less physically demanding than the construction of incubation mounds. The heat sources, solar radiation or geothermal heat, are associated with specific locations such as a suitably exposed beach or an area of volcanic activity. Birds laying in such sites usually excavate a burrow or hole into the substrate, depositing their eggs in material assessed to be of an appropriate incubation temperature. The egg is then covered and the burrow (partially) refilled. There are often great differences between species in the amount of work necessary to gain access to material of suitable temperature for incubation. Melanesian Megapodes *Megapodius eremita*, for instance, lay in the loose sand at the end of extensive and virtually permanent burrows at the huge communal nesting grounds (Bishop 1980). Maleos *Macrocephalon maleo*, however, may simply deposit eggs in shallow depressions in beach sands (Dekker 1990*a*). In general, then, burrow-nesting species need to do very little in terms of providing an incubation environment for their eggs. For the mound builders, however, considerable work may be necessary.

Heat production and thermoregulation of incubation mounds

The source of heat harnessed by most mound-building megapodes (the Malleefowl is a notable exception, see below) is primarily microbial respiration, a product of the active decomposition of organic matter by a vast array of minute organisms. These decomposers are common in leaf litter and may reach immense numbers when gathered into a mound. In Australian Brush-turkey *Alectura lathami* mounds composed of relatively resistant *Eucalyptus* and *Acacia* leaves, the most common fungi on the forest floor are replaced by an unidentified *Penicillium* species within the mound (Seymour and Bradford 1992). The biology and role of these organisms in the functioning of mounds remains to be investigated.

The rate of heat production by the micro-organisms varies markedly with moisture levels, ambient temperatures, the quality of the organic matter, and its rate of supply. Frith's (1956*a*, 1957) classic studies on the Malleefowl revealed the complexity of the heat production process in this species' mounds. The extremely dry environment of the Malleefowl imposes an overriding necessity for the conservation of moisture within the organic matter of the mound. Malleefowl prepare a new mound by excavating a large depression in the ground, at a site selected in part because of the drainage characteristics of the soil type. Substantial amounts of leaf litter are then gathered into the depression. When the first rains of the season dampen the leaf litter, the male mixes the material thoroughly before covering it with a large amount of dry soil and sand. This appears to act primarily as an insulation layer, conserving both moisture and heat. However, the presence of this layer also greatly hampers the male's ability to check and maintain the internal decomposing core. At the height of the breeding season, this vast amount of material is partially or fully removed by the male each day in order to mix or add to the decomposing core or to enable the female to lay.

Frith was the first worker to show experimentally that a megapode was capable of actively manipulating the internal temperatures of the incubation mound. He was able to show that the mound's internal temperatures remained remarkably constant throughout the Malleefowl's prolonged breeding season because of the male's ability to control the processes of heat production within the mound. These manipulations were performed by opening and closing the mound, and adding or removing material in response to changes in internal temperatures detected by the bird.

During spring (September–November), the mound is heated entirely from within and it is opened as briefly as possible to prevent the leakage of heat. By late spring and early summer the combination of internal heat and the increasing ambient temperatures may pose a threat of overheating the eggs. At this time the mound is piled high with sand to minimize solar input (see Fig. 5.2). As summer advances, however, the heat production of the micro-organisms begins to wane as the organic matter dries out. The Malleefowl has solved this problem at this time of the year by spreading the insulation layer of sand on the ground around the mound and allowing the sun to heat the material before it is placed back on the mound in the late afternoon. The relationship between the bird's assessment of the mound's temperature and the required action, indicates a level of apparent judgement that greatly surprised the scientific world of the day. Frith further tested this ability by devising experiments that enabled the bird's responses to changing mound temperatures to be carefully monitored. In this work, artificial heaters were placed inside mounds, allowing the experimenters to control mound temperatures remotely. The reaction of the males to these manipulations showed that the birds were capable of detecting temperature changes of 0.5 °C and that assessments could be made and acted upon very promptly.

Since Frith's work, studies on other megapodes have shown the Malleefowl to be the most specialized member of the group in terms of the activities associated with mound construction and maintenance. While mound building is the commonest method of incubation among megapodes (see Table 5.1), no other species show the level of behavioural sophistication of the Malleefowl. Certainly this correlates directly to the particular environmental challenges of this species' relatively hot, dry habitat. Apart from the Malleefowl, only a few virtually unknown populations of the Australian Brush-turkey live in areas other than the humid tropics and subtropics.

Although Frith's work provided essential information on the gross mechanisms of mound construction and the behaviour associated with its maintenance, little was known of the actual processes of heat production of the mound itself. More recently Seymour and his co-workers have made a number of major breakthroughs in our understanding of temperature regulation in mounds. These findings were obtained as part of wider interest in the ecophysiology of megapodes in general. Because of the complex interactions between the different heat sources utilized by Malleefowl, the relatively simpler mounds of the Australian Brush-turkey were chosen for study on Kangaroo Island off South Australia (Seymour and Bradford 1992). Although mounds from Mount Tamborine in southeast Queensland (Jones 1988b) were similar, the translocated population on Kangaroo Island occurs in an environment much drier than that in much of its natural range.

Australian Brush-turkey mounds pass through two phases during the breeding season. During the initial 'construction' phase internal temperatures are highly unstable, often rising from close to ambient to more than 40 °C (Jones 1988b). During this phase males spend much of the day adding material to the growing pile, and defending it against other males (Jones 1990b). The transition to the 'maintenance' phase is marked by the stabilizing of internal temperatures and a gradual lessening of the males' work rate. At both Kangaroo Island and Mount Tamborine the mean mound temperature was close to 33 °C, with remarkably little variation for individual mounds. Different mounds, however, may stabilize at temperatures between 30.8 and 35.8 °C (Jones 1987a).

The shape of the mound changes from a characteristic bell shape to that of a plateau during the transition from construction to maintenance phases. This is due principally to the reorientation of the male's activities from gathering material to digging, mixing, and attending the mound itself. With the stabilization of internal temperatures the structure may potentially function as an incubator. Females seeking a place to lay their eggs test the temperatures themselves prior to laying, and their digging as well as that of the male leads to the flattening of the mound top.

Male Australian Brush-turkeys also alter the general shape of the mound during certain climatic extremes. Saturation of the mound due to heavy rainfalls may cause a sudden fall in mound temperature, significantly reducing the activities of the micro-organisms. During wet weather, the mound top may be piled high, presumably to aid the run-off of rain (Fleay 1937) and the mound is often opened up soon after the rain stops to facilitate drying (D. Jones, personal observation). Conversely, during periods of extended dry weather, mound tops are flattened or cupped, possibly to aid the retention of whatever precipitation occurs.

These activities of Australian Brush-turkeys do not, however, compare to the regular labours of the Malleefowl. One reason is that the Australian Brush-turkey, like all of the other mound builders, relies entirely upon the heat produced by decomposition, whereas the Malleefowl must be able to switch to solar radiation when required.

While it is certain that these activities are directly associated with mound temperature regulation in most mound-building species, it has now been shown that physiological characteristics of the mound itself fundamentally influence temperature stability. Seymour (1985) applied the concepts of thermal physics in an attempt to understand mound thermoregulation. He proposed that, in the relatively simple brush-turkey mound at least, the structure behaved as a stable homeotherm. In this model, the core temperature of an appropriately constructed and maintained mound constantly returns to some stable equilibrium temperature. This process occurs because of two major characteristics of incubation mounds.

Firstly, the considerable mass of the mound gives it great thermal inertia. Heat production due to microbial respiration occurs throughout the mound, but is maximized in the core where substrate temperatures are highest and are buffered from outside climatic influences by the thick outer layers of soil and litter. Secondly, the relationsip between the production and loss of heat automatically returns the mound's temperature to an equilibrium after opening. For example, when heat is lost during the routine opening required for egg laying, the rate at which heat is lost after the mound is refilled becomes less than the rate of heat production and the mound rewarms. Should the temperatures rise much higher than the equilibrium, heat loss exceeds heat production and the mound cools. Of course, the actual equilibrium temperature achieved will be determined by physical features of the mound and may not be suitable for incubation. It has been the adaptive achievement of the mound builders to harness this process and evolve manipulations of the mound that bring the material close to an optimal temperature around 33 °C.

This model for mound homeothermy has now been tested by comparisons with mounds in nature and by a series of experiments using natural and artificial mounds conducted on Kangaroo Island. One of the predictions of the model was that mounds that had already reached thermal stability should exhibit marked homeothermy over an extended period in the absence of male manipulations. This was confirmed for a series of Australian Brush-turkey mounds at Mount Tamborine where the males responsible for construction had been expelled by competing males. Although these mounds received no attention they remained thermally stable for between 3 and 7 weeks after being abandoned (Jones 1988a). Data obtained from mounds on Kangaroo Island were even more remarkable: a number of mounds remained close to equilibrium for several months (Seymour and Bradford 1992). An abandoned Malleefowl mound in South Australia maintained a temperature of 35 °C for 7 weeks without attention (Weathers et al. 1990).

The experimental manipulation of natural mounds conducted by Seymour and Bradford (1992) involved either adding or removing material to or from active mounds. In each case, internal temperatures rose or fell significantly. In mounds where material had been added, the temperatures reached levels lethal to any eggs that may have been present. The reaction of the males tending these mounds was either to open the overheated mound or to add further material to the cooling mound. These actions eventually reduced the temperatures of the overheated mounds to close to their original temperatures, but the mounds that had decomposing material removed did not rewarm, despite the males' attentions.

The artificial mounds constructed by these workers had been carefully designed to resemble natural mounds, and were moistened using appropriate amounts of water. It was found that only mounds of a certain minimal size, and with adequate moisture content, were able to sustain their heat production.

Moreover, as the experimenters did not replicate the regular maintenance activities of the birds, the artificial mounds did not show temperature stability at levels that would have been suitable for incubation.

These studies have provided the most detailed assessments of the requirements necessary for a functional mound. These are a combination of three factors. Firstly, the mound must consist of a critical mass of leaf litter. For Australian Brush-turkeys on Kangaroo Island, the smallest natural mound consisted of about 3.3 tonnes or 6 m³ in volume (Seymour and Bradford 1992). The average size of functional mounds there was 1.2 m in height and about 4.9 m in diameter, significantly larger than mounds at Mount Tamborine (which averaged 0.85 m height, 3.8 m diameter; Jones 1988b). Seymour and Bradford suggested that this difference in size may be associated with the types of leaf litter used by the respective populations. Mount Tamborine birds used leaf litter derived from various rainforest trees and clearly avoided using the relatively decomposition-resistant *Eucalyptus* leaves (Jones 1988a), whereas the Kangaroo Island birds were forced to use only *Eucalyptus* and other tough leaves. The difference in decomposition rates (and subsequently heat production) between these litter types may be reflected in the mound sizes.

Secondly, the material must have sufficient water content. A critical minimum of 0.2 ml of water per gram of dry material is necessary for significant heat production; most mounds on Kangaroo Island were in the range of 0.25 to 0.35 ml/gram. Although a decline of moisture content leads to reduced heat production, drier mounds also have a decreased thermal conductivity and lose less heat (Fig. 6.1). Thus, there is no simple correlation between water content and temperature. Another advantage of drier mounds is a better diffusion of gases throughout the mound (Seymour *et al.* 1986). Mounds in more humid climates therefore face the problems associated with consistently high moisture content. This raises

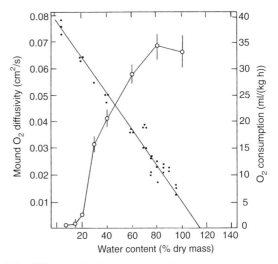

6.1 The relationship between the water content of mound material and oxygen consumption (○) (due to microbial activity) and gas diffusivity (•) in an Australian Brush-turkey mound on Kangaroo Island. (From Seymour 1985; used with permission.)

the possibility of the birds regulating the moisture content of their mounds by altering mound shape, as mentioned above. Such behavioural responses have yet to be studied.

The third factor influencing mound function is the necessity for additions and thorough mixing of fresh mound material. As decomposition proceeds, the maintenance of microbial population densities (and, hence, heat production) depends on the replacement of organic matter. The addition of fresh material may not need to be continual, however. At Mount Tamborine, males added progressively smaller amounts to their mounds as the season advanced and only spent an average of 20 min per day on mixing (Jones 1988b). The larger Kangaroo Island mounds required even less attention and were often left for extended periods with little effect on temperature.

Seymour's model of mound homeothermy has clearly distinguished the key features influencing heat production in the mounds of Australian Brush-turkeys. These features almost certainly operate in the mounds of other species; this model will greatly aid in discerning the critical environmental characteristics that dictate the use of this form of incubation.

Energetics of mound incubation

Megapode incubation techniques raise many questions about their relative energetic costs compared to other methods of incubation. This is particularly pertinent to mound building; it seems likely that most burrow-nesting species incur significantly less energetic costs in having their eggs incubated. For mound incubation, two main questions may be considered: what are the energetic costs for the birds associated with constructing and tending a mound, and what are the energy dynamics associated with organic decomposition within the mound?

Although Malleefowl mounds are not the largest megapode mounds, they do represent the greatest amount of labour. This is because they are completely reconstructed every season and because their maintenance requires the regular removal and replacement of very large amounts of material. An average mound requires the collection of about 3.4 tonnes of material, while opening the mound involves moving about 850 kg, equivalent to about 500 times the bird's body weight! Weathers *et al.* (1993) estimated the energetics of these activities indirectly by determining the relationship between the time a male Malleefowl spent working and the onset of gular flutter. This behaviour (the avian equivalent of panting where the animal loses heat through evaporative cooling) provides a visible indication of the start of heat stress. After determining the relationship between energy expenditure of an adult Malleefowl on a treadmill and the time of start of gular flutter, this was converted into an equivalent work rate for mound tending in the wild. Surprisingly, these workers estimated that the energy expended in mound tending was moderate (about 12 W/kg body weight) or about three times the basal metabolic rate (BMR). This is well below the maximum sustainable aerobic work rate of about 5.7 times BMR of other birds (Weathers and Sullivan 1989). The calculated total work of incubation was similar to the total incubation energy expenditure of other birds. This is probably due to the relatively brief periods of mound work required to maintain functioning mounds; even when tending regularly the male only works between 3 and 5 h per day (Frith 1962*a*). However, these findings relate only to daily energy expenditure, and not to the impact on these birds of sustained work over the extremely long breeding seasons. Furthermore, the major energetic costs associated with the initial construction of the mound need to be accounted for before a more complete picture of mound building energetics is possible.

Heat production by micro-organisms within the mounds represents the release of considerable energy from the organic matter being decomposed. Seymour and Bradford (1992) estimated that an Australian Brush-turkey mound of about 3 tonnes, to which was being added about 800 g of new material each day (Jones 1988*b*), generated substantially more energy per unit time than expected from their model of mound homeothermy. This model was, however, based on Kangaroo Island mounds, where only a proportion of the material (mainly tough *Eucalyptus* leaves) is used in the heat production. *Eucalyptus* leaves make up very little of the mound material at Mount Tamborine, and experiments by Jones (1987*b*) have shown that they are significantly more resistant to decomposition than the rainforest leaves normally used. None the less, the rate of heat produced by a mound is in excess of 20 times that of a resting bird of a similar size (Booth 1985). Thus, in terms of heat production temperature at least, a mound should be capable of incubating many more eggs than could be achieved by brooding under an adult in the usual fashion.

Ecophysiology and adaptations

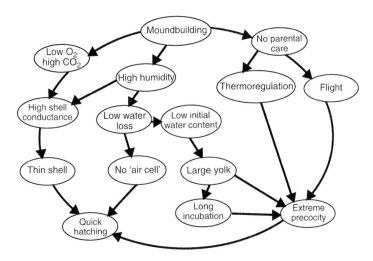

6.2 A possible schema of the numerous interactions among factors associated with the evolution of the megapode incubation method. (From Seymour 1985; used with permission.)

Adaptations of eggs and embryos

Our understanding of the complexities associated with the evolution of the extraordinary adaptations necessary for the success of the megapode incubation method is far from complete (see Fig. 6.2). The following sections provide an overview of current knowledge.

Eggs: size and composition

Megapode eggs range in mass from 231 g in the Maleo to 75 g in the Polynesian Megapode *Megapodius pritchardii* (see Table 4.1). In general these eggs are up to five times the size expected for the body weight of an equivalent galliform (Vleck *et al.* 1984). Moreover, with clutch sizes ranging, on average, from 12 to 24 (Frith 1956*b*), female megapodes produce an overall yearly egg mass greater than any other bird of similar size (see Chapter 4). The rate of egg tissue formation is, however, similar to other galliforms, because the eggs are produced individually at intervals varying between two and thirteen days (Seymour and Ackerman 1980).

As mentioned previously, megapode eggs are very rich in yolk (Dekker and Brom 1990). Indeed, the proportion of yolk is exceeded only by the Brown Kiwi *Apteryx australis* (Dekker and Brom 1992). The high proportion of yolk provides megapode eggs with a high energy to water ratio, a condition necessary for maintaining the embryo throughout the prolonged incubation period. High energy contents are also highly correlated with the extreme precocity of the hatchlings.

Shell structure

Megapode eggshells are surprisingly thin compared to most eggs of similar size, being 31 per cent thinner than predicted for galliforms of equivalent size (Booth 1988*b*). The resulting fragility of megapode eggs is not, however, a serious liability, because when laid they are normally deep within the substrate of the incubation site. Following laying they are relatively free from the hazards of mechanical damage, except during subsequent laying activities. In most birds' eggs, thin eggshells facilitate the loss of water from the egg, a process critical to the normal development of the embryo. Normally, a water loss of about 15 per cent of the initial weight of the egg occurs, and is associated with the formation of a fixed airspace at the blunt end of the egg needed by the embryo to breathe immediately prior to hatching (Seymour 1985). This gas exchange is regulated by the difference in gas tension across the shell and by the conductance of the pores in the shell. In megapodes, this important loss of water would appear to be jeopardized by the very moist conditions in which the eggs are incubated. However, water transfer in megapode eggs is, unexpectedly, very similar to other birds (Seymour et al. 1986). This is due to the remarkably high conductance of the eggshells, which is about twice that predicted for birds of equivalent size (Seymour and Rahn 1978). Furthermore, despite the apparently unsuitable environments in which they develop, the water loss of megapode eggs during incubation is 10–12 per cent of their initial mass, and at a rate that increases 3-fold over the incubation period (Seymour et al. 1987). However, despite the loss of significant amounts of moisture no fixed airspace is formed. These unexpected results are due, not only to the thin shell, but also to the high degree of shell thinning throughout incubation and the nature of the eggshell's pores (Booth and Seymour 1987).

In most birds' eggs the shell becomes 4–8 per cent thinner during incubation. This is due to embryos removing calcium from the inside of the shell for skeletal formation. However, small amounts of thinning have little effect on the pore structure or length and as the density of pores is fixed at the time of shell formation, conductance and water loss generally do not change significantly in most birds (Booth and Seymour 1987). The major change in conductance noted in megapode eggs results from the removal of calcium from the narrow end of the cone-shaped pores. This steadily opens the main region of resistance to gas exchange and leads to greatly increased shell conductance.

Gas exchange

Adequate diffusion of oxygen and carbon dioxide through the shell is also essential for the successful growth of avian embryos. As with water loss, the effectiveness of respiratory gas exchange is a product of shell conductance and the partial pressure difference across the shell. In contrast to most birds' eggs, megapode eggs are surrounded by remarkably high levels of carbon dioxide and low levels of oxygen. Such conditions would be expected to inhibit the passage of gases in and out of the egg.

Studies of gas exchange in megapode eggs (Seymour et al. 1986) have shown that the gas tensions within the shells of late-stage embryos are very similar to those found in eggs incubated in normal nests. Again, this is attributed to increased eggshell conductance. None the less, gas exchange between the embryo and the atmosphere is greatly influenced by features of the material in which the egg is being incubated. Although it is now known that no parts of a mound are oxygen deficient, the source of atmospheric oxygen may be separated from the egg by 60 cm of leaf litter or sand. The movement of gases through the mound is primarily by diffusion through spaces in the material (Seymour 1985). The regular raking, digging and general mixing of the mound material practised by many mound builders therefore assists in the diffusion of gases between the air and the mound. In particular, the addition of fresh material onto the mound by the birds kicking it into the air is likely to maximize the porosity of the mound. The excavation of small, deep holes for the purposes of temperature assessment (see Chapter 4) has also been found to facilitate gas exchange, as well as increasing microbial activities in the vicinity of the holes (Baltin 1969).

Samples of the respiratory gas tensions within mounds of Australian Brush-turkeys and Malleefowl show that the oxygen pressure decreases while carbon dioxide increases with depth to the level of the eggs, where both reach gas tensions that could be dangerous to the eggs (Seymour et al. 1986; see Fig. 6.3). These gas tensions also change throughout the incubation period, especially in relation to

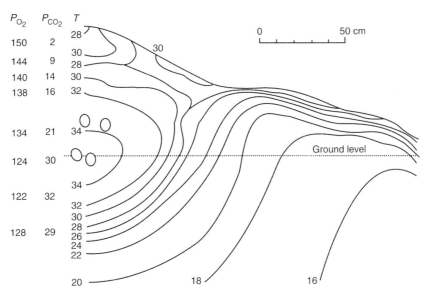

6.3 Gas tensions of oxygen and carbon dioxide, and temperatures, within a Malleefowl mound. (From Seymour and Ackerman 1979; used with permission.)

moisture levels. Measurements were very similar in both Malleefowl and Australian Brush-turkey mounds, despite many structural differences. The gas tensions are due primarily to microbial respiration; these organisms consume oxygen at a rate of about 20 l/h, about 60 times that of all the eggs a mound is likely to contain. Overall, the eggs are responsible for less than a fifth of the total reduction in oxygen pressures at the level of the eggs (Seymour et al. 1986).

By the end of the latter part of the incubation period, the difference in gas tensions between the atmosphere surrounding the eggs and that inside would be expected to have become intolerably high, if gas conductance through the shell had remained the same throughout incubation. However, because of the increases in conductance through shell thinning, gas tensions inside the eggs are virtually identical to measures for open-nesting species (Seymour et al. 1986). This change in conductance allows the embryos to cope with the conflicting demands of minimizing dehydration while increasing gas exchange as metabolic demands increase toward the end of incubation.

Although very well adapted to cope with these physiological challenges, the embryo in its egg still depends upon the activities of the mound tender to ensure that the mound maintains a high porosity to facilitate diffusion of gases. This means balancing the requirements of the microbial populations with those of the developing embryo. In general, mound-building species need to conserve some level of moisture and avoid the possibility of saturation. This observation may also explain why tropical megapodes avoid the wet season for breeding (Frith 1956a).

Unfortunately, very little is known about the processes of gas exchange in incubation sources other than mounds constructed each year, such as those tended by Malleefowl and Australian Brush-turkeys. Moreover, virtually all the work conducted over more than a decade by Seymour and his colleagues has been limited to locations with a rainfall pattern very different from that of the natural distribution of all other megapodes. Although it is likely that the principles provided by these workers will greatly assist in understanding the ostensibly similar mounds of numerous other species in the tropics, it is wise to caution against a simplistic and generalized extrapolation to other climatic zones. This is especially important in relation to non-mound incubation sources, about which little more than the temperature regimes are known.

The eggs of species that utilize heat sources such as geothermal areas and sun-warmed sand may not be subject to gas tensions as severe as those experienced by mound builders, because of the absence of vast populations of respiring micro-organisms. However, such incubation sites may be prone to waterlogging, especially during spells of sustained rainfall. This could seriously reduce the diffusion of gases through the material, as well as reduce the substrate temperatures. Although self-heating sites appear not to require the regular attention of the birds, many communal sites are thoroughly excavated during laying visits. In the nest-burrows of Maleos, moisture content of the substrate varied greatly (from 1.6 to 45.0 per cent), depending on the composition of the material and recent rainfall patterns, though temperatures near the eggs remained at about 32–35 °C (Dekker 1988*d*). The eggs of this and other burrow nesters are deposited at depths varying according to substrate temperatures, an assessment made by the female at the time of laying (see Chapter 4). Subsequent climatic events, especially heavy rains, may seriously alter the suitability of the site for incubation, but of course cannot be manipulated by the birds.

Hatching and the initiation of breathing

All bird embryos exchange gases with the atmosphere through a specialized membrane known as the chorioallantois. This double structure grows out from the embryo which supplies it with blood, eventually completely surrounding the developing chick. This organ supplies the embryo with all its oxygen and removes its carbon dioxide. This poses a potential problem for the chick because hatching necessitates rupturing the sole organ of respiration. Furthermore, having previously been entirely reliant on chorioallantoic respiration, the chick's lungs remain fluid filled and non-functional. The solution almost all birds have found to this problem has been to commence ventilation of the lungs prior to hatching. This is achieved by the chick thrusting its beak into the air cell at the blunt end of the egg, a process known as 'internal pipping'. Air cells form in eggs as a result of the loss of water, and increase in size throughout the development of the embryo. After some hours, the chick uses its egg-tooth, which is absent in megapodes at the time of hatching, to make a small hole in the shell and take its first breaths of external air ('external pipping'). Following the piercing of the chorioallantois, its blood supply slowly diminishes as the lung is drained of fluid and becomes functional. This is a slow process in birds because bird lungs contain constant-volume, blind-ended alveoli. In general, there is a considerable functional overlap of the two forms of respiration, ending with the now blood-free chorioallantois being safely torn open as the chick progressively chips its way out of the shell.

Megapodes, however, undergo hatching very differently. One of the consequences of their unusually moist incubation environment is that no appreciable fixed air cell is formed between the inner and outer shell membranes. The lack of a fixed airspace precludes megapode chicks from internal pipping, so that the lungs can only be ventilated after the chorioallantois and the shell have been ruptured. The process of hatching would therefore appear to involve major risks of blood loss and asphyxiation, as the lungs remain filled with fluid until hatching.

Uniquely among birds, megapodes hatch very rapidly. Using their legs and back the chicks break violently through the membrane and shell within a few minutes. Blood flow through the chorioallantois stops soon after it is torn by the feet, about 25 ml of fluid drains from the egg, and the chick begins to breathe deeply immediately (Seymour 1984). Its lungs, having contained no oxygen immediately prior to hatching, lose fluid and aerate rapidly (Seymour 1984). There is, therefore, virtually no overlap between chorioallantoic and pulmonary respiration. Seymour believes that this unique process of initiating breathing is made possible by the relative ease of hatching. This is facilitated by the relatively thin shell, which has been progressively thinned throughout incubation. None the less, the initially small air volume within the chick's lungs delivers relatively little oxygen to the blood, and the chick is unable to perform any real exertion for perhaps several hours.

Emergence from the mound

Megapode chicks hatch at distances varying between about 20 cm and 1 m or more beneath the surface of the incubation site. In order to emerge from the incubator they must dig their way to the surface, a process they perform without any assistance from the adults. The only indirect influence that adults may have upon the emergence of the chicks is in preventing the compaction of material around the eggs through regular diggings. The many early accounts of adults digging tunnels into the site to assist in the emergence of the young are now interpreted as part of the normal maintenance procedures.

Following a period of recovery after hatching, the hatchling commences to excavate a tunnel through the material toward the surface. The time taken in moving from the level of hatching to the surface varies according to the depth, the compaction of the substrate, the nature of the substrate, and the energy reserves of the individual hatchling. Estimates of the time spent in emergence for mound builders are 2–15 h in Malleefowl (Frith 1959b) and 1–2.5 days in Australian Brush-turkeys (Baltin 1969; Vleck et al. 1984). Little is known of the emergence time for burrow nesters, although Dekker (personal observation) found that Maleo hatchlings dig upwards very slowly, often remaining within the material for several days.

Observations of Australian Brush-turkey and Maleo hatchlings placed in transparent cylinders of mound material (Vleck et al. 1984; R. Dekker, personal observation) have revealed brief bouts of intense activity interspersed with long periods of recovery during which chicks breathe heavily. The chick lies on its back to scrape at the material above it, then compresses the fallen material around its body with its back. This process is likely to be most efficient when the substrate is compressed somewhat firmly; material that is too friable may leave the chick in a hole and unable to reach the material above it. On the other hand, very compacted material may simply be impossible to dislodge. Similarly, the moisture content of the material may greatly affect the chick's digging efficiency, as well as influencing the gas environment in which it has to work.

Obviously, the process of emergence is energetically expensive for most megapode hatchlings. This is a cost that hatchlings in most other species do not have to endure. Estimates of the cost of emergence, based on the 'residence time' of the chick within the mound, indicated that Malleefowl hatchlings use 8 per cent, and brush-turkeys 33 per cent, of the energy expended during the entire preceding incubation period (Vleck et al. 1984). The difference between these two species may be attributed to the relative ease of emergence from the friable sandy Malleefowl mound compared to that of the matted sticks and plant detritus of the brush-turkey mound.

Life after emergence

Having reached the surface of the incubation site, megapode hatchlings do not receive the parental care, protection, brooding, and social environment that would appear to be universal among higher animals. Instead, they must survive without assistance of any kind from their parents. They must find food and shelter, and evade predators. These conditions have selected for the most precocial of all hatchlings among birds.

The life led by megapode hatchlings is probably the least known aspect of megapode biology. This early period is certainly the most dangerous for the chicks; not only are they most vulnerable to predation, they are also faced with finding suitable and adequate food, while coping with the extremes of ambient temperatures. Virtually nothing is known of the foraging abilities of the chicks, although they do emerge with some energy reserves in the form of fats built up during incubation that may be metabolized for a brief period following hatching.

Thermoregulation

One of the most important functions of the brooding of hatchlings in most birds is that of protecting the young from variations in temperature. The hatchlings of many species, including galliforms, are poor thermoregulators for some time after hatching. This is particularly important at lower temperatures, most young birds being able to cope with heat-stress though only in the short term (Dawson and Hudson 1970). Booth (1984, 1985) studied the thermoregulation of Malleefowl and Australian Brush-turkey hatchlings and has shown that both are able to maintain a stable body temperature over considerable ranges of ambient temperatures.

Malleefowl chicks are excellent thermoregulators over temperatures ranging from 3 to 46 °C, the seasonal variation typical of their arid, inland environment. This is achieved by their ability to raise their metabolic rate by up to three times above standard when exposed to cold stress. Booth attributed this to the chicks' substantial energy stores, present as extensive subcutaneous fat and yolk stock. At higher temperatures, the birds lose heat by panting and gular fluttering, which commences at 42 °C, as in other birds. The Malleefowl hatchlings exhibit a thermoneutral zone of 32 to 39 °C. In Australian Brush-turkey hatchlings this zone is 29–38 °C.

Although the mass of neonates of these two species is almost identical (both weigh about 114 g; Booth 1985), Australian Brush-turkeys have a higher metabolic rate, lose water at a greater rate, and have a higher body temperature than Malleefowl. These differences probably reflect the contrasting environments to which the two species are adapted. For example, Australian Brush-turkeys, which have evolved in more equitable climates, are unable to cope with temperatures below 10 °C during their first day, and are less tolerant of thermal stress.

Both species are, however, remarkably adept at coping with the natural ambient temperature variation they are likely to experience following hatching. This is due not only to their relatively large body size and the ability to alter their metabolic rate, but also to the excellent insulative properties of their well-developed plumage (Booth 1984).

7

Reproductive behaviour and mating systems

Introduction

There can be little doubt that the relatively long history of interest in the megapodes has been due to their unique incubation activities. Most published studies (see Diamond (1983) and Jones and Birks (1992) for reviews) concentrate on aspects of the birds' use of their incubation sites. It is surprising, then, that so little of the reproductive behaviour of the birds themselves has been studied. For many species, details of sexual behaviour and mating systems were published for the first time during the last decade. Again, as is evident in other subject areas covered in this book, most information comes from only a few species.

The modern approach to reproduction in animals is strongly evolutionary in emphasis, focusing on individuals rather than on species as a whole. This has led to a re-examination of many previously held ideas. The approach is of particular relevance to megapodes because of the great influence that their incubation method has had, not only on reproduction, but on all aspects of their ecology and behaviour.

This chapter describes behaviour associated directly with reproduction and some evolutionary consequences of the megapode incubation method, and gives a classification of the mating systems exhibited by the family.

Reproductive behaviour

As in most other aspects of megapode biology, interest in behaviour has centred upon reproductive activities. One obvious reason for this is the fact that most species are observable mainly while attending incubation sites; away from these locations they are much more difficult to detect and observe.

The most detailed studies on behaviour have been conducted in captivity (for example Fleay 1937; Baltin 1969; Kloska and Nicolai 1988), but several recent studies of wild populations (for example Coates 1985; Jones 1990a; Birks 1992) have added valuable detail to the general picture. As outlined later in this chapter, some species of megapode are monogamous, others are non-monogamous.

Courtship behaviour

Very little has been published on courtship or pair-formation displays among megapodes. Although species which form apparently permanent pair bonds presumably undertake some process of mate selection and pair formation, this has yet to be described in detail. Where pair bonding does exist it may occur quite early.

In most monogamous megapodes, paired birds remain in close and permanent contact. Among Malleefowl *Leipoa ocellata*, however, paired birds are separated for extended periods due to mound defence by males and the foraging movements of females away from the mound. In this species, Immelmann and Böhner (1984a) described a 'greeting ceremony' which was given during morning reunions of a captive pair and involved the two birds circling, ruffling their feathers and spreading their wings. This display, which was interpreted as helping to maintain the pair bond, has not been observed in wild Malleefowl. Malleefowl are the only monogamous megapodes in which paired birds spend long periods, very often several days, apart. In some monogamous species, however, pairs are separated when females visit egg grounds, leaving their mates in the forest nearby (for example, Polynesian Megapode *Megapodius pritchardii*); it is not known whether there are 'reunion' displays among these species. In all other species, as far as is known, either the paired birds remain together permanently, or no pair bonds exist.

The monogamous megapodes, including the Malleefowl, exhibit a number of behavioural characteristics that strongly suggest closely pair-bonded monogamy. For example, many species show highly synchronized behaviour, especially noticeable during routine activities such as foraging or mound maintenance. Also associated with close pair bonds in birds in general is a lack of obvious or even discernible pre-copulation behaviour, and this has been noted frequently in Malleefowl (for example, Immelmann and Böhner 1984a). Moreover, copulations outside the breeding season are common in this species, behaviour without reproductive value but possibly related to pair bond maintenance. In contrast, males of the monogamous Maleo *Macrocephalon maleo* have been seen vigorously scratching material into the air and running rapidly around in circles in the presence of the female immediately prior to copulation. Similar spirited behaviour has also been described in Orange-footed Megapodes *Megapodius reinwardt* (Banfield 1913). The only function previously suggested for such behaviour has been that of play; it is possible that these activities may be associated with pre-mating behaviour.

Courtship feeding has also been detected among monogamous megapodes. Males in three species (the Maleo, Polynesian Megapode, and Nicobar Megapode *Megapodius nicobariensis*) have been observed bringing food items to feed their mates. In the Maleo, this occurred immediately after copulation (R. Dekker, personal observation). It is probable that this behaviour also occurs in other monogamous species.

Copulations

Megapodes apparently copulate throughout the breeding season. In part this relates to the very long periods during which egg laying can take place; repeated inseminations may be necessary to ensure that all eggs are fertilized. Unfortunately, nothing is known about the relationship between the female's fertile period and egg laying.

Several patterns of copulation behaviour are evident within the family. These patterns differ according to the type of mating system. Copulations are rarely seen among those monogamous species in which males remain close to their mates throughout the breeding period, rather than to the incubation site. It is probable that the female is actively defended by the male in these species; the two birds are rarely more than a few metres apart and any single birds that approach are vigorously expelled. Studies of the behaviour of such species do not mention copulation or even solicitation (Crome and Brown 1979; Coates 1985). In most cases, visits to incubation sites appear to be as brief as possible: female Maleos and communally nesting *Megapodius* species remain only long enough to lay, and leave the area promptly (Weir 1973; Dekker 1990c). Some monogamous species using undefended mounds spend long periods at their incubation sites, but they show no evidence of sexual interactions during these periods (Crome and Brown 1979). Very little is known of the frequency or timing of copulations among these species but apparently unsolicited copulations, involving sudden pursuits by males of their mates around the forest floor, are often reported (for example, Barrett and Crandall 1931; Coates 1985).

In contrast, copulations among non-monogamous species, all of which defend mounds rather than females, appear to occur almost exclusively at or near the mound. Although there are sound theoretical reasons for expecting this to be normally the case (such as the apparent importance to a female of choosing a male known to be successful as a mound owner as opposed to a mound-less male), Dow (1988b) has noted many copulations away from mounds in one population of Australian Brush-turkeys *Alectura lathami*. This has not been detected or even suspected in other mound-defending species, including other populations of this species (Jones 1990a). None the less, the possibility of successful copulations apart from those detected at the mound must be taken into account during future studies. In the majority of cases, however, it is likely that most copulations among these species occur during visits by receptive or laying females to males at their mounds.

Australian Brush-turkeys exhibit an apparently unique display that appears to be directly associated with attracting females. This display, performed away from the mound, involves the male lowering his body to the substrate, spreading the wings and tail and extending the neck forward while pecking at the substrate. The context of this display appears to be the presence of a reluctant or poorly motivated female. Baltin (1969) suggested that females responded to this display because of its resemblance to a laying female. In adopting this posture a male may be attempting to sexually motivate a female by performing behaviour usually associated only with the mound, the normal site for copulations (Jones 1990a). Such a deception may be used by males attempting opportunistic copulations with females normally expected to make mate choice decisions at mounds (Jones 1990b). Baltin (1969) regarded this display as a normal component of the reproductive repertoire of his captive birds. Among wild populations, however, the incidence of the display is varied. The display was rare in one rainforest study site (Jones 1987b, 1990a), but was common in less dense populations where copulations away from the mound were often seen (Dow 1988b).

Evolutionary consequences of the megapode incubation technique

As explained in previous chapters, the evolution of the megapode incubation method was accompanied by a large number of physiological and behavioural adaptations. Perhaps the most profound of these was the development of extremely precocial chicks that are able to live completely independently of their parents from the time of hatching. These features, which distinguish megapode hatchlings as the most precocial of all birds, appear to have been selected to compensate for the reduction and finally the complete absence of parental care in megapodes. Even among other highly precocial species, one or more adults must provide some protection against predators and climatic extremes. Megapode hatchlings, however, receive no adult assistance whatsoever from the time they leave the egg.

This severing of the usually close relationship between chicks and parents is a direct consequence of the use of external sources of heat for incubation. Many bird species attempt to synchronize the hatching of young by delaying the commencement of incubation until the clutch is complete. This has obvious advantages for parents seeking to keep a group of precocial hatchlings together and is critical for species whose young leave the nest soon after they hatch. Megapode embryos, however, begin their development immediately after the egg has been laid into the warm incubation site. Subsequently, each egg hatches separately and at varying intervals throughout a breeding season that may extend for months. This complete asynchrony of hatching prevents a parent from gathering the brood together. The immediate surroundings of the incubation site are often unsuitable habitat (especially where beaches or geothermal areas are used), which necessitates the immediate movement of chicks into the surrounding forests. This also prevents them from grouping, further promoting their solitariness. As would be expected, the mortality rate of young megapodes appears to be extremely high although few figures have been published; in one study, Jones (1988c) estimated a loss of 90–97 per cent for Australian Brush-turkey hatchlings.

Although the steps involved in the evolution of reduced parental attention are difficult to determine (see Chapter 8), the evolutionary influences of this process have been far-reaching. In particular, the abandonment of parental care of hatchlings has released both sexes from what is, for most birds, a primary constraint on reproductive output. It is generally held that, while individuals from any species should seek to produce the largest number of offspring, clutch sizes are a product of a trade-off between various physiological and ecological features, especially the number of offspring that can be raised; the preponderance of monogamy as a mating system suggests that for most species this has proved to be the best way for both males and females to optimize their reproductive success (for example, Lack 1968).

The fact that megapodes are freed from many of the costs associated with looking after their young has provided an opportunity for females to greatly increase their potential fecundity through sustained egg production. All megapode species appear to lay their eggs over extended periods, often lasting for several months and are limited principally by the environmental constraints on the availability of food or the condition of incubation sites.

Increased egg production by females is, however, dependent upon the accumulation of appropriate qualities and quantities of energy and nutrients. This has required females to maximize their foraging effort, often resulting in significantly larger home ranges and increased food intake during the breeding season (Booth 1987*b*). Furthermore, the relatively large size and high yolk content of megapode eggs also necessitate prolonged foraging by egg-laying females. The particularly high amounts of energy, protein, and calcium required for egg production may also lead to feeding excursions into better quality habitat some distance away from the incubation site.

These features, which are probably common to all breeding females megapodes, confront the male with a fundamental problem: how to benefit from the female's reproductive potential without restricting her capacity to produce eggs. In many animals, males simply prevent their mates from moving out of their territories, while keeping other males away. For female megapodes, however, their nutritional needs may require them to move over a wide area. There appear to have been two approaches to this problem among megapode species. Given the evolutionary principle that each male should seek to maximize his own reproductive success, male megapodes have either monopolized the reproductive output of a single female by remaining permanently with her, or monopolized access to an incubation source in order to mate with females visiting the site to lay. These two contrasting strategies among males allow megapode mating systems to be divided initially into two main types based on the particular key reproductive feature that is defended by males (see Oring (1982) for further details), namely, female defence, where a single female is defended by a male (designated type I) and resource defence, where incubation sites (always mounds), rather than females, are monopolized by males (designated type II). Using this schema, all species exhibiting female defence will have, by definition, a monogamous mating system, whereas resource defending species are most likely to be polygynous, with males competing for females through their control of access to the incubation sites.

To some extent, these two strategies appear to have been influenced by the type of incubation site used. On the one hand, naturally occurring sites that function independently of the birds (such as geothermal areas and solar-heated beaches) usually exist on a scale that precludes monopolization by any one male. In species using such sites, males almost always remain permanently with one female. This arrangement allows the female unrestricted foraging opportunities while allowing the male to guard her, ensuring that no other males gain access to his mate. In this situation, the male defends a single female and the incubation site they use may be shared with other pairs.

On the other hand, an incubation site that has been constructed (that is, a mound) is often defended vigorously by the male that has invested his time and energy in its construction. By controlling access to the incubation site he may be able to increase his reproductive success by mating with females seeking a site in which to lay. His success in this strategy will depend on a number of factors, including the number of alternative incubation sites available within the locality, the availability of breeding females within the population, and the relative attractiveness of the male and/or his mound site to the breeding females. As in many other polygynous bird species, these features may lead to considerable variation in the reproductive success of competing males. Furthermore, where environmental conditions enforce very low population densities, mound-building birds may form pairs and mate monogamously, having few other options, as appears to be the case for Malleefowl (but see below).

The identification of these primary influences on the form of the reproductive behaviour of megapodes may provide a useful comparative approach to describing the actual mating system of each species. It needs to be emphasized that there are many variations of the types of interactions between the sexes, even within the traditional categories used to describe mating systems. Rather than simply label each species as 'monogamous' or 'polygynous', it may be much more instructive to detail the overall social organization of which patterns of sexual behaviour form a part.

While recognizing that the breeding activities of only a few species have been studied in detail, it is none the less possible to identify features in most species that suggest strong affinities to the mating system of the better-known species. These features relate to the general form of the species' social organization, as well as the more overt sexual activities. Of particular relevance is the persistence of pair bonds, the temporal and spatial interactions of the sexes, and sexual differences in effort in reproductive activities.

At the current state of knowledge, megapodes exhibit three types of mating systems, as follows.

Type I: male defence of female

Mating system: female-defence monogamy

This type of mating system is exhibited by the largest number of species within the family: all of the *Megapodius* species, including both mound builders and burrow nesters, the Moluccan Megapode *Eulipoa wallacei*, the Maleo, and probably the three *Talegalla* species.

The social organization of species of this type is characterized by close and probably permanent pair bonds. In virtually all contexts, paired birds usually keep close together and where work is required for the maintenance, activation, or assessment of incubation sites, they share duties about evenly. The sexes are monomorphic and there is a high degree of behavioural synchrony (Lincoln 1974; Crome and Brown 1979).

Paired birds remain close together throughout the year; although home ranges often overlap, there may be little interaction between pairs. The often complex co-ordination of avoidance (and in some species, the sharing of the incubation site) by pairs seems to be effected by vocalizations (Crome and Brown 1979). All species are noted for their loud and repetitious calling, especially during visits to incubation sites, at roosts, and often throughout the night (Bishop 1980; Coates 1985). Away from incubation sites, pairs avoid each other and aggressive interactions are rare. Fighting has only been noted commonly where single males have approached a pair or where breeding birds become concentrated at incubation sites (Bishop 1978; R. Dekker, unpublished data).

Despite this pair-dominated social system, these species often assemble in small groups. The nature of the relationships between these individuals is unknown but may relate to the communal use of incubation sites. Species with this type of social organization use all of the incubation heat sources: geothermal areas, solar-heated beaches, and incubation mounds. Of incubation mounds, however, the large, perennial ones (used each year and not defended) appear to be the most common. In contrast to those type II species that defend an incubation site, males of type I species appear to defend a single female. Certainly, it is typical of these species to have pairs remaining together almost permanently. The only exceptions appear to be in those species where laying females visit egg grounds alone while their mates remain in the forest nearby.

The inclusion of the *Talegalla* species in this category must be regarded as tentative. Apart from the observations of Coates (1985), very little is known about any of the three species. None the less, they do resemble the other species included in this type in being monomorphic, highly vocal, without any apparent secondary sexual ornaments, and in appearing to be closely pair bonded. Confirming field data are needed.

Type II: male defence of incubation sites

The only incubation sites that are defended by any megapode are mounds. There is no evidence from any burrow-nesting species of males defending all or part of an incubation site, apart from single burrows during each laying visit. The primary male strategy is one of providing an incubation site and then controlling the access of females to it. Two groups are evident within this type: three apparently polygynous species, and a single monogamous species.

Mating system (type IIa): resource-defence polygyny

The features described in this section are based primarily on studies of wild populations of Australian Brush-turkeys (Jones 1987*b*, 1990*a*, *b*). Recent work on captive Wattled Brush-turkeys *Aepypodius arfakianus* (Kloska and Nicolai 1988) indicates that there are many behavioural and morphological similarities between these two species, and it is probable that the similar but virtually unknown Bruijn's Brush-turkey *Aepypodius bruijnii* also belongs to this group. Thus, the polygynous megapodes all belong to the two genera now known collectively as the brush-turkeys.

The social organization of species within this group is characterized by a pronounced defence of incubation mounds in males and a high degree of independence in females. These species exhibit no pair bonding at all and interact only during the brief though repeated mound visits by females for copulation and egg laying.

Incubation mounds are sited, constructed, maintained, and defended solely by males. Competition over mound possession is intense and males may be expelled from their mound by more dominant individuals. Some males may acquire more than one mound either by construction or by usurping a mound constructed by another male. These interactions force males to remain close to their mounds and all trespassing males are quickly challenged. Females are apparently not restricted in their movements and, in areas where mound densities are high, may be able to visit all or most of the available mounds on a regular basis. Certainly, females make many non-contact visits to males at their mounds throughout the breeding season. Males typically only allow females to lay in their mounds if they copulate immediately beforehand. Both sexes mate promiscuously and it is probable that many males tend mounds containing eggs fertilized by other males (Jones 1990*a*, *b*).

All the species included in this group show distinct sexual dimorphism. Males are slightly larger than females and usually develop or enlarge brightly coloured wattles and combs during the breeding season. These may also be extended or may increase in intensity of colour during sexual interactions. During the non-breeding season such morphological traits diminish, intraspecific aggression decreases, and groups of both sexes are common.

While the classification of the mating system has been designated as resource-defence *polygyny*, this describes only the males' influence within the mating system. In the Australian Brush-turkey, however, a striking feature is the virtual independence of the sexes (Jones 1990a, b). It is clear that males use their control of access to the incubation site as a strategy to meet females. On the other hand, the females are completely independent of males and are able to choose and mate with any and as many males at they may wish. Thus, both sexes are polygamous. This suggests a combined classification to describe the differing male and female strategies evident within this species, as has been used elsewhere for species with similar characteristics: resource-defence *polyandry* (see Oring (1982) for further explanation). Whether such a classification is appropriate for any of the other species remains to be determined.

Mating system (type IIb): resource-defence monogamy

The Malleefowl is the only species in this category. Its social organization is characterized by close and apparently permanent pair bonds but with a high degree of independence between paired individuals. During the construction of incubation mounds the pair keep closely together, sharing much of the work involved. Once the female begins egg production, however, the pair lead mainly solitary lives, with the male remaining near his mound for long periods of time (Frith 1959b) while his mate forages over a wide area away from the mound; relatively little time is spent together and the birds even roost apart. With the cessation of breeding activities the pair reunites.

This type of social organization is strongly influenced by the acquisition and defence of the incubation mound. During the early stage of mound site selection, competition among males frequently leads to fights and may even result in death, but these interactions usually cease when males concentrate on the labours associated with the construction of their mounds. None the less, both sexes do move well away from the mound, males occasionally and females perhaps daily. Indeed, the breeding season home range of females may include the mounds of other males (Booth 1987b).

Mound ownership and perhaps occupation appears to be advertised by a loud booming call, normally delivered near the mound (Frith 1959b). Territorial behaviour breaks down at the end of the breeding season and small groups of pairs may assemble at locally rich feeding locations (Frith 1962a).

This type of mating system is confined to the one megapode living primarily in an arid environment. Although the relationship between the ecological conditions and the bird's mating system remains poorly understood, it is certain that these conditions impose significant constraints on all aspects of the ecology of this species. For example, the relatively low density of birds and mounds must reduce the number of possible social interactions to a few individuals. None the less, in some populations at least, females are able to visit some other males at their mounds during solitary movements.

Despite the high degree of spatial and temporal independence of the sexes, the mating system is distinctly monogamous. Pair bonds persist throughout the year and are probably life-long. However, the potential for extra-pair copulations is enhanced by the absence of mate guarding by males, especially where the population density is sufficiently high to minimize intermound distances. The only case of polygyny reported to date involved one male tending two mounds which were used by separate females (Weathers *et al.* 1990). One explanation for this apparently unusual situation is that the artificial provision of mound materials and foods enabled this male (a wild bird, but living in a national park) to overcome the normal constraints to breeding in the area, thereby allowing him to construct an additional mound. It is noteworthy that the mating system of this bird very closely resembles that of the other resource-defending species (type IIa).

These features highlight the sources of variability in reproductive behaviour that exist for males of species that exhibit resource defence as opposed to mate defence. In particular, the inability of resource-defending males to prevent possible matings between females and other males indicates that males, in receiving eggs laid by potentially promiscuous females, may be subject to a high level of cuckoldry. This is a risk that is impossible for males to control because of the delay of several days between a copulation and the resulting egg being laid. Thus, these males may be forced to incubate eggs fertilized by other males as an unavoidable part of their reproductive strategy. Therefore, as Jones (1990a) has argued, the dangers of cuckoldry for these mound-building species relate primarily to the costs they may incur as a result of caring for another male's eggs. However, in these species, the only parental care provided is the provision and maintenance of the mound: a form of care fully shareable among all of the eggs.

Moreover, Jones (1990b) has suggested that, for Australian Brush-turkeys and possibly for all mound-defending species, mounds are used by males to attract females, as well as simply to provide an incubator. Males seeking to obtain the maximum number of copulations must remain at their mounds for as long as possible. For Australian Brush-turkeys, the males most successful in receiving eggs were those that maintained viable mounds for the longest period (Jones 1987b). Other mating tactics employed by males of this species involved increasing the proportion of mounds under their control by expelling some males from their mounds, constructing additional mounds, and even usurping the mounds of other males (Jones 1990b).

8

Evolution of megapode incubation strategies

Introduction

Superficially, the breeding strategy of megapodes shows more resemblance to the breeding behaviour of crocodiles and other reptiles than to that of birds (see Chapter 6). Some scientists have therefore assumed that megapodes are primitive birds which have inherited their breeding behaviour directly from their reptilian ancestors. Others have suggested that mound building and burrow nesting are strategies which are derived from 'normal' avian incubation and, thus, that ancestral megapodes incubated their eggs with body heat, as do other galliforms. In this chapter we shall review the various arguments about the origin of the megapode incubation strategy.

Megapode incubation: reptilian or avian?

The most detailed study of megapode relationships and the evolution of their incubation strategy was conducted by Clark (1960, 1964a, b). Clark rejected the theory proposed by Portmann (1938, 1955) that the megapode breeding strategy is a primitive trait. Portmann's ideas were based on several reptile-like characters in megapodes, such as the large number of eggs, the long incubation period, lack of an egg-tooth at hatching, lack of natal down, precocity of young, and lack of parental care. Clark pointed out that in fact megapode embryos do have an egg-tooth during the initial stage of their development and chicks do not lack natal down, while the other reptile-like characters could be explained as adaptations to their particular breeding strategy. He therefore concluded that the megapode incubation strategy was not a reptilian trait, but evolved from the typical avian behaviour of incubating eggs with body heat.

Seymour and Ackerman (1980) also compared the breeding biology of megapodes with that of reptiles and they classified the reptile-like features put forward by Portmann as convergent adaptations to their particular nesting habits, thus supporting Clark's conclusion.

Recently, Dekker (1990c) and Dekker and Brom (1992) have tried to reconstruct the behavioural changes with regard to breeding which must have taken place during the evolution of megapodes (see phylogenetic tree, Fig. 8.1). On the basis of the presumed relationships between megapodes, other birds, and reptiles put forward by Cracraft (1973) and Cracraft and Mindell (1989), they deduced that megapodes did not inherit their incubation strategy directly from reptiles, and that any similarity between their respective incubation strategies was due to convergent evolution. They therefore reached the same conclusion as Clark, and Seymour and Ackerman, though from a different standpoint: incubation by means of body heat is a typical avian character, exhibited in all birds except megapodes, but absent in reptiles. In the course of evolution, the reptilian method of incubation was already lost, and the typical

74 The Megapodes

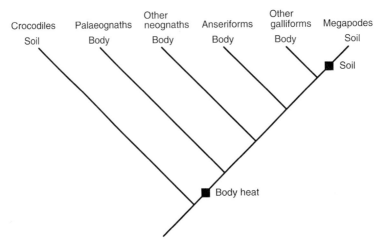

8.1 The onset of body heat incubation in birds and soil incubation in megapodes. (After Dekker 1990c; phylogenetic tree according to Cracraft 1973 and Cracraft and Mindell 1989.)

avian incubation strategy developed, before the first avian offshoot, the palaeognaths (the ostriches and allies) branched off. As megapodes and other galliforms did not branch off until later, and as all galliforms except megapodes incubate their eggs by means of body heat, their common ancestor must also have incubated its eggs by means of body heat, just as the palaeognaths did and still do. Thus, the common ancestor of all megapodes originally incubated its eggs by means of body heat, and developed the unique megapode breeding strategy only later; although very similar to that of reptiles, it could never have been inherited directly from them, as Fig. 8.1 illustrates.

From mound building to burrow nesting

With a substantial body of evidence suggesting that the megapode incubation strategy has been derived from the typical way of incubating eggs with body heat, the question of which strategy was developed first, mound building or burrow nesting, can now be tackled.

Mound building is the more commonly used strategy. It is exhibited by 19 species, representing all genera except *Eulipoa* and *Macrocephalon* (see Table 5.1). The genera *Aepypodius*, *Alectura*, *Leipoa*, and *Talegalla* consist entirely of mound builders, whereas the Moluccan Megapode *Eulipoa wallacei* and the Maleo *Macrocephalon maleo* are invariably burrow nesters. *Megapodius* is the only genus which is represented by mound builders as well as burrow nesters. Most *Megapodius* species build mounds or lay their eggs between decaying roots of trees. In certain species, some populations build mounds and others dig burrows. Only the Polynesian Megapode *Megapodius pritchardii* is an obligate burrow nester, laying its eggs at volcanically heated sites, though it is possible that it also uses solar radiation heat for incubation.

Some ornithologists, such as Meyer and Wiglesworth (1898), Frith (1962a), and Immelmann and Sossinka (1986), have considered that the more simple method, burrow nesting, must have been the initial strategy and that mound building was derived from it. Clark (1960, 1964a, b), on the other hand, speculated that mound building developed first and was later modified to burrow nesting in the Maleo and some *Megapodius* species; he felt that it was easier to conceive of mounds evolving from a simple nest than from the burying and abandoning of eggs in the ground. Clark supposed that the megapodes were derived from an ancestral jungle-nesting group rather than from birds nesting in sandy, open areas. His scenario was as follows: as an initial step in becoming a mound builder, ancestral megapodes covered their eggs with organic matter whenever they left the nest, as do grebes (Podicipedidae), for example. Gradually they added more and more organic matter and returned less and less frequently to check the nest temperature. With less attention paid to the nest, eggs would hatch in the absence of the adults and selection would favour the most precocial chicks; this in turn would have led to larger eggs. To date, however, no clear advantages of the megapode breeding strategy relative to that of other galliforms have been suggested.

Evidence from phylogeny

Dekker (1990c) and Dekker and Brom (1992) reconstructed the changes in breeding behaviour within the megapode family itself, in the same way as they had reconstructed the changes during the evolution of megapodes (described above). For this purpose, they plotted the incubation strategies of the species of each genus over the phylogenetic tree illustrating the relationships of megapodes presented in Chapter 2 (Fig. 2.2). Step by step, while going back along 'the branches of the tree', the breeding strategy used by common ancestors could be determined (Fig. 8.2). The common ancestor of *Macrocephalon*, *Eulipoa*, and *Megapodius* incubated its eggs in mounds as well as in burrows, as is still encountered nowadays in some *Megapodius* species, such as the Melanesian Megapode *M. eremita*, Micronesian Megapode *M. laperouse*, and Philippine Megapode *M. cumingii*. Others, such as the Polynesian Megapode, lost the ability to incubate eggs in mounds, while several species, such as the Nicobar Megapode *M. nicobariensis* and Orange-footed Megapode *M. reinwardt*, lost the ability to incubate their eggs in burrows. Both the Maleo and Moluccan Megapode entirely lost the ability to build mounds and became true burrow nesters.

Figure 8.2 also illustrates that burrow nesting was developed after the remaining four extant genera, *Alectura*, *Aepypodius*, *Talegalla*, and *Leipoa*, which are all mound builders, had branched off and that the common ancestor of all extant megapodes must therefore have been a mound builder. Thus mound building represents the more primitive (plesiomorphic) condi-

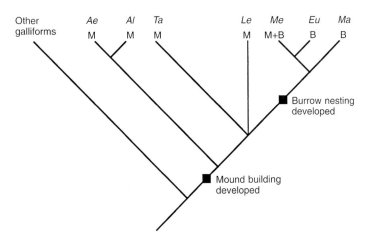

8.2 The onset of mound building and burrow nesting in megapodes. (After Dekker 1990c and Dekker and Brom 1992; phylogenetic tree according to Fig. 2.2.) *Ae* = *Aepypodius*, *Al* = *Alectura*, *Ta* = *Talegalla*, *Le* = *Leipoa*, *Me* = *Megapodius*, *Eu* = *Eulipoa*, *Ma* = *Macrocephalon*, M = mound building, B = burrow nesting.

tion and burrow nesting has been derived from it, as Clark (1960, 1964a, b) has already speculated. The development of the burrow-nesting strategy by the common ancestor of *Macrocephalon*, *Eulipoa*, and *Megapodius* may have been triggered by the discovery of new heat sources suitable for incubation, coinciding with the occupation of new habitats in new locations.

If Dekker and Brom had plotted the same incubation strategies over the phylogenetic tree presented by Clark (see Fig. 2.1), in which *Aepypodius*, *Alectura*, *Leipoa*, and *Talegalla* are presumed to form a monophyletic group, no definite answer would have been obtained: mound building, burrow nesting, or a combination of both strategies could each have represented the ancestral condition. However, of these three possibilities, the second and third are less likely than the first, because the number of evolutionary steps necessary to explain the process would have been greater or their direction much less probable. If, for instance, the megapode incubation strategy was presumed to have started as burrow nesting, then mound building would have developed twice independently: once in the *Aepypodius*, *Alectura*, *Leipoa*, *Talegalla* group, and once in *Megapodius*.

Evidence from egg yolk

Further support for the conclusion that burrow nesting has been derived from mound building is given by the yolk content of eggs of species of both groups (Dekker 1990c; Dekker and Brom 1992). Eggs of precocial avian species contain a larger proportion of yolk than eggs of less precocial or altricial species. Ancestral megapodes which developed mound building must have raised chicks which were less precocial than those of extant megapodes, and their eggs must have contained a smaller proportion of yolk, resembling the condition in other galliforms. The proportion of yolk in eggs of representatives of all galliforms except megapodes ranges from 32 to 49 per cent of the egg contents weight. Eggs of species in the mound-building *Aepypodius*, *Alectura*, *Leipoa* group contain 48–55 per cent yolk, while it ranges from 61 to 69 per cent in the *Macrocephalon*, *Eulipoa*, *Megapodius* group (see Table 2.1). This is considered as an indication of the direction in which the megapode breeding strategy developed: from a large amount of yolk and precocial chicks in galliforms which incubate their eggs with body heat and take care of their young, to a larger amount of yolk and super-precocial chicks which are independent in mound-building megapodes, to a still higher amount of yolk in burrow-nesting megapodes (see also Figs 2.2 and 6.2). From this point of view it would be interesting to compare chicks of mound-building megapodes with those of burrow nesters, as it would be expected that the latter were slightly further developed at hatching. Similarly, the incubation period at the same constant temperature would be expected to be shorter for eggs of, for example, the Australian Brush-turkey *Alectura lathami* which contains 48–52 per cent yolk, than for eggs of the Maleo with 61–64 per cent yolk. Descriptions of the incubation period and of maturity at hatching in the literature seem to lend support to both these hypotheses; however, no direct comparisons have yet been made, partly due to the fact that Australia, where most of the research is conducted, lacks the presence of burrow nesters.

Evolution of the burrow-nesting strategy

Burrow nesting in volcanic soils, and burrow nesting at sun-exposed beaches, are in fact two distinct strategies, as different heat sources are utilized. The question of whether volcanic or solar heat was exploited first cannot be answered by plotting both strategies over a phylogenetic tree of the genus *Megapodius*, the only genus in which all strategies are exhibited, because the relationships within this genus have not yet been unravelled. However, Dekker (1990c) and Dekker and Brom (1992) have argued that burrow nesting at geothermal sites developed first. Some burrow nesters, such as the Maleo and possibly the Moluccan Megapode, bury their eggs at sun-exposed beaches as well as at volcanically heated sites. There is also one species, the Melanesian Megapode, which burrows in beaches and volcanic soils, and also builds mounds. However, there are no megapodes which burrow at sun-exposed beaches and build mounds, but do not incubate their eggs at volcanic soils. This may mean that burrow nesting in volcanic soils is a necessary intermediate step between mound building and burrow nesting on sun-exposed beaches. The situation on the main island of New Guinea lends support to this; vast stretches of apparently suitable tropical beaches are available there, yet none of the six species of megapode living in New Guinea (representing three different genera, *Aepypodius*, *Megapodius*, and *Talegalla*) utilize the beaches; all are mound-builders. Volcanic soils are, however, rare or absent, and perhaps the reason that the New Guinea birds do not use the beaches is that they have had no opportunity to take the intermediate step of burrow nesting in volcanic soil. This would suggest that beach nesting is derived from burrow nesting in volcanic soils and not directly from mound building.

Place of origin of megapode breeding

The fact that megapodes are restricted to Indo-Australia, and that their particular breeding behaviour is not exhibited by any other avian family elsewhere in the world, is probably related to the absence of carnivores in that region at the time that mound building developed (Dekker 1989c).

Some female galliforms minimize predation by carnivores during incubation by the use of adaptations such as cryptic plumage and minimal loss of body heat and body odour. Other galliforms, such as the neotropical curassows and guans (Cracidae), nest in trees and thus avoid confrontations with ground-living predators. Mound-building megapodes on the other hand are very conspicuous while at their 'nest' where they spend day after day, noisily scratching leaves and soil from the forest floor. Both male and female frequently disappear with their head or even their entire body deep inside their nesting holes while digging in the mound, thus losing the opportunity to observe what is going on in the immediate surroundings or to escape if a predator attacks. This makes them highly prone to predation by territorial carnivores roaming the area, and may lead from the extermination of small, local populations to the disappearance of megapodes from much larger areas. The failure of mound-building *Megapodius* species to settle on the mainland of north Borneo and on the largest island in the Kangean Archipelago, despite their presence on small (but carnivore-free) islands nearby, has been attributed to the presence of various species of cats (Felidae) and civet-cats (Viverridae) (Dekker 1989c; see Chapter 3). Indo-Australia, the only area which was originally devoid of mammalian carnivores, seems therefore to be the most likely place for the megapode breeding strategy to have developed.

9

Conservation

Introduction

Megapodes are increasingly threatened by over-exploitation of their eggs, predation by humans and the carnivores they have introduced, competition for food with introduced herbivores, and destruction of their habitat. Whether on the Nicobars, on Niuafo'ou, in the Moluccas, or in the Philippines, everywhere the situation gives cause for concern. Megapode eggs are considered a delicacy since they are large and rich in yolk, a natural resource which is quite important for some local economies. This chapter summarizes the status of the various species, threats to their survival, conservation projects, and measures for their protection.

Human exploitation of eggs

Large-scale exploitation of megapode eggs has been described for the Maleo *Macrocephalon maleo*, the Moluccan Megapode *Eulipoa wallacei*, the Melanesian Megapode *Megapodius eremita*, and the Polynesian Megapode *Megapodius pritchardii*, all of which are burrow nesters. In these species the harvested eggs form an intricate part of village economy and can be such a lucrative business that nesting grounds are divided into plots and exploited according to strict rules. Similarly, in New Guinea the eggs of mound builders are exploited: a mound may become the property of the person who finds it and he retains the right to harvest the eggs in subsequent years. Even the ownership of entire nesting grounds is sometimes subject to tribal law and custom. Serious fights, and even deaths, resulting from disputes over nesting grounds are still remembered by older people on New Britain.

This system of controlled exploitation worked well in the past, as only a part of the daily egg-production was harvested, leaving enough eggs to hatch. It still sometimes works quite well at the present day. In the Moluccas and some parts of Sulawesi, tenants can buy the sole rights of exploitation for a period of one or more years. A financial deal is made between the tenant and the village community and is respected by all villagers. As recently as 1991 the exploitation rights to one major nesting ground of the Moluccan Megapode on Haruku were sold for 2 750 000 Rupiah (*c.* $US 1500) for 1 year; an enormous amount by local standards. Eggs were sold for only RP 250 each (*c.* 15 US cents). However, the yearly income from the egg sales is estimated to have approached RP 10 000 000. A large amount of the income from egg sales is refunded to the community and in this particular case it was used for maintenance of the village mosque (Dekker 1991). Thus, the nesting ground of the Moluccan Megapode so vividly described by Martin as early as 1894 and again in 1953 by Wiljes-Hissink still seems to flourish under these strict rules.

In many instances, however, the loss of traditions and the mixing of cultures has turned the tide considerably in recent years. Tenants disappear and nesting sites are invaded by an uncontrolled number of egg collectors; this has happened, for example, on the majority of Maleo nesting grounds in Sulawesi, where the species is rapidly declining (Dekker 1990a).

The number of eggs that are harvested is often tremendous. Bishop (1978, 1980) reported that inhabitants of a single village collected 18 000 eggs per month at Pokili, a communal nesting ground of the Melanesian Megapode in western New Britain. The total number of eggs produced monthly at that particular site must have been several times higher, because people from other villages collected there as well. Kisokau (1976) estimated that 30 000 eggs per season were laid at Garu, another breeding ground of the Melanesian Megapode in western New Britain. Downes (1972) mentioned that there are indications of average minimum monthly collections of 5000 eggs, but did not indicate at which nesting ground. At the nesting ground of the Moluccan Megapode at Kailolo, Haruku, at the time of writing an estimated 40 000 eggs are harvested each year (R. Dekker, personal observation). The number of eggs collected at nesting grounds of other species such as the Maleo has probably never been as high as that.

In 1947, 9705 Maleo eggs were collected at the Panua nesting ground in North Sulawesi (Uno 1949). According to MacKinnon (1981), the daily production there did not exceed two or three eggs 30 years later.

Mound-building megapodes suffer heavy egg losses as well, but because their mounds are scattered throughout often dense forests some do remain undiscovered if the forested area is large enough. Endemics of small islands, such as the Nicobar Megapode *Megapodius nicobariensis*, are particularly vulnerable because of the limited habitat available in which to hide mounds.

Human impact on individual species and their conservation status

Brush-turkeys and Talegallas

The Australian and Wattled Brush-turkeys, *Alectura lathami* and *Aepypodius arfakianus*, as well as two of the three *Talegalla* species, are considered stable at present according to criteria and threat categories described by Mace and Lande (1991) (see Table 9.1). All are

Table 9.1 Conservation status of megapodes according to Mace–Lande threat categories (Mace and Lande 1991)

Australian Brush-turkey	S	Micronesian Megapode	V
Wattled Brush-turkey	S	Nicobar Megapode	V
Black-billed Talegalla	S	Philippine Megapode	V
Brown-collared Talegalla	S	Sula Megapode	V
Dusky Megapode	S	Tanimbar Megapode	V
Forsten's Megapode	S	Vanuatu Megapode	V
Melanesian Megapode	S	Maleo	V/E
New Guinea Megapode	S	Bruijn's Brush-turkey	E?
Orange-footed Megapode	S	Polynesian Megapode	E
Red-billed Talegalla	V?	Biak Megapode	E
Malleefowl	V	Moluccan Megapode	E/C

Mace–Lande criteria: S, Stable; V, Vulnerable; E, Endangered; C, Critical; ? status uncertain due to lack of data.
Definitions according to the Mace–Lande criteria: critical: 50% probability of extinction within 5 years or two generations, whichever is longer; endangered: 20% probability of extinction within 20 years, or 10 generations, whichever is longer; vulnerable: 10% probability of extinction within 100 years.

mound builders with populations which are widely distributed or which are living in areas without a high human population. However, the status of the Red-billed Talegalla *Talegalla cuvieri* from the Vogelkop peninsula of New Guinea, and various isolated subspecies such as the blue-wattled race of the Wattled Brush-turkey, *A. arfakianus misoliensis*, is less well understood, but appears to be vulnerable.

The status of the Bruijn's Brush-turkey *Aepypodius bruijnii* from Waigeu is unclear. It is known from 13 or 14 specimens which were collected in the nineteenth century by native hunters employed by A. A. Bruijn, a resident of Ternate, Indonesia who was engaged in the bird-skin and feather trade; also from one other specimen which was collected in 1938 by Joseph Kakiaij. The species has never been observed in the wild by outsiders on any other occasion. However, it was known to residents of at least four villages on the island in 1986 (J. M. Diamond, personal communication). *Aepypodius* species are shy and quiet and the population of *A. bruijnii* is probably small. The terrain on Waigeu is extremely rugged karst and its inaccessibility probably affords a great deal of protection for the birds and their eggs.

The Malleefowl

The Malleefowl *Leipoa ocellata* of the semi-arid environment of southern mainland Australia has undergone a severe decline in numbers and area of distribution during the twentieth century. By 1916 the species was already disappearing from many parts of New South Wales and the decline has increased tremendously; in 1985, the population in New South Wales was estimated at 750 breeding pairs. It may also be gone from the arid inland and from areas of central and south western Victoria; the population in Victoria is now estimated at less than 1000 pairs. The total population of this species may not exceed 2000 pairs.

The Malleefowl is threatened primarily because its habitat, the mallee, is being cleared and fragmented or modified for the cultivation of wheat, grazed upon by domestic stock, or frequently burned. Fire makes the quality of the habitat low or unsuitable for the Malleefowl for the first 10–20 years; it improves slightly during the next decades to become truly suitable only after 40–60 years. However, the typical fire frequency prescribed for optimizing forage production for domestic stock is only 20 years (Benshemesh 1990*b*). Food shortage through competition with domestic stock, goats, and rabbits adds to the problem, as has been shown by Priddel and Wheeler (1990*a*).

Red Foxes *Vulpes vulpes* impose another serious threat, especially in small reserves and in habitats that are of marginal quality for Malleefowl, causing a high rate of predation of eggs, chicks, and adults. Locally, foxes may take one-third of all the eggs and in one instance are known to have killed half of a total of 100 chicks within 4 months after they were released; up to 40 per cent of these were dead within the first 2 weeks. By destroying eggs and killing chicks, the fox alone may reduce the reproductive output by up to 75 per cent (Priddel 1990).

The following conservation measures have been proposed:

(1) protection and improvement of Malleefowl habitat;
(2) fire management to keep the quality of the habitat adequate for viable populations;
(3) removal of competing grazers to guarantee the availability of food;
(4) control of foxes to reduce predation;
(5) captive breeding to provide a source of animals for reintroduction into areas where the negative impacts on the species have been ameliorated by concerted management action;
(6) monitoring of breeding densities (Frith 1962*b*; Brickhill 1987*b*; Benshemesh 1990*b*; Priddel 1990).

Various state governments are now supporting conservation-oriented research and protection of some remaining patches of suitable Mallee-fowl habitat. A captive breeding project aimed at reintroducing birds into the wild has started in New South Wales with some positive results so far, while fox eradication programmes are running simultaneously. The Malleefowl is thus the subject of an extraordinary conservation effort.

The Maleo

The Maleo is a Red Data species classified as vulnerable because its eggs are collected by humans, its nesting grounds and habitat destroyed for human expansion and commerce, and the birds themselves killed and disturbed by feral dogs. Conservation of the Maleo requires a different approach from that proposed for the Malleefowl. Changing traditions in Sulawesi due to the influence of transmigrants from Java and Bali have altered rules and laws which are said to have been obeyed by the original, native egg collectors. In the past, egg harvesting was strictly forbidden during certain periods, which allowed at least a percentage to hatch, adding new birds to the population. Nowadays, however, there is no control on the number of Maleo eggs harvested, and the resulting competition for eggs means few are left to hatch. A further threat lies in the destruction of entire nesting grounds, which causes a total loss of the source of recruitment for individual populations. A list of 48 nesting grounds for the whole of Sulawesi (Dekker 1990a) showed that only 18 were known along the coast where the sun provides the incubation heat; the remaining 30 are situated inland in areas heated by geothermal activity or along rivers where the sun warms the ground. Eight of the 48 nesting grounds have been abandoned during this century, and several others are expected to be abandoned soon; 21 are more or less severely threatened; only four were considered safe for the time being, while data were insufficient for the remaining 15. It is predicted that by the turn of this century all coastal nesting grounds will have been abandoned. Since that publication, 37 new nesting grounds have been reported in several papers and through a survey conducted by Marc Argeloo along the coast of North Sulawesi. This brings the total number of nesting grounds to 85 (M. Argeloo, in preparation; see Species accounts for details). Unfortunately, these new sites face the inevitable threats reported for most of the other nesting grounds.

Conservation projects for Maleos in Sulawesi have been initiated and eggs have been incubated in predator-proof hatcheries built at the nesting ground to guarantee more offspring (MacKinnon 1981; Dekker and Wattel 1987). So far, this has produced some satisfactory results. Hatching success in the Dumoga-Bone National Park, North Sulawesi, in 1985–1986 was 55 and 75 per cent in hatcheries at two nesting grounds, giving rise to more than 1200 chicks. A similar hatching percentage was obtained in 1991. Large number of eggs must be incubated to give a significant increase in the number of adult birds, as chick mortality in megapodes is high; no quantitative data are available for the Maleo, but research in Australia has indicated a mortality of more than 90 per cent for both the Australian Brush-turkey and the Malleefowl (Jones 1988c; Priddel 1990). Nesting grounds are often not much larger than one or two hectares and can easily be protected against egg collectors and predators. Thick ground-covering secondary vegetation, especially *Lantana camara* and *Imperata cylindrica*, which covers the nesting ground almost everywhere on disturbed soil in Sulawesi, hinders the Maleos and must be cleared. The local people are often ignorant of the impact of their harvest and need to be better informed about the species, its status, and its prospects; during the 1991 conservation project, much attention was paid to education through the distribution of posters and information leaflets to schools and public buildings. Finally, small-scale national and international nature tourism visits to nesting grounds might help to increase the awareness and stimulate the economy of the people living alongside megapodes. A conservation project such as the one in the Dumoga-Bone National Park might even aim at re-establishing a controlled harvesting of eggs for the inhabitants of nearby villages, once the Maleo population has been brought back to a reasonable number. With an adequately controlled system, the number of surplus eggs that could be harvested in the future would not only be much larger than the number of eggs collected today, but would also be guaranteed for a prolonged period of time. Such a strategy would not only be beneficial for the local community, it might also guarantee the survival of local populations of the Maleo.

The Moluccan Megapode

The situation with regard to the burrow-nesting Moluccan Megapode is unclear, but there is little reason for optimism, despite the rediscovery of a magnificent nesting ground at Kailolo on Haruku, Central Moluccas, in 1991 (R. Dekker, personal observation). The present status of the Moluccan Megapode falls within the category 'endangered' or possibly even 'critical'. If data from local people on the number of eggs collected at Kailolo are correct, the megapode population using this site might be as large as 4000 pairs. However, the birds dependent on this single site for their reproduction are drawn not only from Haruku (which is too small and too deforested to carry such a large population), but also from other places, mainly the south coast of Ceram; this is indicated by birds flying in from the sea and landing on the beach prior to egg laying (R. Dekker, personal observation). Unfortunately, in their descriptions of the Kailolo nesting ground, neither Martin (1894) nor de Wiljes-Hissink (1953) gave data on the number of birds involved, other than that they were very common, so it is not possible to estimate the impact of nearly 100 years of controlled egg collecting. In 1991, 60 eggs from Kailolo were seen in the market of Masohi, south Ceram, 3 hours by boat from Haruku and close to another nesting ground of the Moluccan Megapode. This site, on the margins of the airstrip of Amahai, was used by only a small number of birds and villagers from Amahai not only collected all the eggs but also excavated sand and gravel from the site for road construction. Information on nesting grounds elsewhere is poor and outdated (see Species account).

Megapodius

Several burrow-nesting *Megapodius* species whose eggs are harvested face problems similar to those described for the Maleo and the Moluccan Megapode (see Table 9.1). Conservation projects and surveys have been initiated for the Polynesian Megapode, which is considered endangered, and the Micronesian Megapode *Megapodius laperouse*, which is regarded as vulnerable. Plans have been made to investigate the possibility of introducing birds to other islands by the translocation of eggs or chicks. As part of a conservation project which started in 1991, eggs of the Polynesian Megapode (endemic to the small volcanic island of Niuafo'ou) were transported in 1992 to the island of Late, where suitable hot soils or beaches are available in the absence of cats and humans and they subsequently hatched (Curio 1992b). In June 1993, D. Rinke transferred 37 eggs and seven chicks to the rugged and uninhabited island of Fonualei in northern Tonga. Surprisingly, nearly one year later Rinke observed a fully grown megapode on one of his trips to the island (Rinke 1994). Reintroduction has also been suggested for the Maleo, the Malleefowl, and the Micronesian Megapode (which is monitored in the Marianas by the United States Fish and Wildlife Service). None of these plans has yet been put into practice, however.

Mound-building *Megapodius* species face slightly different problems, although mounds on small islands are often easily accessible and are thus visited and exploited on a regular basis. The Philippine Megapode *M. cumingii* seems to have vanished recently from several islands in the Philippines, raising concern for the survival of some of its races. The status of the Sula Megapode *M. bernsteinii* from the Sula and Banggai Islands between Sulawesi and the Moluccas was completely unknown until 1991. In that year, two separate expeditions encountered it regularly in both the Sula (Davidson *et al.* 1994) and Banggai archipelagos (Indrawan *et al.* 1992). The status of the species in the Banggai archipelago was considered to be vulnerable, and the snaring of adult birds was regarded as a greater threat than egg collecting. Habitat destruction probably also causes problems, despite the ability of the birds to adapt to secondary vegetation. The Nicobar Megapode was found to be more common than expected on Great Nicobar, the largest island of the Nicobar group; on the basis of local information it had been thought that only 50–400 birds survived. The breeding population in a narrow strip along the entire coast of Great Nicobar was estimated in March 1992 at 780 pairs. Egg collecting and trapping of adult birds took place only on a very small and local scale, while the habitat of the species seems safe due to the establishment of two reserves on Great Nicobar in 1992, Campbell Bay and Galathea National Park. Owing to lack of knowledge about the occurrence of the species on the remaining islands, its status is considered vulnerable (Dekker 1992). The Biak Megapode *M. geelvinkianus* was observed in reasonable numbers on the very small island of Supiori in 1991 (D. Gibbs, *in litt.*); nothing further is known about this species. It is considered to be endangered according to the Mace–Lande threat categories. Little is known about the status of the Tanimbar Megapode *M. tenimberensis*, endemic to Tanimbar Island; owing to its restricted range and threats to its habitat it is regarded as being vulnerable. It was regularly recorded along a trail in southeast Tanimbar in December 1991 (K. Monk, *in litt.*). Even less is known about the Vanuatu Megapode *M. layardi*, endemic to Vanuatu. Owing to its restricted range it is also considered vulnerable. Some species, such as the Orange-footed Megapode *M. reinwardt*, are still widely distributed (though recent and reliable data are few), but seem to be disappearing from an increasing number of small islands.

Captive breeding

When most conservation measures fail, captive breeding programmes can be initiated, as in the Malleefowl. It is difficult to re-create artificially the conditions under which eggs of mound builders are incubated, so one alternative is to attempt to incubate them in hatcheries on volcanic soils or sun-exposed beaches. However, often this is not possible, owing to the absence of suitable sites. In addition, the effects that this method of incubation and hatching might have on breeding behaviour, when the chicks (mound builders by origin) reach maturity, have not yet been studied. All that is known is that Malleefowl hatched in incubators have successfully bred when released back into the wild (D. Priddel, personal communication).

Megapodes have several advantages over other birds in conservation management: the chicks do not receive any parental care and can thus be released immediately after hatching without being hand reared. Eggs of mound builders, and recently of the burrow-nesting Maleo also, have successfully hatched under special conditions in electric incubators. Experiments with artificial mounds, however, even if provided with additional heat generated by heating elements inserted in the mound, have yielded unsatisfactory results in most cases, since the temperature inside the mounds did not reach the optimum of 33–34 °C, or did not stabilize at this temperature for a prolonged period. Mounds will not warm to incubation temperatures unless they are of certain minimum dimensions (about 0.75 m in height and 2.0 m in diameter in the case of the Australian Brush-turkey).

For successful incubation of megapode eggs, certain conditions have to be sustained (for details see Dekker 1990c; Winn 1992). Eggs must be incubated in soil in a vertical position with the blunt end upwards and should not be turned during the long incubation period of 2–3 months. The optimal incubation temperature is 34 °C, which results in 80 per cent hatching success for Malleefowl eggs; at 36 °C hatching success was 44 per cent, at 38 °C, 38 per cent, and at 32 °C only 22 per cent (Booth 1987a). These data are likely to have general validity for all megapodes. The incubation period is negatively correlated with the incubation temperature: the higher the temperature the shorter the incubation period. Temperatures as low as 28 °C can be tolerated for at least 4 days, while 38 °C is considered to be the highest temperature at which the embryo can survive, especially shortly before hatching. The humidity of the soil is also a critical factor. Both dry and saturated substrate conditions may stop the development of the embryo due to dehydration, changes in O_2 and CO_2 tensions, and/or unsuitable temperatures. A moisture content of 20 per cent, calculated as 100 × (wet mass-dry mass)/wet mass, was used successfully in captive breeding efforts of the Maleo in the collection of the New York Zoological Society in the USA. A relative air humidity near 100 per cent in the incubator in which eggs buried in sand were brooded gave similar good results. Both temperature and humidity determine the depth at which the egg should be buried. From a practical point of view this need not be deeper than 20–25 cm.

Ackerman and Seagrave (1987) reported that water loss from megapode eggs during incubation is negatively correlated with thermal conductivity (TC). TC of sand is greater than that of compost and TC is greater in a wet than in a dry medium. Water loss from the egg is therefore smaller in (wet) sand (or burrows) and greater in (dry) compost (or mounds). The temperature increase of the egg and embryo shortly before hatching depends on TC as well and is greater in compost than in sand. It is therefore expected that megapode eggs incubated in mounds will heat up more during incubation than eggs incubated in burrows. Thus, eggs incubated in sand have less chance of overheating.

Chicks of all 22 species of megapode are fully independent of their parents; they can run as soon as they have hatched and fly shortly after and they fend for themselves, so they do not need any special care. This makes them 'ideal' birds for reintroduction programmes and they can even be released on the day of hatching. The hatchlings subsist on their subcutaneous yolk reserve for the first few days, requiring very little other food. Their weight decreases initially, by up to 12 g on the first day (average approximately 10 g) and by an average of 5 g on the second day. After a few days (up to a week) it begins to increase. This pattern has been recorded for both the Malleefowl and the Australian Brush-turkey. Provided with the appropriate food it is fairly simple to raise the chicks, as long as they are kept in a quiet place. Malleefowl chicks have been observed to feed on ground-living insects, mainly ants (Formicidae) and beetles (Chrysomelidae and Carabidae) for the first 2 weeks and subsequently they shifted to seeds of several species of *Acacia*. Maleo chicks have been raised successfully on a diet of grasshoppers, termites, a variety of other insects, peanuts, and maize. Both species obtained a body mass of *c.* 300 g after 70 days. Captive Malleefowl weighed only 75 per cent of their adult body weight after 1.5 years, indicating that maturity is not reached within the first year. Australian Brush-turkey chicks grow much faster, reaching a weight of *c.* 800 g after 70 days and approaching adult body mass after 220 days (Booth 1989a; Dekker 1989a). Both Malleefowl and Maleo reach maturity in their second or third year, while the Australian Brush-turkey reaches maturity in its first year, at least under captive conditions but probably also under natural conditions. Although the data are scarce, chick mortality seems to be high under natural conditions; in a study conducted by Jones (1988c), the estimated post-emergence mortality in the Australian Brush-turkey was 90 and 97 per cent in two consecutive breeding seasons; feral cats were mainly responsible for this.

Captive breeding alone is unlikely to stop the recent decline of megapode populations. Active protection in their natural environment and eradication of introduced predators must always have priority; otherwise within a few decades more megapodes may be added to the list of extinct species.

Conflict with humans

The Australian Brush-turkey has been declining in numbers throughout the period of European occupation of Australia, due primarily to habitat destruction and hunting. Since its inclusion on the protected fauna list in Australia in 1974, however, and the subsequent cessation of most hunting, the species has recovered its previous population densities in many locations in southeast Queensland. Indeed, it has become a major problem to people living adjacent to its normal habitat by invading house-yards in order to construct mounds. The attendant destruction of gardens, lawns, and landscaping is currently causing significant amounts of damage (Jones and Everding 1991). The species is also responsible for consistent depredations of bananas, pine seedlings in plantations, sugar shoots and other crops in northern New South Wales and south and central Queensland (Barker 1949; Keys 1990). Thus, while the majority of megapodes are under serious or increasing threats due to interactions with humans, the Australian Brush-turkey has not only recovered from previous population losses, but has even become a 'pest'. This example clearly shows that at least one megapode species is able to adapt, accept, and even use for its own benefit the presence of human beings. As for the other species, several have shown signs of a similar acceptance of or adaptation to the presence of human beings, so there are some grounds for hope. However, until man's hunger for eggs abates, there is little likelihood that the fortunes of these other species will also be reversed.

PART II

Species accounts

Genus *Alectura* Latham

Alectura Latham, 1824. *General History of Birds*, **10**, p. 455.

Synonym: *Catheturus* Swainson, 1837.

One species, two subspecies, endemic to E Australia.

Large dark small-headed long-tailed megapode. Head and neck mostly bare, boldly coloured; ad ♂ in breeding season has wrinkled and swollen vascular skin extending from middle of neck downwards, hanging from rear and side of neck down into broad flap-like folded wattle over upper chest when bird at rest, inflated to large pouch round lower neck in display; wattle shrunken and inconspicuous outside breeding season, inconspicuous in ♀ and imm. Bill short, two-thirds to three-quarters of length of head; deep at base, strongly compressed laterally; tip of culmen strongly decurved. Nostril large, rounded. Wing short, broad, tip bluntly rounded; 10 primaries, p4–p5 longest, outermost feather (p10) 7–9 cm shorter than p4–p5, p9 3.5–5 cm shorter, p8 1.5–3 cm shorter, p7 1–1.5 cm shorter, p6 and p3 0–1 cm shorter, p2 1–2 cm shorter, innermost (p1) 2–4 cm shorter. Tail relatively (and absolutely) longest of all megapodes; 18 feathers, each markedly broad, up to 7 cm; fourth central pair (t4) longest, central pair (t1) 4–8 cm shorter, outer pair (t9) 8–11 cm shorter; tail laterally compressed, each tail-half fan-shaped as seen from the side. Upper tail-coverts short. Tarsus and toes relatively short for a megapode, though very strong; tibia fully feathered, feathering extending over upper front of tarsus; front of tarsus with two rows of large hexagonal scutes. Claws long and strong, relatively longest of all large megapodes. Basal half of first phalanges of middle and inner toe connected by small web; a trace of a web between middle and outer toe also.

Australian Brush-turkey *Alectura lathami* J. E. Gray, 1831

Alectura Lathami J. E. Gray, 1831, *Zoological Miscellany*, **1**, p. 4.

PLATE 1

Polytypic. Two subspecies. *Alectura lathami lathami* J. E. Gray, 1831, E Australia from Mt. Amos, just south of Cooktown (Queensland), south to New South Wales. (Synonyms: *Meliagris lindesayii* Jameson, 1835; *Catheturus australis* Swainson, 1837; *Catheturus novaehollandiae* Bonaparte, 1856; *A. l. robinsoni* Mathews, 1912, Cairns). *Alectura lathami purpureicollis* (Le Souëf, 1898), NE Australia from Cooktown north to Cape York.

Description
PLUMAGES
ADULT MALE: head and neck mostly bare, except for sparse short black bristles on forehead and crown and very scanty hair-like bristles on nape, cheek, chin, throat, and neck; bristles on forehead pointed forward and upward, recurved, those on crown upward and backward, straight. Mantle and scapulars dark brownish-black, slightly tinged plumbeous when plumage fresh; fringes along feather tips slightly paler, brown-grey. Back and rump grey-brown, feathering dense and woolly; upper tail coverts dark fuscous-brown with indistinctly paler grey- or olive-brown tips. Chest and side of breast brownish-black, feather tips with rather distinct paler grey-brown or buff-brown fringes; breast, belly, and flank fuscous-brown with broad and contrasting off-white fringes, giving underparts a scalloped appearance; width and contrast of pale scalloping somewhat dependent on abrasion and bleaching, and also some individual vari-

Australian Brush-turkey *Alectura lathami*

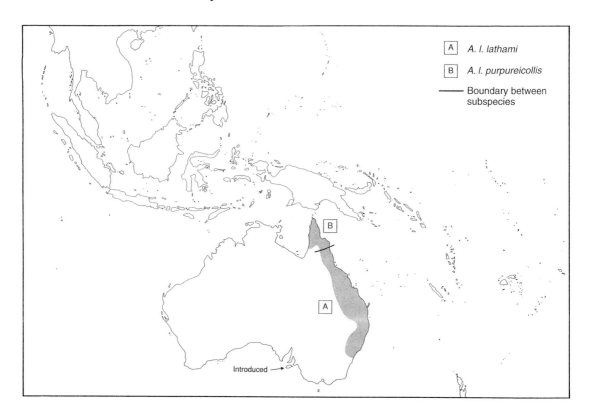

ation: in some birds (perhaps especially in *purpureicollis*) fringes broad and white, coalescent, almost hiding dark feather centres on belly. Vent grey-brown, feathers tipped dirty white or isabelline; rear flank and under tail coverts more or less uniform dark fuscous-brown; feathering of thigh fuscous-brown with ill-defined pale brown or isabelline feather tips. Tail black-brown, almost black on terminal halves of tail feathers. Flight feathers, tertials, and upper wing-coverts dark fuscous-brown, almost black on inner webs of tertials and flight feathers, slightly paler and more grey-brown on shorter coverts and outer fringe of flight feathers. Under wing-coverts and axillaries blackish-brown.

ADULT FEMALE: like ad ♂, but head and neck somewhat less bare, forehead, crown, cheek, and area round ear more densely covered with bristle-like feathers, neck covered with sparse hair-like plumes, only lore and region round eye fully bare, though naked skin well visible elsewhere. Differs from ad ♂ mainly in colour and development of wattle on neck (see 'Bare parts') and in size (see 'Measurements').

CHICK: crown, hindneck, and upper mantle brown-buff, sometimes lightly tinged rufous; some darker brown of feather bases partly visible. Remainder of upperparts, whole upperwing, and flight feathers sooty brown, upperparts with fuscous-brown or pure brown feather tips. Lore and region round eye bare; a small patch of dark brown feathers round ear opening; remaining side of head as well as side and front of neck, chin, and throat buff. Chest buff-brown or rufous-brown, breast and flank greyish-buff, mid-belly and vent pale buff. In some birds, lower mantle to back and upperwing and chest to belly very faintly barred dusky grey, but general appearance of chick mainly uniform brown.

IMMATURE: rather like ad ♀ and sometimes difficult to separate. In first imm plumage, at age of 30–100 days, feathering of head and neck less bristly, forming lanceolate black-brown plumes (forming feathered cap over forehead, crown, and nape), barbs of each feather less reduced than in ad, leaving less of bare skin visible, but lore and region round eye fully bare; outer primaries still juv, p10 rather sharply pointed, length 145–151 mm to insertion in skin ($n = 4$; in ad 156–172 mm, mean 163.6 ± 5.42, $n = 14$); wing and tail rather short (in nominate *lathami* wing 253, 259, and 259 mm, tail 223, 232, and 245 mm, $n = 3$); tail feathers narrow, 4–5 cm (in ad, 6–8 cm), less stiff than in ad. In second imm plumage, at age of c. 3–10 months, even more closely similar to ad ♀, feathering of head and neck more bristly and scanty; outer primaries and tail still in first imm plumage, p10 often somewhat pointed, length 146–153 mm (149.1 ± 2.56, $n = 5$); wing (to first imm wing-tip; nominate *lathami*) 267–304 mm (285.6 ± 14.75, $n = 7$), tail 226–237 mm (232 ± 4.34, $n = 5$). Some ♂♂ have a fully developed wattle but show more extensive bristles on forehead, crown, and neck than ad ♂ and have wing- and tail-length below average of ad ♂; these perhaps in third imm body plumage with second imm outer primaries and tail.

BARE PARTS

ADULT MALE: *A. l. lathami*. Iris light or pale brown, olive, blue, blue-grey, pale grey, greyish-white, brownish-white, or dull brown-grey. Eyelids sometimes yellow. Skin of head and upper half of neck deep red or bright flesh-red; lower half of neck including wattle bright saturated cadmium-yellow. Wattle conspicuous and strongly vascularized in breeding season, forming thick pouch round base of neck, but much reduced in non-breeding season, when colour of head, neck, and wattle less bright, more pinkish-red or pale red on head and upper neck, more lemon- or buffish-yellow on lower neck; skin of lower neck still furrowed and loose. Bill black-brown to black; base tinged green or pale olive. Leg and foot dark horn-brown to black-brown, side and rear of tarsus and sometimes undersurface of foot dull greenish-yellow, olive, or grey; sometimes stained with reddish soil.

ADULT FEMALE: *A. l. lathami*. Like ad ♂, but colour of head and neck less bright and wattle scarcely developed, much as in ad ♂ in non-breeding season. Iris also recorded as ochre and brown.

CHICK: *A. l. lathami*. Iris light brown. Bare skin of lore and region round eye dull yellow. Bill black. Leg and foot dark olive-brown or flesh-brown.

IMMATURE: *A. l. lathami*. Like ad ♀, but no trace of wattle; iris brown to pale brown, skin of head and upper neck and dull reddish or pale maroon, lower neck yellow.

A. l. purpureicollis: like nominate *lathami*, including red head and upper half of neck, but lower half of neck and (if present) wattle mauve, ivory-grey, purple-white, whitish-purple, or pink instead of yellow. In ad, iris brownish- or greyish-white, but also recorded as 'dark yellow' (BMNH); bill recorded as grey-green or green-grey (D. P. Vernon, *in litt.*); leg and foot brown or dark brown, but also recorded as 'yellow' and 'red' (BMNH).

MOULTS

Primary moult serially descendant. Two moulting series in wing present in older imm, two to four in ad. Moult starts with p1 and with suspended previous series (about midway in wing) between late Feb and early June; in ad suspended again early July to early Aug until next moulting season Feb–June, but in some imm moult continues Aug–Jan. In juv, p1 of first moulting series shed about May–June, before juv outer primaries fully grown; second moulting series starts with p1 when first series reaches p6–p8 (on average, 6.8, $n = 4$) and outermost primaries still juv; first series not suspended in six birds examined, but suspension of second series frequent when it reaches outer primaries and a new third series has started on innermost primaries. In ad with two moulting series, distance between inner and outer series four to nine primaries (on average, 5.7, $n = 6$), in those with three or four moulting series, distance between series three to five primaries (on average, 3.9, $n = 9$). In ad, sometimes only one or two primaries of each series replaced in single moulting season, perhaps especially in older birds, either because season short or because moult slow. In SE Queensland, 80% of all ad moult body- and flight-feathers between Feb and May in non-breeding season (Rogers et al. 1990).

MEASUREMENTS
ADULT: *A. l. lathami*. S and C Queensland, skins of 15 ♂♂ and 14 ♀♀ (BMNH, RMNH, SMTD, ZFMK, ZMA, ZMB).

MALES	n	mean	s.d.	range
wing	9	325.1	7.13	313–334
tail	9	258.6	13.58	236–274
tarsus	12	100.9	3.21	97.0–106.0
mid-toe	12	53.2	3.37	49.7–59.1
mid-claw	10	28.0	2.11	25.5–31.1
bill (skull)	11	43.5	1.80	40.7–45.7
bill (nostril)	11	22.4	1.12	20.3–24.3
bill depth	10	18.9	0.77	17.4–20.2

FEMALES	n	mean	s.d.	range
wing	9	310.6	8.00	298–322
tail	8	247.2	9.48	226–260
tarsus	12	90.1	4.71	84.0–99.2
mid-toe	13	48.6	2.26	45.8–52.3
mid-claw	10	25.8	2.61	21.6–29.2
bill (skull)	9	40.9	1.97	38.2–43.8
bill (nostril)	9	20.6	0.93	19.5–22.0
bill depth	8	17.3	1.21	15.6–18.7

Wing, ♂ 321.8 ± 14.1 mm ($n = 42$), ♀ 304.7 ± 12.4 mm ($n = 34$); tail, ♂ 265.7 ± 15.3 mm ($n = 42$), ♀ 259.9 ± 12.6 mm ($n = 31$); tarsus, ♂ 123.1 ± 4.6 mm ($n = 41$), ♀ 110.4 ± 3.8 mm ($n = 32$); bill (exposed culmen), ♂ 39.1 ± 1.63 mm ($n = 43$), ♀ 36.2 ± 1.82 mm ($n = 32$) (Mt. Tamborine, New South Wales, Rogers et al. 1990).

ADULT: *A. l. purpureicollis*. Cape York area, skins of 6 ♂♂ and 2 ♀♀ (BMNH, ZFMK).

MALES	n	mean	s.d.	range
wing	4	309.8	11.90	298–326
tail	5	235.0	7.84	228–247
tarsus	6	102.4	3.06	99.0–106.5
mid-toe	6	54.1	2.50	51.8–57.8
mid-claw	6	25.9	0.78	24.3–26.4
bill (skull)	5	43.1	1.27	41.5–44.4
bill (nostril)	5	21.6	0.92	20.5–22.3
bill depth	5	19.1	0.97	17.4–19.8

FEMALES	n	mean	range
wing	2	304.5	298–311
tail	2	234.0	232–236
tarsus	2	92.2	90.7–93.7
mid-toe	2	50.2	50.0–50.4
mid-claw	2	24.0	23.8–24.3
bill (skull)	2	39.5	39.3–39.7
bill (nostril)	2	18.9	18.2–19.6
bill depth	2	17.1	16.5–17.7

CHICK: probably just fledged: wing 92, 101, 101 mm ($n = 3$); tarsus 31.9, 34.8 mm ($n = 2$) (BMNH, ZFMK, ZMB).

WEIGHTS
ADULT: *A. l. lathami* (SE Queensland). ♂ 2120–2950 g (2520 ± 180, $n = 37$), ♀ 1980–2510 g (2160 ± 250, $n = 23$) (Rogers et al. 1990).

CHICK: c. 80 g at hatching; reaches full size in captivity after approximately 4 months.

GEOGRAPHICAL VARIATION
Marked in coloration of bare parts: nominate *lathami*, occurring south from Mt. Amos, just south of Cooktown (D. P. Vernon, *in litt.*), has lower neck and (in ad) wattle yellow, *purpureicollis* north from Cooktown has lower neck mauve to purple-white; head and upper neck red in both. Slight geographical variation in size, but probably clinal, without abrupt change: nominate *lathami* from southern part of range larger than *purpureicollis* in north, especially tail, but leg and foot more or less

similar; tail/tarsus ratio of nominate *lathami* south of 25°S 2.35–2.85 (2.67 ± 0.16, $n = 10$), between 25°S and 17°S 2.34–2.66 (2.47 ± 0.10, $n = 10$); in *purpureicollis*, north of 15°S, 2.14–2.47 (2.31 ± 0.13, $n = 5$). Yellow-wattled birds of intermediate size from central Queensland sometimes separated as *robinsoni* Mathews, 1912, but difference from typical nominate *lathami* from further south too small and overlap too large to warrant recognition.

Range and status
A. l. lathami: From Mt. Amos, just south of Cooktown (Queensland), south to New South Wales. One pair introduced on Kangaroo I. (S Australia) in 1936; species still present. *A. l. purpureicollis*: Cape York District of N Queensland, north of Cooktown. However, a red-wattled Brush-turkey (nominate *lathami*) has been reported from Mt. Cook near Cooktown, thus within the range of *purpureicollis* (Ford 1988). See also Blakers *et al.* (1984).

Locally common, even invading urban environments.

Field characters
Length 60–70 cm. Unmistakable. Large greyish-black turkey-like galliform with small bare head red and neck also red. Adult ♂ has yellow wattle at base of neck, especially obvious during the breeding season. Black tail large and laterally compressed.

Voice
Jones (1987*b*) recognized four calls. The most characteristic is a deep, resonant call of one ('oom') or two ('oo-oom') syllables, descending in pitch at the end (see sonogram). This booming sound is produced by inflating the neck sac and forcing air out through the nostrils, and is given by the adult ♂ only (see Fig. 4.3). Volume and pitch are dependent upon the size of the sac which develops each year during the breeding season. Booms may be repeated up to several times per minute. Rarely would one ♂ give more than 10–20 booms in a morning. Male booming vocalization are highly variable, and individuals can sometimes be identified by the boom they

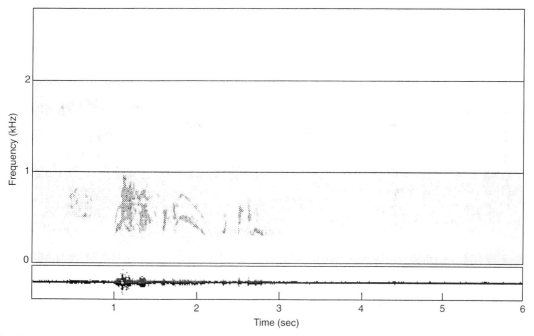

A. l. lathami
S. Birks and K. Davis, Mt. Tamborine, Queensland.

make. Older ♂♂ produce longer booms (i.e. five or six syllables), with syllables varying in length but the first one always longest. One-year-old ♂♂ seem incapable of booming (S. Birks, *in litt.*). Booms are produced in several contexts: (1) when the ♂ is at the mound, to advertise the location to ♀♀ or to deter rival ♂♂; ♂♂ never boom when a ♀ is present at the mound; (2) during aggressive contests over food when the ♂ rushes at a rival, stops abruptly, arches its neck, and inflates its wattle to give the vocalization; (3) at the roost during the evening or at dawn when birds displace one another on the branches; (4) when ♂♂ encounter each other away from mounds, with the aggressive ♂ booming before or after chasing a rival (S. Birks, *in litt.*). After the breeding season, the ♂ becomes rather silent.

The other calls recognized by Jones (1987b) are: (1) a series of deep, very low, frequently repeated gulping sounds, uttered slowly (0.5–1 per sec) or rapidly (2–6 per sec) by both sexes, and indicative of disturbance, especially by predators; (2) the main contact call, similar to the previous call but louder and deeper in pitch, uttered only infrequently and most typically by the ♂ on returning to its mound; (3) a very loud, high-pitched scream given by birds upon capture by a predator.

Juvenile (first-year) birds are apparently unable to make any vocalizations other than soft grunts given, for example, when prevented from getting to a food source by more dominant birds (S. Birks, *in litt.*). A grunting nasal sound produced by a chick was first heard at the age of 68 days, while a chick only 2 days old made a single, quiet, squawking 'anck'. This was the only chick out of about 50 handled ever heard to vocalize (S. Birks, *in litt.*). Coles (1937) reported a call described as 'moonik' by a chick at the age of 16 days.

Habitat and general habits
Humid rainforests, scrubby creek margins, scrub (*Lantana camara*), and urban environments. Also found in dry closed forest types inland from the coast, including dry sclerophyll *Eucalyptus* forest (Dow 1980), *Acacia* forest, mangroves, and isolated limestone hills (Barrett and Crandall 1931; Blakers *et al.* 1984), and even plantations of Hoop Pine *Araucaria* (Keys 1990).

All birds are primarily solitary while foraging, although large numbers will congregate at localized rich food sources. During the non-breeding season, birds of any age and either sex often perch together during the midday rest period, and spend the night at large communal roosts about 5–15 m above the ground, usually at a different site from that used at midday. ♂♂ often interact aggressively, displacing each other from roost spots. During the breeding season, however, mound-tending ♂♂ remain near their mound for most of the day, interacting with other birds only when they approach the mound. In this period, birds of both sexes usually use the communal roost closest to their mound, with ♀♀ often roosting close to a mound they are currently visiting. Such roosts, some of which are used for several years, can, for example, consist of one ♂ and two or three ♀♀ (S. Birks, *in litt.*).

Throughout the breeding season, adult ♂♂ are continually involved in dominance interactions, which include chasing, booming, but only rarely fighting. The inflatable neck sac appears to be used both as a signal of status and as a means for producing the booming vocalization. ♂♂ defend their mounds vigorously, though they are often expelled from them. They will also temporarily defend local rich food sources from other birds. There are no defended territories; home ranges of both ♂ and ♀ are significantly larger in the breeding than in the non-breeding season although there is no difference between sexes (Jones 1990a).

♀♀ are often aggressive to other ♀♀ at mounds, with some tolerating each other's presence, even when egg laying, while others fight violently and repeatedly over access to mounds. When a ♀ fights, she may leap upon her offender and use her legs and feet to scratch. Fights are much less common than chases, however. ♂♂ do not usually participate in any way in these fights (S. Birks, *in litt.*).

Chicks are entirely solitary at first, but join social groups within a few weeks. Social structure among adults is largely determined by direct agonistic interactions leading to a relatively linear hierarchy among mound-tending ♂♂, but little other discernible structure. However, there is evidence that associations between similarly-aged individuals develop among juveniles during their first few months. The duration of these associations diminishes progressively with age, especially in ♂♂; group membership among ♀♀ may persist for several years (Jones 1987*b*, 1990*a*).

Food
Wide range of seeds, fruits, berries, and other vegetable materials including young sprouts of sugar-cane and potatoes; also invertebrates such as beetle larvae, and even frogs which birds may find when working at the mound as well as when foraging. Reported to eat seed of grasses and *Geijera* sp. (Cleland 1912) and berries of *Hodgkinsonia* (Bravery 1970). Label on skin in collection BMNH mentioned 'granules, fruits, grubs, beetles, etc.' In captivity fond of mice, tadpoles, and snails (Coles 1937). (See 'Feeding ecology', p. 38.)

Displays and breeding behaviour
Promiscuous, no pair bond. ♂ arrives at the mound early each day of the breeding season and awaits ♀. Some ♂♂ produce regular booming vocalizations (see 'Voice'), apparently to advertise their location or availability. ♀ approaches ♂ in a crouched position with neck withdrawn, tail folded, and walking slowly and silently. Upon detecting the approaching ♀, the ♂ walks slowly around the top of the mound, frequently pecking at the substrate. On some occasions, especially if the ♀ does not advance or if she is encountered away from the mound, the ♂ will approach her giving 'Flattened' display with body lowered to the ground, wings spread widely, tail broad, and neck outstretched towards her. Less commonly the feet may be stretched out before the body, the tail trailing behind and head well back.

The ♀ solicits by fluffing her body feathers, drooping her wings, and spreading her tail over her back. Copulation, lasting about 5 sec (range 3–11 sec) (Jones 1990*a*), involves the ♂ walking onto the ♀'s back, grasping her neck skin and lowering his tail. The ♂'s neck sac and head are visibly bright, with the neck sac fully extended. Copulation normally occurs on the mound, sometimes nearby or further away (Dow 1988*b*). S. Birks (*in litt.*) has recognized three types of copulations: those solicited by the ♀, unsolicited but tolerated by the ♀, and forced with the ♀ trying to escape. Solicited and unsolicited copulations are common during the period immediately preceding egg laying; also, ♀♀ make special visits to the mound in between egg-laying visits to copulate with the ♂ and to dig in his mound. ♂♂ are nearly always aggressive to some extend when ♀♀ visit their mounds. This aggression takes the form of pecks at and rushes toward the ♀, and varies dramatically in intensity. Some pecks barely make contact, while other are violent and may even lead to feather loss. ♀♀ usually shield themselves with a wing when this occurs.

Egg laying involves the ♀ in excavating a deep conical hole into the centre of the mound; this usually takes 30–45 min. The ♂ does not assist at all at this stage, although he may remove some material before copulation, but rather commonly disrupts ♀'s work, frequently pecking at her head. Both ♀ and ♂ check the temperature of the mound by taking billfulls of warm substrate, and the ♂ also routinely digs small holes into the mound in order to assess the temperature. At the completion of laying, which takes 2–6 min, the ♀ treads material around the egg, but she leaves within 5–15 min, typically because the ♂ chases her away. The ♂ normally completes the filling of the hole. ♀♀ usually lay at one mound and mate with that ♂ for several weeks, then switch to a different mound and repeat this behaviour.

Because a varying number of ♀♀ lay in a single mound, and individual ♀♀ lay in more than one mound, clutch size is difficult to estimate. In captivity, single ♀♀ produce a large number of eggs (in one case 56: Coles 1937), but 18–24 eggs (Frith 1956*b*) is more probable in the wild. The interval between eggs is usually 2–5 days but may be more, especially when the ♀'s reserves are depleted (Frith 1956*b*; Vleck *et al.* 1984). Individual mounds may contain between 0 and 58 eggs, though the average is about 15 (Jones 1987*b*). ♀♀ do not have a full breeding season during their first year of adulthood.

Breeding

NEST: constructed by ♂ only. A large mound of soil and leaf litter, placed in deep shade in forest. Construction takes 3–6 weeks, and mounds may be tended continuously for up to 8 months. Initially they are piled high (1–1.3 m), but they flatten out to 0.8–1 m with use. On average, mounds in SE Queensland are 3.6 by 4.1 m in diameter (with maximum and minimum diameter ranging 3 to 7 m), and they weigh 2–4 tonnes (Jones 1988*a*; S. Birks, *in litt.*). Although traditional mound-sites are often used, individual ♂♂ do not normally use the same site 2 years following. A small percentage of ♂♂ build mounds on new sites.

EGGS: different from other megapode eggs, except those of the Wattled Brush-turkey *Aepypodius arfakianus* and possibly Bruijn's Brush-turkey *Aepypodius bruijnii*, in being white instead of pinkish-brown. Similar in shape to eggs of *Aepypodius*, *Leipoa* and probably *Talegalla*, but on average slightly less elongated than eggs of *Eulipoa*, *Macrocephalon*, and *Megapodius*. SIZE: *A. l. lathami*. 93.8 ± 3.2 × 59.1 ± 1.9 mm ($n = 25$, Vleck *et al.* 1984); 83.5–97.6 × 58.4–62.2 mm (91.3 × 61.5, $n = 54$, Schönwetter 1961). *A. l. purpureicollis*: 81.0–98.5 × 54.1–61.8 mm (91.1 × 59.6, $n = 23$, Schönwetter 1961). WEIGHTS: 151–221 g (180 g, $n = 25$, Vleck *et al.* 1984); 211 ± 3.0 g ($n = 6$, Seymour and Rahn 1978). Other data within these ranges.

BREEDING SEASON: *A. l. lathami*. From May or June to Jan or Feb in SE Queensland. *A. l. purpureicollis*: eggs in Feb, chick in mound in Mar in Cape York.

Local names

Wee-lah or gweela (Gray 1861, Sorenson 1920).

References

Gray (1861), Le Souëf (1898), Mathews (1910–1911), Cleland (1912), Macgillivray (1914), Sorenson (1920), Ashby (1922), Barrett and Crandall (1931), Jerrard (1933), Hyem (1936), Coles (1937), Frith (1956*b*), Schönwetter (1961), Sutter (1966), Baltin (1969), Bravery (1970), Abbott (1974), Seymour and Rahn (1978), Dow (1980, 1988 *a*, *b*), Blakers *et al.* (1984), Vleck *et al.* (1984), Jones (1987*b*, 1988*a*, *b*, *c*, 1990*a*, *b*), Ford (1988), Keys (1990), Rogers *et al.* (1990), Beruldsen (1991), Jones and Everding (1991), Seymour (1991), Jones and Birks (1992).

Genus *Aepypodius* Oustalet

Aepypodius Oustalet, 1880. *Comptes rendus hebdomadaires des séances de l'académie des sciences, Paris,* **90**, p. 907.

Two species: one in New Guinea, Japen I., and Misol I., the other restricted to Waigeu I.

Rather large to large mainly black megapodes. Head and neck mostly naked, skin brightly coloured. A fleshy crest on forehead and crown, a flat fleshy shield on nape, extending laterally in a pair of short lobes in one species, in long wattles in the other. Bare fore-neck ending in long pendulous wattle below. Bill relatively short, about three-quarters of length of head; deep at base, culmen and cutting edges strongly decurved; nostrils rounded. Wing short, broad; tip strongly rounded; 10 primaries, p5–p6 longest, p10 (outermost) 5–9 cm shorter than p5–p6, p9 3–5 cm shorter, p8 and p2 1–2.5 cm shorter, p7

0–1 cm shorter, p3 0.5–1.5 cm shorter, p1 (innermost) 2–3.5 cm shorter. Tail moderately long, composed of 14–18 feathers; middle pair shorter than second and third pairs; tail laterally compressed, each tail half fan-shaped.

Upper tail coverts short, deep chestnut or maroon. Tarsi and toes relatively largest of all large megapodes, claws relatively short; front of tarsus covered with single row of broad scutes; tibia feathered down to tibio-tarsal joint.

Wattled Brush-turkey *Aepypodius arfakianus* (Salvadori, 1877)

Talegallus arfakianus Salvadori, 1877. *Annali del Museo Civico di Storia Naturale di Genova*, **9**, p. 333.

PLATE 1

Polytypic. Two subspecies. *Aepypodius arfakianus arfakianus* (Salvadori, 1877), mainland New Guinea and Japen I. (Synonyms: *pyrrhopygius* Schlegel, 1879, Vogelkop peninsula; *arfaki* Schlegel, 1879, Arfak Mts.) *Aepypodius arfakianus misoliensis* Ripley, 1957, Misol I.

Description
PLUMAGES
ADULT MALE: head and neck mostly bare. Forehead and crown with fleshy chicken-like comb, highest on forehead, widening and flattening towards hindcrown; nape covered with flattened fleshy shield; bare foreneck ending in single pendulous wattle. Surface of comb, foreneck, and wattle somewhat wrinkled and furrowed, not warty. Top and side of head as well as side and front of neck covered with scanty short hair-like black bristles, often tending to concentrate to form a patch between comb and nape shield, a stripe from upper lore over eye backwards, and a ring round ear opening. Hindneck densely covered

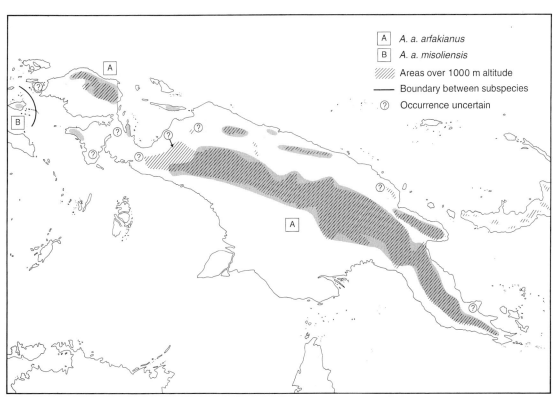

with stiff black feathers, chin and throat often with traces of scanty light grey feathers. When sexually inactive, front part of comb forms narrow ridge c. 1 cm high, rear part flattened out into small lobe on each side of head, pendulous wattle on foreneck rather thin and not longer than a few centimetres. When sexually excited, comb swollen, slightly protruding over base of upper mandible and over upper lore, covering entire forehead and crown as a wrinkled fleshy mass, 2.5–3 cm high; nape-shield slightly swollen, smooth, somethings extending into a short lobe at each side of nape; pendulous wattle thick and swollen, up to c. 6 cm long, surface covered with bead-like rounded vesicles. Upperparts and tail glossy black, except for chestnut-brown tips to feathers of rump and uniform deep chestnut or maroon upper tail-coverts. Wing black, like upperparts, but upper surface sometimes tinged brown and undersurface usually slightly tinged grey. Underparts sooty-black with slight grey tone, more sooty-grey on vent; feathers of breast and belly fringed paler or darker grey to variable extent, underparts in some birds appearing distinctly scalloped light grey (but less so than in Australian Brush-turkey *Alectura lathami*), others virtually uniform black.

ADULT FEMALE: like adult ♂, but top of head more densely covered by bristle-like feathers, comb just indicated by narrow low ridge; shield on nape inconspicuous or absent; chin and throat covered with short feathers or with traces of feathers, usually black rather than light grey; foreneck virtually unswollen, pendulous wattle thin and short (two short wattles in ♀ from Mt. Giluwe—BMNH); size generally smaller, especially tarsus (see 'Measurements').

CHICK: upperparts largely uniform fuscous-brown, faintly barred black on back and tertials; underparts mainly grey-brown, faintly barred with darker brown. Chin and cheek ochre-yellow. Flight feathers uniform black-brown to sooty-black. Bill much deeper at base than in *Megapodius*, nostril more rounded than in *Talegalla*.

IMMATURE: like adult ♀, sometimes hard to distinguish. In first imm plumage, outer primaries still juv (length of p10 c. 120 mm to insertion in skin; in adult, 132–150, $n = 14$), tail juv (120–130 mm long); head mostly though thinly covered with black hair-like bristles, only lores and skin round eye and ear bare; nape, hindneck, and side of neck densely feathered; comb restricted to narrow low ridge just visible between bristles; chin and throat densely covered with fragmentary light grey feathers. Foreneck not swollen, no pendulous wattle. Body browner than in ad, less deep black. Sexes similar. Subsequent imm plumages hard to separate from ad ♀, but second generation p10 more pointed than in ad, 117–127 mm long ($n = 5$).

BARE PARTS
ADULT MALE: *A. a. arfakianus*. Comb pink-red. Shield on nape bluish-white or pale blue-grey. Side of head and neck, foreneck, and wattle bluish-white, pale greenish-blue, or pale blue-grey, sometimes more greenish-yellow on upper side of head; lower end of wattle pink-red or blood-red. Colour of head and neck somewhat dependent on sexual activity of bird, comb and lower end of wattle deeper purplish-red and remainder of head and neck more bluish-white when excited, red paler and blue duller and more greenish when inactive. Iris greenish-yellow, pale yellowish-grey, grey, greyish-green, yellow-brown, tan, hazel, or brown; eyelids pink-red; a strip of skin at base of bill pink-red or light purple-red. Bill yellowish green-horn, greenish with grey tip, dark greenish-slate, brown with pale green lower mandible, olive-black, or blackish-grey. Leg dull yellowish-green, light green, sage-green, greenish-yellow-grey, dull yellowish-olive, olive-green, olive-brown, or greenish-brown, rear or front of tarsus and whole foot darker olive-grey, greenish-slate, or dull olive-brown. On Japen, race unknown (see 'Geographical variation'), ad ♂ had skin of head and neck dull red, comb and wattle bright red (Rothschild and Hartert 1901).

ADULT FEMALE: *A. a. arfakianus*. Like ad ♂, but colours less bright, red of comb and wattle reddish-pink, blue-grey of remainder of head and neck duller greenish-grey, especially on side of head.

CHICK: *A. a. arfakianus*. Bill horn-brown, foot dark.

IMMATURE: *A. a. arfakianus*. Like ad ♀, but iris also recorded as dark brown, bill as greyish-black or black-brown, tarsus as grey, clear brown, or slate-brown. Skin of head and neck dull greenish-grey or olive.

ADULT: *A. a. misoliensis*. Like nominate *arfakianus*, but comb bright light blue in both sexes, instead of red.

CHICK: *A. a. misoliensis*. Iris grey-brown, bare skin round eye and on lore bright blue, bill horn-black with cutting edges greyish-horn, tarsus pink-flesh, toes dull flesh-grey. At 7 weeks, iris dark-brown, bare skin round eye caerulean-blue, on lore and round ear violet-blue, bill slate-black with horn cutting edges, legs greyish-blue, scutes on front of tarsus and upper surface of toes horn-brown, rear of tarsus and soles with shade of pink-flesh.

MOULTS

Of 15 skins of imm and ad birds checked for primary moult, all showed serially descendant moult with generally two moulting series in each wing (once, three series); moult centres four ($n = 3$), five ($n = 3$), or six ($n = 6$) feathers apart. Moult completely suspended in both series in four birds; moult suspended in one series but one or two feathers growing in other series in six birds; one or two feathers growing in both series in remaining five birds. Only a few skins are dated; hence difficult to find out whether a clear moulting season exists. In captivity in Europe, both ♂ and ♀ started to moult one week after last egg was laid.

MEASUREMENTS

ADULT: *A. a. arfakianus*. Western mainland New Guinea east to Snow Mts; data of skins of 1 ♂, 2 ♀♀, and 5 unsexed adults combined (BMNH, RMNH, ZMA).

	n	mean	s.d.	range
wing	8	263.1	5.25	255–271
tail	8	129.9	6.10	120–138
tarsus	7	89.5	4.77	82.4–95.0
mid-toe	8	50.2	2.10	47.5–52.6
mid-claw	7	17.8	1.38	15.8–19.6
bill (skull)	6	37.5	1.66	35.8–39.3
bill (nostril)	7	19.4	1.54	17.4–21.9
bill depth	6	17.4	1.47	15.6–19.4

Japen, skin of 1 ♂: wing 257 mm, tail 121 mm, tarsus 90.8 mm, mid-toe 50.3 mm, mid-claw 18.9 mm, bill to skull 37.1 mm, to nostril 18.9 mm, bill depth *c*. 18.5 mm (FMNH).

Papua New Guinea, skins (BMNH, RMNH, SMTD, ZMB).

MALES	n	mean	s.d.	range
wing	9	275.6	6.52	268–288
tail	8	132.3	5.98	125–138
tarsus	8	96.9	5.32	88.9–101.7
mid-toe	8	51.6	2.45	48.0–54.5
mid-claw	9	20.7	1.27	18.9–23.1
bill (skull)	8	39.9	1.73	36.5–41.3
bill (nostril)	9	20.6	0.86	19.4–21.9
bill depth	9	18.7	0.63	17.9–19.6

FEMALES	n	mean	s.d.	range
wing	9	269.6	4.50	261–274
tail	8	138.4	5.71	129–146
tarsus	9	86.1	3.43	79.8–91.5
mid-toe	8	49.3	2.14	46.0–52.5
mid-claw	8	18.6	0.78	17.1–19.8
bill (skull)	8	37.1	2.01	35.4–41.0
bill (nostril)	8	17.9	1.38	16.0–20.1
bill depth	9	16.9	1.46	15.9–18.8

Due to comb, bill to skull often difficult to measure; length to frontal base of comb on average 6.8 mm less than to skull. Note marked difference between sexes in tail and tarsus length.

A. a. misoliensis: skin of ♂ (ZMA): wing 263 mm, tail 134 mm, tarsus 95 mm, mid-toe 52.5 mm, mid-claw 19.4 mm, bill to nostril 21.1 mm.

CHICK: wing 102, 107 mm; tarsus 35, 37 mm (Salvadori 1882).

WEIGHTS
A. a. arfakianus: ♂ 1450, 1525 g ($n = 2$), ♀ 1350, 1530 g, ($n = 2$) (AMNH). *A. a. misoliensis:* in captivity, ♂ *c.* 1600 g, ♀ *c.* 1200 g. Chicks newly hatched in captivity 115–125 g (118.8 ± 4.79, $n = 4$) (Robiller *et al.* 1985*a*).

GEOGRAPHICAL VARIATION
Perhaps slight variation in size on mainland New Guinea, eastern birds from sample cited in Measurements averaging slightly larger than western ones, especially in wing and tail, but samples small. However, this is not fully supported by data from literature: birds from northern Snow Mts (western Central Range) small indeed, wing of ♂ ($n = 5$) 260–268 mm (263.6), ♀ ($n = 6$) 252–261 mm (256.2) (Rand 1942*b*), but those of Arfak Mts (northwest) had wing (both sexes, $n = 9$) 260–272 mm (267.8); tail ($n = 9$) 130–149 mm (141.9) (Ripley 1957), one ♂ had wing of 290 mm (Oustalet 1881).

Isolated population from Misol described as *misoliensis* by Ripley (1957); said to have bill more slender than in mainland birds, less highly arched; chestnut of upper tail-coverts duller, less rich; feathers of vent fringed with paler slate grey; size generally smaller: wing of one ♂ and two ♀♀ 243, 261.5, 264 mm; tail 131.5, 141.5 mm; bill to nostril 19, 20, 21.5 mm. However, colour of upper tail-coverts, colour and extent of grey of vent (and belly), and bill shape somewhat variable on mainland New Guinea, and all measurements fall within data of mainland population except for single bird with wing of 243 mm, which may have been not fully mature. A paratype from Misol examined by Mees (1965) did not differ in colour and size from mainland birds, the grey appearance of the belly being mainly due to the absence of many feathers, revealing grey of bases on remaining feathers. Colour of comb of *misoliensis* quite different from nominate *arfakianus*, however.

Isolated population of Japen perhaps a valid subspecies also, but only one bird examined (FMNH). Size small, like smallest birds from Vogelkop; colour of comb and wattle bright red (as in nominate *arfakianus*, comb not blue as in *misoliensis*), but bare skin of head and neck red, duller than red of comb and wattle, not pale blue as in nominate *arfakianus* and *misoliensis*.

Variation in grey fringes of feathers on breast and belly individual, not geographical, not sex- or age-related: both ad ♂ and imm ♀ may have uniform black underparts without grey, or both may show conspicuous grey scalloping, somewhat resembling similar markings of *Alectura lathami*.

Range and status
A. a. arfakianus: main mountain range of New Guinea from Nassau Mts to Owen Stanley Mts; isolated in Tamrau and Arfak Mts (Vogelkop peninsula) in north, Wandammen Mts and Bombarai peninsula (in west), Gauttier, Cyclops, Torricelli, and Sepik Mts (northern mainland), and Saruwaged Mts (Huon peninsula); apparently not in Adelbert Mts. Also on Japen (race unknown). *A. a. misoliensis:* restricted to Misol.

Widely distributed though uncommon throughout its range, rare on Misol. Japen population probably small and restricted to higher altitudes.

Field characters
Length *c.* 50 cm. A blackish medium-sized megapode with dark maroon rump and upper tail coverts and laterally compressed tail. Head, throat, and neck bluish-white and mostly naked. Comb on crown red (blue in *misoliensis*), wattle on central lower neck long and pendulous in breeding ♂, whitish with red tip. ♀ slightly smaller with duller head coloration and much smaller wattle. Can be distinguished from sympatric *Talegalla* spp. by its pale head and neck, presence of comb and wattle, maroon rump, rounded nostrils, and smaller size. Usually silent (in contrast with *Talegalla* spp.) and therefore easily overlooked.

Voice
Based on description of captive *misoliensis*: at the beginning of the breeding season the ♂

produces a series of seven, often similar, rhythmic crowing notes from or near the mound 'hja hja hjahjahjahja hja', lasting for 3 sec (Kloska 1986). Later in the season the frequency decreases, and the call is less rhythmic and consists of five notes only. At the end of the season it consists of two notes (1.3 sec) or a single note. While calling the ♂ stretches its body, neck and head upwards (see Fig. 4.2). Both call and posture differ from those of *Alectura lathami* (to which it is closely related).

The ♀ produces a brief (1-syllable) call at the mound during egg laying. No contact calls between ♂ and ♀ have been reported. Voice of nominate *arfakianus* from New Guinea described as a harsh, explosive crowing, consisting of six down-slurred notes 'kyew kyew-kyew-kyew-kyew kyew'.

Five different vocalizations have been described for imm *misoliensis*, two of which are related to or modifications of the calls described above. The other three are a short and harsh distress cry made by an immature ♀ when chased by a ♂, a relatively deep threatening-call 'gock-gock' made by both sexes, and a powerful defence cry only observed in ♂♂, when defending themselves against other ♂♂.

Habitat and general habits

Appears to prefer steep forested mountain sides at 750–2700 m altitude. On the mainland of New Guinea normally above 1000 m. Sympatric with *Talegalla* and *Megapodius* spp. between 750 and 1500 m. On Misol it appears to have a very local distribution in the central limestone hills at about 300–840 m altitude. On Japen, mounds have been found at 1250 m (Dwyer 1981; Ripley 1964). Terrestrial, walks in typical brush-turkey posture with head and broad tail elevated. Shy and inconspicuous. When disturbed, it flies up into trees (whereas *Talegalla* spp. prefer to remain on the ground and run away). Roosts in trees. Imm suspected to live in small groups as in *Alectura*.

Food

Little information; reported to eat fallen fruits, pieces of hard fruits, seeds, and probably also insects, thus largely omnivorous as other megapodes.

Displays and breeding behaviour

The following account is based on Kloska and Nicolai (1988). ♂♂ normally solitary at the incubation mound, and responsible for its building and maintenance. Breeding behaviour resembles that of *Alectura lathami* and can probably be characterized as 'resource-defence polygyny plus polyandry'. When displaying, ♂♂ have wattles grotesquely enlarged, tail cocked up over the back and sometimes hidden under the wings (making them appear tail-less), and feathers bristled on upperback. Mound-building is regularly interrupted by display posture at the top of the mound. While displaying, the ♂ circles round the top of the mound. During morning hours, the ♂ tests the temperature of the mound by making 10–20 cm deep 'test-holes', and probing inside them with head and neck and bumping with his body while the wings are spread. After having repeated this two or three times, the 'test hole' is refilled with mound material. Depending upon weather conditions, up to 15 such holes can be made in a single morning.

The ♀ usually visits the mound on the day of egg laying only. The ♂ reacts to her presence by running towards her with wings raised, stopping at a short distance from her, and flapping both wings briefly, but without attacking. He then immediately returns to the top of the mound and runs down to her again. The ♀ reacts by moving back a few steps and shaking her feathers. Eventually, the ♂ takes up the display posture described above on top of his mound. The ♀ then raises her body feathers and lets her wings hang down slightly. After running across the top, the ♂ leaves the mound, allowing the ♀ to ascend with bristled feathers and to enter a 'test hole'. In a stretched posture and with comb fully expanded, the ♂ then approaches the ♀ and chases her off. She leaves the mound slowly with her head and tail lowered, and wings spread down. This sequence may be repeated as much as 20 times in an hour. Copulation normally takes place at the top of the mound and occurs when the ♀ stays at the mound while the ♂ approaches her. Copulation lasts for about 5–10 sec, during which time the ♂ holds the skin at the back of the ♀'s head with his bill (as in the Australian Brush-turkey), and this is repeated once or twice, sometimes as often as five times, on a single morning. 65% of all copulations ($n = 52$) of a pair observed in captivity (in an enclosure of 21 m^2) took place on the day of egg laying (Kloska and Nicolai 1988). Following copulation (in 94% of the cases, $n = 17$), the ♀ starts digging a hole for her egg. At first the ♂ reacts aggressively by pecking her head and neck and trying to push her off his mound. Suddenly he stops, and allows the ♀ to dig a hole while he stands close behind her, slanting his tail and spreading his wings down slightly, and following her movements carefully. In captivity, the ♀ needed between 10 and 75 min to dig the 25–50 cm deep hole. During digging, the bluish-white coloration of her head became more intense, and her wattle doubled in size. When the temperature inside the hole has been found suitable after testing it several times by taking mound-material in her bill, the ♀ pushes herself inside the hole. Only head, neck and tail are visible and when the egg is actually laid her head is stretched forward and her wings spread over the surface of the mound. Laying may last for 4–8 min, during which time the ♀'s bill is open and here eyes flicker. Immediately afterwards, she starts closing the egg hole; at this stage the ♂ gets more aggressive and after a few minutes chases her away. In captivity, a ♀ *misoliensis* produced 20 eggs at intervals of 3–11 days (average 5.9 days) over a period of 4 months. The comb of the ♂ decreased in size and lost its bright colour a week after the last egg was laid. Later in the season, when ♂ territoriality and activity decrease, both birds can be seen digging at the mound simultaneously.

Breeding

NEST: *A. a. arfakianus*. Mound different from that of *Talegalla*, somewhat similar to a small *Megapodius* mound in form though with a much greater leaf and twig content, about 1.5–2.0 m high and 3.0–3.5 m in diameter. *Megapodius* is reported to lay its eggs sometimes in mounds of this species.

EGGS: different from those of other megapodes, except *Alectura*, in being matt white with a few sparse small buff markings; elongated oblong-ovate, with one end slightly more pointed than the other. SIZE: *A. a. arfakianus*. 82.2–98.8 × 51.0–64.2 mm (91.1 ± 5.1 × 59.4 ± 2.8, $n = 27$, Mt. Sisa, S Highland Province, Dwyer 1981); 80–101 × 58.5–63.3 mm (93.2 × 60.8, $n = 15$, Astrolabe Mts, Parker 1967*b*); 88–95.8 × 58 – 62.5 mm (92.1 ± 3.00 × 60.5 ± 1.47, $n = 7$), Hydrographer Mts, Hartert 1930). WEIGHTS: *A. a. arfakianus*. 120–213 g (192.0 ± 22.3, $n = 37$, Mt. Sisa, S Highland Province, Dwyer 1981).

BREEDING SEASON: *A. a. arfakianus*. Probably depending upon local climate, and thus varying between locations. Eggs for sale near Mt. Sisa, S Highland Province, nearly year round, in Sept, Oct, Dec, Jan, Feb, Mar, Apr, May, and June (Dwyer 1981). Eggs from the Hydrographer Mts in Feb, the Arfak Mts in May and June (Hartert 1930), from Saiko, NE New Guinea, at 1800–2000 m, in Sept (Parker 1967*b*), from Enaena, Mt. Simpson, E New Guinea, at 1500 m, in Sept, and from Mafulu, SE New Guinea, in Oct (Mayr and Rand 1937). A ♀ collected with egg in oviduct on the northern slope of Mt. Giluwe, Central Highlands, in July. Birds in breeding condition were taken in Feb, Mar, and June at Bernhard Camp, N Snow Mts (Rand 1942*b*). Chick from Cyclops Mts in Aug–Sept (Hartert 1930). Incubation period unknown under natural conditions, but likely to be as variable as in other megapodes, depending on incubation temperature.

Local names
Gi:e (Mt. Sisa, Highland Province), alonga (W Dani), gnok (Misol), ajinda (Japen), dasiári (Daribi), cya (Fore, E Highlands).

References
Salvadori (1877, 1882), Schlegel (1879, 1880a), Oustalet (1880, 1881), Ogilvie-Grant (1893, 1897), Rothschild and Hartert (1901, 1913), Hartert (1930), Mayr (1930, 1941), Rothschild et al. (1932), Junge (1937), Mayr and Rand (1937), Rand (1942b), Gilliard (1950), Sims (1956), Ripley (1957, 1960, 1964), Mees (1965), Parker (1967b), Harrison and Frith (1970), Diamond (1972, 1985), Dwyer (1981), Coates (1985), Robiller et al. (1985), Anon (1986), Beehler et al. (1986), Kloska (1986), Kloska and Nicolai (1988).

Bruijn's Brush-turkey *Aepypodius bruijnii* (Oustalet, 1880)

Talegallus Bruijnii Oustalet, 1880. *Comptes rendus hebdomadaires des séances de l'académie des sciences, Paris*, **90**, p. 906.

PLATE 1

Other name: Waigeo Brush-turkey.

Monotypic.

Description
PLUMAGES

ADULT MALE: head and neck bare, except for sparse black bristles; latter mainly confined to strip at side of forehead and above eye, but sparsely present also on lower neck, side of neck, and hindneck, soft and very sparse on remainder of head and neck; fine pale-grey hair-like bristles on throat. A chicken-like fleshy comb on central forehead and crown; as in the Wattled Brush-turkey *A. arfakianus*, comb wider and more flattened on forehead (probably extending over base of culmen when bird sexually excited), narrower and higher on forecrown above eye, widening and flattening out in a wide shield on nape. In contrast to *A. arfakianus*, flattened nape shield extends into long pendulous wattle on each side, and surface of entire comb, nape shield, and upper surface of wattles covered with dense, *c.* 5 mm long wart-like papillae. Surface of swollen foreneck smooth; lower foreneck with long smooth or furrowed wattle ending in horny knob. Upperparts, wing, and tail deep black, but back and rump more sooty, upper tail-coverts deep chestnut or maroon, and longer upper wing-coverts tinged brown. Chest deep dull chestnut, some dark grey of feather bases sometimes partly visible; remainder of underparts dark olive-grey, feathers of breast and belly suffused chestnut-brown on tips to variable extent, feather tips of flanks, lower belly, and vent washed olive-grey.

ADULT FEMALE: like ad ♂, but hindneck and side of neck covered with sparse black feathers; bristle-like feathers on forehead and crown

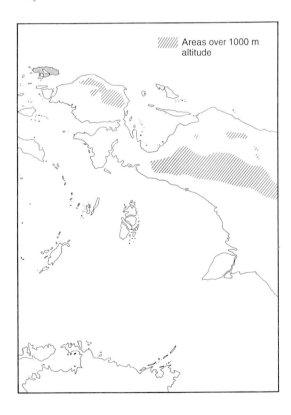

well developed. Comb as in ♂, but small and rather inconspicuous, shield on nape just indicated or virtually absent, without a wattle on each side. Foreneck hardly swollen, wattle on lower foreneck small or absent.

CHICK: unknown to science.

IMMATURE: no wattles (Rand and Gilliard 1967). A near-ad ♂ has comb narrow and low, extending from forehead to central hind crown, closely covered with warty papillae, bordered by numerous fine bristles at sides; nape shield indicated, but not yet attached to comb, ending in small lobe at each side but without wattles; wattle on lower foreneck c. 18 mm long (Oustalet 1881). Ageing probably as in *A. arfakianus*.

BARE PARTS
ADULT MALE: bare skin of head coral-red according to Rand and Gilliard (1967), but illustrated orange with reddish comb in Oustalet (1880); in skins, tinged orange. Iris hazel. Bill and foot fuscous, blackish-brown, or black.

ADULT FEMALE: in bird collected by J. Kakiaij in 1938, iris orange, bill black, foot 'half orange' (De Schauensee 1940b).

MOULTS
Primary moult serially descendant. Of three undated birds examined, moult suspended in one, active in both others; in actively moulting birds, two and three moulting series appear to be present, in bird with moult suspended three or four series, asymmetrical between wings.

MEASUREMENTS
Waigeu, ♂. Data combined for skins of three examined (BMNH, RMNH, SMTD) and data from literature (Oustalet 1881; Salvadori 1882; both with tail and tarsus recalculated due to different measuring techniques; Rand and Gilliard 1967). Exposed culmen measured to front of comb.

	n	mean	s.d.	range
wing	6	303.5	5.47	296–310
tail	6	144.7	4.63	139–150
tarsus	6	105.9	4.67	100–111
mid-toe	4	58.9	1.04	57–60
mid-claw	5	20.8	1.78	18–22.5
exposed culmen	6	32.7	2.88	27–35
bill depth	5	20.0	0.71	19–21

♀ (n = 1; perhaps imm): wing 251 mm; tail 141 mm; tarsus 93 mm; exposed culmen 27 mm; bill depth (nostril) 17 mm (De Schauensee 1940b). Imm ♂ (n = 1): wing 270 mm (Oustalet 1881). Bill to nostril on average 10.1 mm less than exposed culmen.

WEIGHTS
No information.

GEOGRAPHICAL VARIATION
None reported.

Range and status
Endemic to Waigeu I. So far, only 15 specimens have been obtained, 13 by A. A. Bruijn's native collectors between 1879 and 1885, and one in 1938 from Jeimon on the east side of Majalibit Bay, a few miles inland from Siam; the 15th specimen, in Senckenberg Museum in Frankfurt labelled 'Aug 1904', may have been another of Bruijn's specimens. Skins now in the collection of AMNH, New York (six), MHN, Paris (two, including type specimen), RMNH, Leiden (one), BMNH, Tring (one), SMTD, Dresden (one), Senckenberg Museum, Frankfurt (one), Turati Collection, Milano (one), the Academy of Natural Sciences (ANS), Philadelphia (one), and the Zoological Institute and Zoological Museum of the University of Hamburg (one).

Status: unknown. Has never been observed in the wild despite extensive search by at least 12 expeditions. Dekker and colleagues (M. Argeloo and Ch. Vermeulen) failed to find the species during a 10-day survey in SE Waigeu in October 1993. Residents of four villages (Selpele, Kabui, Warsambim, and Lupintol) on Waigeu gave detailed descriptions of a megapode larger than the Dusky Megapode *Megapodius freycinet*, which was evidently *A. bruijnii* (J. M. Diamond, *in litt.*).

A brush-turkey observed on Batanta (south of Waigeu) by a member of an expedition in 1986 might have been *A. bruijnii* (J. M. Diamond, *in litt.*). No other reports of *Aepypodius* spp. or any other large megapode exist for this island. Occurrence of *A. bruijnii* on Batanta also reported by local people, but these observations proved to be wrong and referred to the Magnificent Ground Pigeon *Otidiphaps nobilis* (Greenway 1966).

Field characters
Length *c*. 55 cm. The only large brush-turkey on Waigeu. Sympatric with *Megapodius freycinet* from which it can easily be distinguished by its naked red head and neck with red comb, elongated wattles on nape and foreneck, maroon rump, and laterally compressed tail. General coloration brownish-black with chestnut-brown underparts becoming olive-greyish on sides and abdomen.

Voice
Unknown.

Habitat and general habits
Mountain forests on Waigeu. General habits undescribed, but probably similar to *A. arfakianus*.

Displays and breeding behaviour
Unknown. Most probably a mound builder, as *A. arfakianus*.

Breeding
Unknown; neither mounds nor eggs have ever been described.

Local names
Mantankentewp (Kabui), mantankemtup (Warsambim), hakal (Lupintol), mangwap (Urbinasopen).

References
Oustalet (1880, 1881), Salvadori (1882), Ogilvie-Grant (1893, 1897), Rothschild and Hartert (1901), Shufeldt (1919), Rothschild *et al.* (1932), De Schauensee (1940*a*, *b*), Mayr (1941), Greenway (1966), Rand and Gilliard (1967), Beehler *et al.* (1986).

Genus *Talegalla* Lesson

Talegalla Lesson, 1828. *Manuel d'Ornithologie*, **2**, p. 185.

Three species, all endemic to New Guinea region.

Rather large to large black megapodes. Bristle-like feathers of forehead project over base of culmen, curved upward. Feathers of crown bristle-like and depressed in two species (*cuvieri* and *fuscirostris*), lanceolate and forming dense crest in one species (*jobiensis*). Side of head and neck and foreneck mostly naked, skin brightly coloured. No wattles. Bill relatively long, about seven-eighths of length of head; deep at base, terminal half of culmen and cutting edges strongly decurved. Nostrils oval. Wing short, broad; tip strongly rounded; 10 primaries, p5–p6 longest, p10 (outermost) 5–7 cm shorter, p9 2.5–4 cm shorter, p8 1–2.5 cm shorter, p7 and p4 0–1 cm shorter, p3 0.5–1.5 cm shorter, p2 1–2 cm shorter, p1 (innermost) 2–3.5 cm shorter. Tail relatively long, composed of 16 feathers; tip rounded. Upper tail-coverts short, black. Tarsus and toes relatively long, strong; claws short. Extent of feathering on tibia variable, 0.5–1.0 (–1.5) cm of lower tibia bare in *cuvieri* and *jobiensis*, 1.5–3.5 cm in *fuscirostris*. Front of tarsus covered with single row of scutes, but these less broad and more extensively divided towards joints than in *Aepypodius*.

Note on nomenclature

Ever since its first description by Lesson, the name of this genus has caused confusion. In his monograph on megapodes in 1881, Oustalet claimed that *Talegallus* Lesson, 1826 (*Voyage autour du Monde, exécuté par Ordre du Roi, sur la Corvette de sa Majesté, La Coquille, pendant les années 1822, 1823, 1824, et 1825, Zoologie*, p. 715; pl. 38) had priority over *Talegalla* Lesson, 1828 (*Manuel d'Ornithologie*, 2, p. 185). Until about 1930, however, both *Talegalla* and *Talegallus* were used in the literature. Peters (1934) gave way to common use in accepting *Talegalla* in preference to *Talegallus*, but uncertainty remained as no explanation was given for this decision. In addition, Neave (1940, *Nomenclator Zoologicus*, 4, p. 391) stated that *Talegalla* Lesson, 1828, had priority over *Talegallus* Lesson, 1826.

Although 1826 is given as the year of publication of *Voyage autour du Monde...*, the particular part containing the text describing the genus on p. 715 was published on 1 May 1830 (see *The Annals and Magazine of Natural History*, (7) 7, 1901, p. 391). Plate 38, showing the species under the name '*Talegallus Cuvieri*', was published on 29 November 1828 (see Ronsil, R., 1948, *Encyclopédie Ornithologique* 8, *Bibliographie Ornithologique Française* 1. Paris, Lechevalier, p. 162). However, Lesson's description of *Talegalla* in *Manuel d'Ornithologie* was published on 28 June 1828, thus earlier than both plate and text in *Voyage autour du Monde...* This means, therefore, that *Talegalla* has priority over *Talegallus*.

Red-billed Talegalla *Talegalla cuvieri* Lesson, 1828

Talegalla Cuvieri Lesson, 1828. *Manuel d'Ornithologie*, **2**, p. 186.   PLATE 2

Other names: Red-billed Brush-turkey, Cuvier's Brush-turkey.

Polytypic. Two subspecies. *Talegalla cuvieri cuvieri* Lesson, 1828, Misol I., Salawati I., and Vogelkop peninsula of W New Guinea. *Talegalla cuvieri granti* Roselaar, 1994, southern foothills of Weyland and western Snow Mts, west-central New Guinea.

Description
PLUMAGES

ADULT: sexes similar. Forehead covered with black bristle-like feathers, each consisting mainly of a shaft with its lateral barbs short and wide apart; these feathers project forward at base, extending above nostrils, but are curled upward and backward at tip. Feathers of crown and hindneck similar to those of forehead, but barbs often even more strongly reduced, shafts showing as glossy black bristles, not or hardly curved but closely depressed to skull instead. Side of head bare, except for short bristles round ear and some very fine bristles on upper cheek. Lower cheek, chin, and throat covered with scanty short and loose grey feathers; neck covered all round with a dense layer of longer but equally loose black feathers. Entire body, tail, and wing uniform black; deepest and somewhat glossy on upperparts, chest, tertials, and tail, usually slightly duller and often browner on belly, vent, wing-coverts, and flight feathers, especially if plumage abraded. Tibia covered with black-brown feathers, leaving 7.5 (3–12) mm of lower tibia bare (measured at side to tibiotarsal joint, $n = 15$).

CHICK: upperparts fuscous-brown, barred and speckled with distinct red-brown to rufous-ochre marks (but marks occasionally faint or almost absent); rump and tail-coverts greyish-black. Sides of head and underparts rufous-brown, paler rufous to isabelline-ochre on cheek, chin, throat, belly, and vent; marked to varying extent with brown or dull black spots or bars. Tertials, upperwing, and flight feathers dull black or brown-black, longer coverts, tertials, and secondaries barred or spotted rufous. Probably inseparable from chick of Brown-collared Talegalla *T. jobiensis* (which

Red-billed Talegalla *Talegalla cuvieri* 107

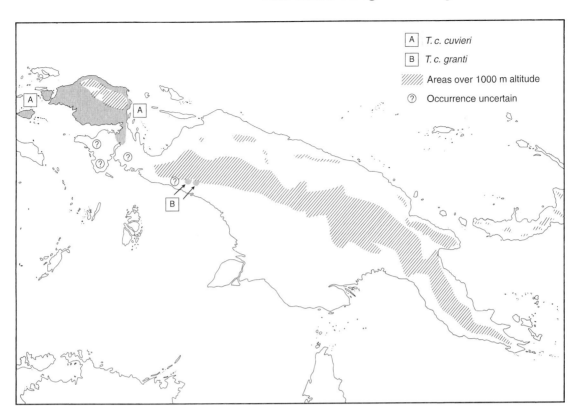

A	*T. c. cuvieri*
B	*T. c. granti*
/////	Areas over 1000 m altitude
?	Occurrence uncertain

see); a chick of *T. cuvieri* from Passim (western Geelvink Bay) hardly separable from one of *T. jobiensis* from Japen (Meyer 1874a).

IMMATURE: in first imm plumage, crown and nape covered with lanceolate feathers, dull grey with black shaft-streaks, feathers rather soft and loose, not as reduced as in ad; lore and area round eye bare; cheek, chin, and neck feathered, bare skin hardly visible, cheek and chin grey, side and rear of neck rufous with faintly sooty feather-centres or grey with indistinct dull rufous feather tips; foreneck dark grey, feathers with faint rufous-grey fringe. Body dull black-brown (feathers grown last are blacker than those grown earlier), some older feathers with faint rufous-grey fringe along tip, on underparts sometimes with faint rufous shaft streak. Some upper wing-coverts and secondaries still juv, barred or spotted rufous and black; outer primaries still juv, p10 strongly pointed at tip; length of full-grown p10 from tip to insertion in skin 92–115 mm (103.0 ± 11.5, $n = 3$); tail still juv, short, 118 mm ($n = 1$). In second imm plumage, similar to ad, but second generation of outer primaries and all tail shorter than ad; p10 somewhat pointed, 141 mm (132–148 mm, $n = 10$) (in ad, 148 mm, range 142–162 mm, $n = 11$), tail 138–159 mm (152.5 ± 8.55, $n = 6$) (in ad, 164 mm, range 155–175 mm, $n = 31$); lower neck with variable amount of chestnut, in some (younger?) birds all chestnut with black spots and bars; in other (older?) birds black prevailing and chestnut restricted to some shaft-streaks or bars at basal side or rear of neck. Chestnut at base of neck in second imm plumage about similar to that of ad *T. jobiensis*; some *T. cuvieri* without imm character of wing show traces of chestnut at base of neck, these perhaps imm in third imm plumage but perhaps occasionally ad.

BARE PARTS
ADULT: iris clear pale yellow, lemon-yellow, dull yellow, yellow with orange inner ring, or

yellow-red; on Misol, pale brown or yellowish (Rothschild and Hartert 1901). Bare skin of head and neck yellowish, yellow-green, yellow-olive, greenish-brown; record of reddish-brown (Ogilvie-Grant 1893) not substantiated; in *granti*, medium olive or olive-green (BMNH). Bill reddish-yellow, yellow-red, orange, orange-red, red, or blood-red; in *granti*, dull tomato-red (BMNH) or brown and yellow (Rothschild and Hartert 1913). Leg and foot yellow, orange-yellow, pale orange, or yellow-red; in *granti* rich or deep orange (BMNH) or orange-yellow (Rothschild and Hartert 1913), on Misol pale chrome or yellowish-olive (Rothschild and Hartert 1913; Ripley 1964).

CHICK: iris grey, brown, or black. Bill dark brown, greyish-brown, or horn-black, lower mandible pink with brown or horn-black tip and yellow-red base. Leg and foot yellow or orange-yellow.

IMMATURE: iris pink-yellow, dull yellow, or pale dirty yellow. Bare skin of head and neck dull yellow. Bill red, or greyish-brown with paler lower mandible. Leg and foot yellow, flesh-yellow, or salmon.

MOULTS

Primary moult serially descendant. Usually two moult series on each wing, rarely one or three; in imm, innermost series (second generation) started with p1 when two juv outer primaries (first generation) still growing, third generation started with p1 when second generation series active on p7 (p6–p8). In ad and older imm, the two series six or seven primary feathers apart ($n = 17$), more rarely, four, five, or eight feathers apart (each in two birds). Perhaps a moulting season in first half of year: during Jan–Mar, moult in nine series active, in five suspended; in Apr–June, five active, two suspended; in July–Sept, two active, five suspended; in Oct–Dec, two active, two suspended. Moult in most birds apparently starting with p1 Nov–Jan (when a previous series may still be active in outer wing), in more advanced birds followed by a next moulting series from p1 in Mar–Apr, with moult suspended from May or June, when first series has reached p6–p8 and second (if any) p1–p3.

MEASUREMENTS

ADULT: *T. c. cuvieri*. Misol; skins of 1 ♂, 2 ♀♀ and 1 unsexed bird (BMNH, RMNH).

	n	mean	s.d.	range
wing	4	273.8	3.10	271–278
tail	4	160.2	5.24	155–167
tarsus	4	86.8	1.59	85.0–88.5
mid-toe	4	49.0	1.91	46.9–51.5
mid-claw	4	18.1	1.00	17.0–18.4
bill (skull)	4	39.6	0.86	38.6–40.7
bill (nostril)	4	22.6	0.90	21.7–23.8
bill depth	4	18.4	1.04	17.3–19.5

T. c. cuvieri: Salawati and Vogelkop peninsula; skins of 5 ♂♂, 11 ♀♀, and 3 unsexed birds (BMNH, RMNH, SMTD, ZMA).

	n	mean	s.d.	range
wing	18	285.8	4.25	279–291
tail	19	164.3	4.78	156–172
tarsus	17	90.7	2.91	85.1–95.5
mid-toe	15	49.8	1.47	47.0–52.0
mid-claw	17	18.5	1.35	15.6–20.4
bill (skull)	17	40.8	1.33	38.5–43.1
bill (nostril)	16	23.6	1.21	21.2–26.1
bill depth	15	18.8	0.85	17.3–19.8

Single ♀ from Salawati (RMNH) is among the larger birds in sample above. Sexed birds from Vogelkop peninsula from sample above: wing, ♂ 280–291 mm (286.8 ± 4.21, $n = 5$), ♀ 280–291 mm (286.0 ± 4.11, $n = 10$); tail, ♂ 156–166 mm (161.8 ± 3.91, $n = 5$), ♀ 157–172 mm (163.7 ± 5.23, $n = 10$); tarsus, ♂ 85.1–91.8 mm (89.6 ± 3.78, $n = 5$), ♀ 88.2–95.5 mm (91.6 ± 2.41, $n = 10$). Literature data from 4 ♂♂ and 1 ♀; wing 274–288 mm (281.8 ± 6.14, $n = 5$); tail 145–177 mm (157.6 ± 11.99, $n = 5$) (Gyldenstolpe 1955).

T. c. granti: Iwaka R. (southern foothills of Snow Mts, west-central New Guinea), skins of 3 ♂♂ (BMNH).

	n	mean	s.d.	range
wing	3	290.3	4.62	285–293
tail	3	169.3	3.51	166–173
tarsus	2	99.2	–	97.8–100.7
mid-toe	3	57.3	2.91	54.0–59.4
mid-claw	3	19.6	0.40	19.1–19.8
bill (skull)	3	45.7	1.46	44.3–47.2
bill (nostril)	3	26.6	0.86	25.8–27.5
bill depth	3	20.9	0.74	20.1–21.5

CHICK: at hatching, wing 120 mm, tarsus 37 mm.

WEIGHTS
T. c. cuvieri: ♀ 1785 g ($n = 1$, Hartert 1930).

GEOGRAPHICAL VARIATION
Slight in colour, considerable in size. Birds from Misol on average smaller than typical nominate *cuvieri* from Vogelkop peninsula, perhaps warranting recognition as a separate race, but number examined small and most measurements still within theoretical size range (mean ± 3 × s.d.) of Vogelkop birds. Though sample of Iwaka R. birds (southern foothills of Snow Mts) examined small also, these birds are much larger than typical nominate *cuvieri* from Vogelkop, especially in tarsus, foot, and bill measurements, requiring recognition as *granti*; large birds occur also on upper Setakwa R. (Rothschild and Hartert 1913). In the foothills of Snow Mts, *granti* occurs side by side with Black-billed Talegalla *T. fuscirostris occidentis*, though living at somewhat higher altitude than the latter, which is mainly a lowland species. Relatively large size of *T. c. granti* compared with local small *T. fuscirostris* probably due to character displacement, large difference in size of bill and foot enabling the two species to overlap in geographical range without competition.

Range and status
T. c. cuvieri: Vogelkop peninsula, NW New Guinea, as well as Misol and Salawati. *T. c. granti*: Iwaka R. (near upper Mimika) and Utakwa R. in the southern foothills of Weyland and western Snow Mts (where it overlaps with *T. fuscirostris*; see that species, and Chapter 3). Locally quite common in the lowland forests on Vogelkop peninsula. Common in lowland forest at 100 m above sea level on Misol. Not on Gilolo (= Halmahera), *contra* Ogilvie-Grant (1893, 1915).

Field characters
Length 52–57 cm. Can best be distinguished from other *Talegalla* spp. by orange-red bill (blackish in *T. fuscirostris* and yellow-brown to red-brown in *T. jobiensis*); also distinguishable from *T. fuscirostris*, with which it overlaps locally, in having tibia almost completely feathered. As other *Talegalla* spp. (and in contrast with *Aepypodius* spp.), notoriously noisy, calling frequently day and night, the sound carrying for over a mile (J. M. Diamond, *in litt.*).

Voice
A loud and penetrating 'kok, kok' or 'wok-wok', with the nasal quality of a donkey's bray. The alarm note is described as a pig-like grunt (Rand and Gilliard 1967; Ripley 1964). Chicks silent.

Habitat and general habits
Mainly a lowland forest species in New Guinea, but up to 1500 m according to Diamond (1972). Occurs up to 1200 m in moss forest on Mt. Bantjiet, Tamrau Mts, with mounds found at 450 and 900 m (Gilliard and LeCroy 1970). On Misol, in wet evergreen forest in all parts of the island (Ripley 1960). Very wary in foothills of southwest Snow Mts (Iwaka R.) (Ogilvie-Grant 1915). When disturbed, it remains on the ground and runs away (whereas the Wattled Brush-turkey *Aepypodius arfakianus* prefers to fly up into trees).

Chicks remain solitary for at least 1 month and avoid others with antagonistic behaviour, such as fluffing up feathers and spreading of wings.

Food
Unknown, but probably omnivorous as other megapodes.

Displays and breeding behaviour
Undescribed.

Breeding
NEST: mound. Two mounds in Tamrau Mts at 450 and 900 m were constructed of leaves and sticks and built near the base of large trees. They measured 1.8 × 3.3 m and 2.4 × 2.7 m respectively, being 45 and 60 cm high (Gilliard and LeCroy 1970). Another mound was reported to be 1.35 m high and 3.6 m wide at the base (Bergman 1963, which see for illustration).

EGGS: similar in coloration and shape to *Megapodius* eggs, being pinkish-buff or pinkish-brown. SIZE: 87.2–98.5 × 60.0–63.5 mm (94.0 × 62.5, $n = 4$, Schönwetter 1961); 95.3 × 63.5 mm ($n = 1$, Shufeldt 1919); 101.4 × 68.6 mm ($n = 1$, coll. ZMA; large for *T. cuvieri*, might be egg of *T. jobiensis*). WEIGHTS: no information.

BREEDING SEASON: one egg and one chick collected from two mounds near Arandai, southern part of the Vogelkop peninsula, late Apr (Bergman 1963). Eggshell fragments in the mound showed that at least one egg had hatched earlier, which indicates that egg laying must have already taken place in Feb. Chick collected on Salawati in July, others elsewhere in Jan (1) and Mar (1). Female from Misol with granular ovaries in Nov.

Local names
Mangoipe (Manokwari), ungwau (Andai), manguab (Mum, near Waren), nang-wo (Mt. Bantjiet, Tamrau).

References
Lesson (1828), Meyer (1874*a*), Salvadori (1877, 1882), Oustalet (1880), Ogilvie-Grant (1893, 1897, 1915), Rothschild and Hartert (1901, 1913), Shufeldt (1919), Hartert (1930), Rothschild *et al.* (1932), Hartert *et al.* (1936), Mayr (1941), Gyldenstolpe (1955), Ripley (1960, 1964), Schönwetter (1961), Bergman (1963), Mees (1965), Rand and Gilliard (1967), Gilliard and LeCroy (1970), Diamond (1972), Holmes (1989), Roselaar (1994).

Black-billed Talegalla *Talegalla fuscirostris* Salvadori, 1877

Talegallus fuscirostris Salvadori, 1877. *Annali del Museo Civico di Storia Naturale di Genova*, **9**, p. 332.

PLATE 2

Other names: Black-billed Brush-turkey, Dark-billed Brush-turkey, Yellow-legged Brush-turkey.

Polytypic. Four subspecies. *Talegalla fuscirostris fuscirostris* Salvadori, 1877, south coast of E New Guinea between *c.* 145 and 148°E; *Talegalla fuscirostris occidentis* White, 1938, lowlands at southern foot of Central Range of W New Guinea, from Etna Bay to about Sabang; *Talegalla fuscirostris aruensis* Roselaar, 1994, Aru Is and southern lowlands of New Guinea in Merauke area and (probably this race) basin of lower Fly R.; *Talegalla fuscirostris meyeri* Roselaar, 1994, southern shore of Geelvink Bay from Wandammen peninsula to about Siriwo basin; unknown what race inhabits remainder of lowlands and foothills of S New Guinea.

Description
PLUMAGES
ADULT: sexes similar. Exceedingly similar to Red-billed Talegalla *T. cuvieri*, differing only in dark bill (see 'Bare parts') and more extensively bare tibia (at sides, 24 mm of tibia bare, range 15–33, $n = 31$; in *T. cuvieri*, 7.5 mm, range 3–12, $n = 15$); size often smaller, but not everywhere in geographical range: see 'Measurements' for both species. Also, feathers on

Black-billed Talegalla *Talegalla fuscirostris*

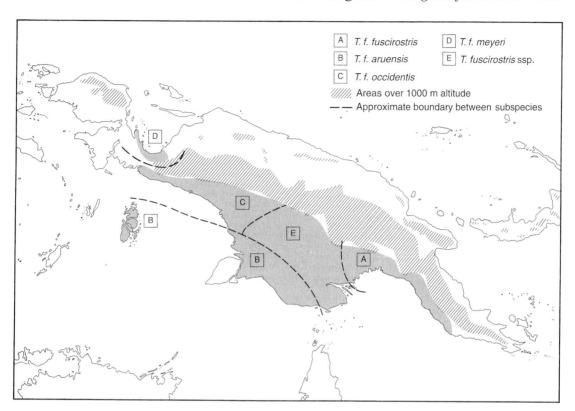

cap on average less reduced than in *T. cuvieri*, short barbs present along shafts, each barb tending to curl upwards, cap resembling barbed-band surface (in *T. cuvieri*, feathers mere bristles). ♀ not markedly smaller than ♂ (*contra* Mayr 1938).

CHICK: upperparts and upper wing-coverts sooty black-brown, grizzled dark rufous-brown from cap to scapulars and back, slightly more rufous on hindneck, but nowhere contrasting. Forehead, side of head and neck, and foreneck buff, finely dappled grey, gradually merging into sooty of remainder of head and body; a sooty black ring round eye, chin and throat pale buff or off-white. Underparts rusty-brown, subterminal black feather-bases showing as irregular dark barring or mottling; rusty tinge deepest on chest, more rufous towards lower belly. Tail-coverts sooty black. Flight feathers, greater upper wing-coverts, and tertials uniform sooty black.

IMMATURE: in first imm plumage, head and neck fully covered with (rather loose) feathers, except for bare lore and region round eye and ear. Feathers of forehead and crown black, lanceolate, less reduced than in ad; feathers of chin and throat soft and short, grey with black shafts, those of side and rear of neck black. Feathers at base of neck either black or, in some birds, red-brown, in latter forming somewhat contrasting ring round neck. Body less deep black than in adult, dark brown of feather centres visible. Outer primaries and tail still juv; p10 strongly pointed, 103.4 mm (90–110) long to insertion in skin ($n = 4$); tail 106.9 mm (83–118, $n = 6$). In second imm plumage, feathers of forehead and crown more reduced, but still less bristle-like than in adult; chin, throat, and neck fairly densely covered with scattered short grey feathers; no red-brown at base of neck; tip of second generation p10 still somewhat pointed, length of p10 134 mm (120–140, $n = 4$); tail 137.1 mm

(125–150, n = 5) (in ad, p10 140.5 mm long, range 132–160, n = 15).

BARE PARTS

ADULT: iris dark sepia, coffee-brown, red-brown, brown, dark brown, or black-brown (on 20 labels of skins examined and also from literature: all these birds from southern Geelvink Bay and SW New Guinea east to Fly R.); occasionally recorded as yellow (Salvadori 1882, E New Guinea). Bare skin of head and neck dusky cobalt-blue, bluish-black, or blackish-grey; in *aruensis*, either blackish-grey or greenish- to yellowish-brown. Bill dark horn-brown to black, sometimes blue (*meyeri*), slate-grey (*aruensis*), yellowish-brown (nominate *fuscirostris*), or dark horn to black with dirty white lower mandible (*occidentis*). Leg and foot chrome- or lemon-yellow, sometimes flesh-pink, orange, or ochre, rarely (once, Merauke area) olive-green with yellow tinge; once yellow and once red in *meyeri*.

CHICK: iris brown, dark sepia, or dark brown. Bill dark horn-brown to brown-black, base of lower mandible pinkish-grey. Leg and foot pale salmon, ochre-yellow, yellow, or orange-yellow.

IMMATURE: like ad, but bill also recorded as grey or dirty white (upper Lorentz R.).

MOULTS

Primary moult serially descendant. Generally, two moult series on each wing, usually separated by five or six (imm) or five to eight (ad) feathers; centres not necessarily both active: in 23 ad, both centres active in 13%, both suspended in 13%, one active and one suspended in 22% (thus in 48% of birds, three generations of primaries in one wing), one active on inner wing while other had just finished with regrowth of p10 in 35%, one suspended on inner wing while other had reached p10 in 17% (in these 52% of birds, two generations of primaries in one wing); in imm, both series usually active or, in small young, one series has started while outer primaries still juv, growing or just fully grown. Rarely (in two of 50 birds examined) three series active, on p1–p2, p6, and p10. Apparently a clear moulting season: in Jan–Mar, moult active on all eight moult centres of four birds examined; in Apr–June, six centres active, one with moult suspended; in July–Sept, ten active, six suspended; in Oct–Dec, 12 active, 14 suspended. Moult apparently usually starts with p1 Aug–Sept, and again with p1 Dec–Jan before previous series completed; birds starting May–June mainly imm; duration of primary moult of one series (if moult not suspended) probably 6–8 months. In E New Guinea, singles in moult in Jan, June, Nov, and Dec, another not (yet) moulting in Aug (Mayr and Rand 1937).

MEASUREMENTS

ADULT: *T. f. meyeri*. Southern shore of Geelvink Bay; skins of 3 ♂♂, 1 ♀, and 3 unsexed birds (MZB, SMTD, ZMB).

	n	mean	s.d.	range
wing	7	278.0	8.27	265–286
tail	7	164.9	12.43	154–191
tarsus	6	88.8	3.05	86.2–94.2
mid-toe	4	49.8	2.55	48.0–53.5
mid-claw	4	19.2	0.64	18.3–19.8
bill (skull)	4	40.2	0.78	39.0–40.6
bill (nostril)	6	22.2	0.85	21.2–23.2
bill depth	5	18.3	0.71	17.4–19.3

T. f. occidentis: lowlands and hills just south of Snow Mts, between Kapare R. and upper Lorentz R.; skins (BMNH, RMNH).

MALE	n	mean	s.d.	range
wing	16	266.1	5.10	260–278
tail	16	147.4	6.02	137–159
tarsus	16	88.3	2.08	85.4–92.7
mid-toe	12	46.8	1.85	43.0–49.3
mid-claw	16	17.4	0.89	16.3–18.9
bill (skull)	16	38.5	1.67	35.7–40.4
bill (nostril)	16	21.8	1.75	19.0–24.6
bill depth	14	17.0	0.73	15.8–18.0

FEMALES	n	mean	s.d.	range
wing	7	269.6	6.24	261–281
tail	7	147.1	4.91	142–152
tarsus	6	87.7	3.09	82.5–91.7
mid-toe	6	47.9	2.16	45.1–49.7
mid-claw	7	17.6	1.30	16.0–19.8
bill (skull)	7	39.6	1.04	38.5–41.0
bill (nostril)	7	22.3	0.78	20.9–23.4
bill depth	5	17.0	0.69	15.9–17.7

Tarsus in 20 of 22 birds 86 mm or longer (in *T. f. aruensis*, 85 mm or shorter in seven of nine birds). Birds from Mimika area in AMNH: wing, 259–274 mm (266.9 ± 4.50, $n = 11$); tail, 142–160 mm (150.0 ± 6.50, $n = 9$) (Mayr 1938).

T. f. aruensis: Aru Is and Merauke area (S New Guinea); skins of 5 ♂♂ and 4 ♀♀ (RMNH, SMTD, ZMA, ZMB).

	n	mean	s.d.	range
wing	9	274.6	8.26	265–286
tail	9	156.6	10.43	144–178
tarsus	9	83.9	2.62	78.0–88.6
mid-toe	9	46.8	1.13	44.6–48.0
mid-claw	9	17.6	1.64	15.8–20.5
bill (skull)	9	38.4	1.07	36.7–40.2
bill (nostril)	9	21.6	0.75	20.5–22.7
bill depth	9	16.5	0.73	15.5–17.6

Fly R. area (S New Guinea), wing, ♂ 256–290 mm (mean 269, $n = 10$), ♀ 252–269 mm (263.2 ± 7.08, $n = 5$) (Rand 1942a).

T. f. fuscirostris: SE New Guinea; skins of 3 ♂♂, 2 ♀♀, and 2 unsexed birds (BMNH, RMNH, ZFMK, ZMB), and data on wing and tail of 5 ♂♂ and 7 ♀♀ from Mayr and Rand (1937) and Mayr (1938).

	n	mean	s.d.	range
wing	19	281.6	8.44	268–299
tail	15	175.4	5.59	168–186
tarsus	7	90.2	3.26	84.3–94.2
mid-toe	7	48.6	2.76	45.0–51.5
mid-claw	7	18.5	1.68	16.1–19.8
bill (skull)	7	38.6	1.23	36.6–40.0
bill (nostril)	7	21.8	0.79	20.8–22.7
bill depth	7	17.1	0.61	16.4–18.0

CHICK: at hatching or shortly thereafter: wing 102–116 mm (109.6 ± 5.80, $n = 6$); tarsus 33.5–36.7 mm (35.6 ± 1.43, $n = 4$) (BMNH, RMNH).

WEIGHTS
ADULT: *T. f. aruensis*. ♂ 1325, 1330, ♀ 1000, 1275 g (Merauke area: Mees 1982; RMNH). *T. f. fuscirostris*. ♂ 1400, ♀ 1560 g (E New Guinea: ZFMK).

GEOGRAPHICAL VARIATION
Marked, but in size only. Nominate *fuscirostris* from SE New Guinea large, with relatively long tail but relatively short tarsus; *aruensis* of Merauke area and Aru Is rather similar in proportions to nominate *fuscirostris*, but generally smaller and tail relatively shorter, especially length of tarsus and tail showing little overlap: tarsus of nominate *fuscirostris* below 88 mm in only one of seven birds (of *aruensis*, over 88 mm in only one of nine birds), tail of nominate *fuscirostris* over 168 mm in all of 15 birds (of 9 *aruensis*, all below 164 mm except for one of 178 mm).

Despite isolated position of population of Aru Is, no difference in size between *aruensis* from Aru and from Merauke. *T. f. occidentis* of lowlands at southern foot of Snow Mts smaller still than *aruensis*, tail relatively shorter (in Mimika area more markedly so than along upper Lorentz R.), but tarsus relatively much longer, virtually always over 85 mm (only one of 22 *occidentis* below 85 mm), while tarsus of *aruensis* over 85 mm in only two of nine birds; tarsus and bill of *occidentis* close in size to nominate *fuscirostris*, but wing and (especially) tail shorter. *T. f. meyeri* of southern shore of Geelvink Bay large; wing and tarsus almost as long as nominate *fuscirostris*, bill longer and distinctly deeper, tail shorter. Position of *meyeri* within *T. fuscirostris* problematical: feathering of head, black neck of imm, large amount of bare tibia (21–26 mm), and (in skins) dusky bill coloration all point to position in *T. fuscirostris*. However, measurements virtually similar to *T. c. cuvieri* from Vogelkop peninsula which occurs slightly further north than *meyeri* along western shore of

Geelvink Bay; their ranges may meet, but apparently no overlap occurs. After *T. cuvieri* had crossed the Central Range southward to invade the range of *T. f. occidentis*, a race of strongly divergent size developed (*T. c. granti*), but when *T. fuscirostris* crossed the Central Range northward towards the range of *T. cuvieri*, it diverged in size towards *T. c. cuvieri*, away from *T. f. occidentis*, though plumage and bare part characters still appear to support inclusion in *T. fuscirostris*. Also, *T. f. meyeri* probably close in size to Brown-collared Talegalla *T. jobiensis jobiensis* from eastern shore of Geelvink Bay, and in fact leg colour of one bird from Ta R. (upper Siriwo basin, SE Geelvink Bay) recorded 'red' on label, like *T. jobiensis*, though another bird from there yellow, and legs of five birds from Rubi and Nappan (SW Geelvink Bay) appear reddish-yellow in skins; other characters do not support a possible intergradation into *T. jobiensis*, as feathering of cap and tibia and neck colour quite different. The species of *Talegalla* occurring on the eastern shore of Geelvink Bay is not yet identified; seen but not collected at Mt. Elephant (mouth of Warenai R.) (Meyer 1874*a*), and no description published of birds occurring between Mt. Elephant and mainland coast opposite Japen I.

Range and status

T. f. meyeri: southern shore of Geelvink Bay from Wandammen peninsula to about Siriwo basin. *T. f. occidentis*: lowlands at southern foot of Central Range of W New Guinea, from Etna Bay to about Sabang. *T. f. aruensis*: Aru Is and southern lowlands of New Guinea in Merauke area and (probably this race) basin of lower Fly R. *T. f. fuscirostris*: south coast of E New Guinea between *c*. 145 and 148°E.

T. f. fuscirostris overlaps with *T. jobiensis longicauda* for at least 90 (perhaps 600) km around 147°E in SE New Guinea, and with *T. cuvieri granti* for probably at least 240 km around 137°E in SW New Guinea. In areas of overlap they segregate altitudinally, with *T. cuvieri* and *T. jobiensis* confined to higher altitudes. Diamond (1972) suggests that the genus *Talegalla* once consisted of a superspecies ring of three allopatric forms, and that the range of the southern form, *T. fuscirostris*, is being invaded by *T. cuvieri* from the west and by *T. jobiensis* from the east, with the forms segregating altitudinally in each overlap zone. (See also Chapter 3.)

Widely distributed and fairly common.

Field characters

Length 51–58 cm. Plumage entirely black except for greyish semi-naked sides of head and brownish-black flight feathers. Differs from *T. cuvieri* (with which it overlaps locally: see Range and status) in having bill usually black to blackish-brown instead of orange-red, and tibia naked for distal 15 mm instead of feathered. Bill and legs proportionately smaller than in *T. cuvieri*. As other *Talegalla* spp. (and in contrast with *Aepypodius* spp.) notoriously noisy, calling frequently day and night, the sound carrying for over a mile (J. M. Diamond, *in litt.*).

Voice

See sonogram. Coates (1985) describes the advertising call as follows: 'a series of three to five very loud raucous honking or braying notes given rapidly on a rising scale followed either by a single short level note or one to three shorter notes uttered more slowly on a falling scale to give a minimum of four notes and a maximum of seven: "wha-wha-wha hah" (duration 2.5 sec); or "wha-wha-wha-wa hah" (3 sec); or "wha-wha-wha-wa hah ha" (3.5 sec); or "wha-wha-wha-wa hah hah hah" (5 sec) or 'whoa-wha-wha-wa-wa hah hah' (5 sec). A phrase may be repeated several or many times, with a pause of a few to several seconds between each series. Then there will be a long interval before the phrases are given again. Sometimes a pair calls together. The advertising call, most commonly given near the mound, can be heard at any time of the day, especially around noon, and sometimes at night. Other calls occasionally heard are a rapidly uttered series of up to 20 loud raucous notes at the same pitch which sounds like laughter, a repeatedly uttered short low guttural rail-like 'ou', which might function as a

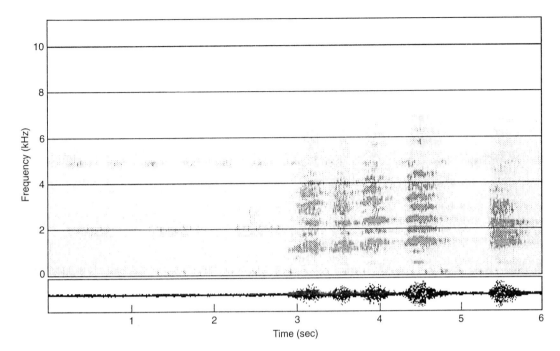

T. f. fuscirostris
K. D. Bishop, Varirata N. P., Papua New Guinea.

contact call, a repeated gulping note given when suspicious, and faint clicking notes (Coates 1985).

Habitat and general habits

Frequents rainforest, monsoon forest and sometimes gallery forest below 100 m, sometimes as high as 800 m. Terrestrial and shy; mostly occurs singly or in pairs. Stately-looking with head held high and broad, laterally compressed tail elevated; tail may be depressed when suspicious (Coates 1985). At a study site in lowland forest, density was estimated to be three birds per 10 ha (Bell 1982). Despite earlier reports that this species and the Orange-footed Megapode *Megapodius reinwardt* do not live in the same area (Rand and Gilliard 1967), Coates (1985) reported it not unusual to find both in the same locality, with active mounds of the two species only 200 m apart.

Food

Omnivorous as other megapodes; diet reported to include insects, small lizards, seeds, grubs, and fallen fruits.

Displays and breeding behaviour

Undescribed. A few observations of ♂ and ♀ at their mound, with one bird digging and the other at close range keeping watch, may suggest monogamy, if this behaviour is regular. If so, breeding behaviour would differ from that of the Australian and Wattled Brush-turkeys, *Alectura lathami* and *Aepypodius arfakianus*.

Breeding

NEST: mound, which is a broad, low heap of forest floor debris, mainly composed of leaves, about 1 m high with a flattish top measuring some 2.4 m across at the top and 3.6 m at the base (Rand and Gilliard 1967). It differs from the mound of *A. arfakianus* and Orange-footed Megapode in being low, flat, irregular, and untidy, rather like a randomly raked up heap of litter. Mounds are often located near the base of a tall tree on well drained, nearly level ground on the forest floor. One mound in the Port Moresby district varied between 5.4 and 7 m in width, and between 60 and 90 cm in height, and covered an area of 30 m^2. A mound may be used for several years before it is abandoned (Coates 1985). A much higher mound is reported by Ogilvie-Grant (1897), measuring 3.3 m in height and 7.5 m in diameter.

EGGS: similar to *Megapodius* eggs, but slightly less elongated; light pinkish to pale or dark buffy brown without gloss, oval in shape, shell slightly rough. During incubation, the pigment on the shell flakes off, giving the egg an irregular dirty brownish-white pattern. SIZE: 95.5–99.0 × 59.0–63.8 mm (97.0 × 61.0, n = 12, Schönwetter 1961); 87–97 × 55–62 mm (93 × 58.7, n = 3); 99 × 59–61 mm (n = 2); 93.5–94.3 × 59.4–59.4 mm (n = 2, RMNH); 90.5–97 × 60–62.4 mm (n = 2, Aru Is, RMNH). WEIGHTS: no information.

BREEDING SEASON: activity at a mound in the Port Moresby area was recorded during the rainy season and early dry season from Oct to May, with occasional visits during the dry season. Two mounds at 700–800 m altitude in hill forest at Varirata National Park near Port Moresby were visited during the period of low rainfall starting in Sept, which ceased suddenly in Nov until at least Jan (at the start of the wet season) (Coates 1985). Eggs and chicks from Utakwa and Mimika R. area, SW New Guinea, in Feb and Mar (Junge 1937); downy chicks in the Trans-Fly region in Feb, May, June, and Aug, from the Aru Is in Feb and May. Birds in breeding condition collected in central S New Guinea in Aug, Oct, Dec, and Jan (Rand 1942a).

References
Meyer (1874a, 1890b), Salvadori (1877, 1882), Oustalet (1881), Ogilvie-Grant (1893, 1897, 1915), Rothschild and Hartert (1901, 1913), Rothschild *et al.* (1932), Hartert *et al.* (1936), Junge (1937), Mayr and Rand (1937), Mayr (1938, 1941), White (1938), Rand (1942a), Van Bemmel (1947), Schönwetter (1961), Rand and Gilliard (1967), Diamond (1972), Mees (1982), Coates (1985), Holmes (1989), Roselaar (1994).

Brown-collared Talegalla   *Talegalla jobiensis* Meyer, 1874

Talegallus jobiensis A. B. Meyer, 1874. *Sitzungsberichte der Mathematisch-Naturwissenschaftlichen Classe der Kaiserlichen Akademie der Wissenschaften, Wien*, **69** (1–5), p. 74.

PLATE 2

Other names: Brown-collared Brush-turkey, Red-legged Brush-turkey, Jobi I. Brush-turkey, Long-tailed Brush-turkey.

Polytypic. Two subspecies. *Talegalla jobiensis jobiensis* A. B. Meyer, 1874, Japen I.; *Talegalla jobiensis longicauda* A. B. Meyer, 1891, N New Guinea from Mamberamo R. east to Milne Bay, and (perhaps this species) locally on southern slope of Central Range in E New Guinea.

Description
PLUMAGES
ADULT: sexes similar. Forehead covered with narrow black forward-projecting feathers with recurved tips, as in *T. cuvieri* and *T. fuscirostris*, but feathering of crown different from these species: black feathers lanceolate, webs not reduced, forming dense bushy crest projecting onto nape, quite unlike glossy black bristles lying closely to skull of both other species; tips of backward- or upward-pointing crown feathers sometimes slightly recurved, giving head a curassow-like appearance, quite different from thin-wet-hair look of *T. fuscirostris* and (especially) *T. cuvieri*. Lore and area round eye and ear naked, apart from ring of short black bristles round ear-opening. Cheek, chin, throat, and upper front and side of neck covered with

Brown-collared Talegalla *Talegalla jobiensis*

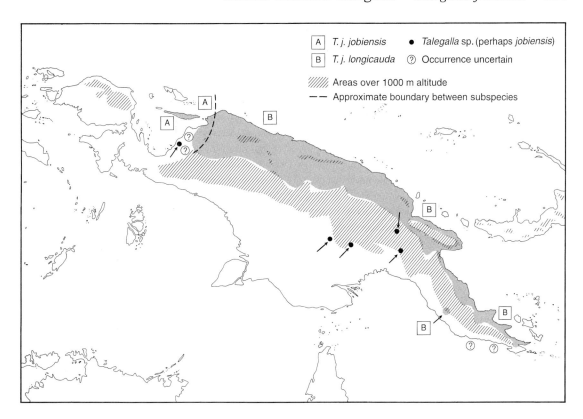

scanty short grey feathers, leaving much of skin visible; rear of neck with short black lanceolate feathers; feathers of lower neck somewhat wider, black with rufous-chestnut spots or bars, chestnut forming indistinct ring round lower neck; in some birds, chestnut restricted to a few fine bars on lower hindneck, in others spots and bars extensive, black on lower neck restricted to some spots or streaks, independent of age or locality. Entire body, wing, and tail black, deepest and glossiest on mantle, scapulars, and chest, sometimes slightly browner on belly and wing, depending on bleaching and abrasion. Lower end of tibia bare for 7.9 mm (3–13, $n = 15$), measured at side. In worn plumage, feathers of crown become narrower through abrasion, but never as bristle-like as in *T. cuvieri*.

CHICK: forehead, side of head, chin, and throat isabel-buff, except for sooty patch round eye extending to ear. Crown to mantle fuscous-brown, finely grizzled red-brown or cinnamon, faintly barred on hindneck. Scapulars and wing-coverts black with contrasting cinnamon (if plumage fresh) or tawny-yellow (if plumage worn) bars, black and pale bars each *c.* 3–4 mm wide. Rump and tail-coverts blackish-grey with some rufous grizzling. Underparts bright rusty-cinnamon, feather centres dark grey subterminally, appearing as dusky bars on chest, breast, and flank; belly almost uniform cinnamon to tawny-yellow. Flight feathers and tertials black with contrasting cinnamon-pink notches along outer webs and narrow irregular cinnamon-pink fringe on tip. Differs from *T. fuscirostris* in more rufous general colour, grizzled rufous rather than uniform dusky cap to mantle, and more extensive and more contrasting markings on rear of body and wing. Probably not separable from chick of *T. cuvieri*. Marks on wing of *T. cuvieri* stated to be more buffish, and underparts of *T. jobiensis* less bright red-brown (Meyer 1874*a*), but colour in both species dependent on abrasion and bleaching, becoming paler with age.

IMMATURE: like ad, but feathering of cheek, chin, throat, and neck less reduced, skin largely hidden. Feathers of cap as in ad, but often somewhat less bushy and dense; lower neck rufous with variable amount of dull black spots and bars. Body browner and duller than in ad. In first imm plumage, some feathers of head and body sometimes with traces of a rufous bar; outer primaries and tail still juv, fully grown p10 pointed, 90–120 mm long from insertion in skin to tip, tail *c.* 120 mm long. In second imm plumage, head and body as in ad, but second generation p10 and tail have not yet attained ad length, tip of p10 slightly pointed, length 130–150 mm (in ad, 140–170 mm), tail 145–160 mm (in ad, 146–203 mm).

BARE PARTS

ADULT: iris pale brown, brown, red-brown, brown-red, chestnut, or red. Bare skin of side of head reddish-, fulvous-, or purplish-brown, dull red, dull brick red, dark red, or coral-red, on neck salmon-pink, dusky pink, cherry-red, or blood-red. Bill dull yellow-brown, brown-horn, dark red-brown, dirty-red, or yellow-red, cutting edges and lower mandible paler horn-colour, pink-horn, or reddish-horn. Leg and foot salmon, pink-red, deep orange, dull magenta, reddish-orange, pale coral-red, light carmine-red, brick-red, bright red, or vermilion, often slightly lilac-grey on scutes of front of tarsus and upper surface of toes.

CHICK: at hatching, iris and upper mandible brown, lower mandible light brown, leg and foot brownish orange; at age of 6 weeks, iris and bill dark brown, leg and foot orange (Ripley 1964).

IMMATURE: iris brown or red-brown. Bare skin of head and neck salmon-pink to reddish-grey. Upper mandible dark brown or red-brown, blackish towards tip; cutting edges and lower mandible dull orange or brownish-red. Leg and foot dull orange to deep red.

MOULTS
Primary moult serially descendant. Of 19 ad examined, five had primary moult suspended (one each in Mar, Apr, May, June, and Sept), 14 in active moult (one Jan, one Apr, two Sept, one Dec, nine undated); two imm (one Mar, one undated) both in active primary moult. Generally (16 birds), two moulting series present in each wing, second series separated by four (in one bird), five (in two), six (in eight), seven (in two), or nine (in one) feathers from first; in three birds, three moulting series present, four (three to five) feathers apart.

MEASUREMENTS
ADULT: *T. j. jobiensis*. Japen; skins of 3 ♂♂, 2 ♀♀, and 1 unsexed bird (RMNH, SMTD, ZMA), and data on wing and tail of 1 ♂ and 1 ♀ from Rothschild *et al.* (1932).

	n	mean	s.d.	range
wing	8	276.8	5.92	268–286
tail	8	160.2	5.39	154–170
tarsus	5	91.3	2.17	88.8–94.2
mid-toe	5	49.2	1.31	47.5–50.5
mid-claw	5	18.6	1.28	16.8–20.4
bill (skull)	6	41.4	1.33	40.4–44.0
bill (nostril)	6	23.3	0.91	22.1–24.8
bill depth	3	18.5	0.26	18.3–18.8

According to Salvadori (1882), measurements of two ♂♂ and one unsexed bird: wing 285–295 mm, tail 160–165 mm, tarsus 90 mm.

T. j. longicauda: W New Guinea from Mamberamo R. east to Humboldt Bay area, southward to northern slope of Snow Mts, skins of 11 ♂♂, 14 ♀♀, and one unsexed bird, including data from BMNH, RMNH, Rand (1942b, wing only), and van Bemmel (1947), excluding literature data of probable imm with wing 255–257 mm.

	n	mean	s.d.	range
wing	26	285.5	9.75	264–307
tail	11	177.7	11.38	162–199
tarsus	5	92.0	4.30	85.3–96.8
mid-toe	4	52.8	2.06	51.0–55.5
mid-claw	4	18.7	1.16	17.6–20.2
bill (skull)	5	40.4	0.59	39.7–41.0
bill (nostril)	5	23.2	1.68	21.5–25.0
bill depth	4	18.7	0.31	18.3–19.0

E New Guinea from Astrolabe Bay to Milne Bay; skins of 9 ♂♂, 2 ♀♀, and 4 unsexed birds (BMNH, RMNH, SMTD, ZFMK, ZMB); wing and tail include data of 3 unsexed birds from Rothschild et al. (1932).

	n	mean	s.d.	range
wing	18	295.1	8.21	277–306
tail	18	188.2	8.40	171–202
tarsus	14	93.7	3.52	88.2–99.8
mid-toe	11	51.8	2.77	48.2–57.3
mid-claw	14	19.6	1.60	16.8–22.8
bill (skull)	15	43.4	2.38	40.5–49.0
bill (nostril)	14	24.7	1.63	23.2–29.3
bill depth	14	20.3	1.31	18.3–23.0

Probably both races, Japen to Humboldt Bay (Mayr 1938): wing, ♂ 261–286 mm (275.9 ± 9.08, $n = 8$); ♀ 257–286 mm (273.7 ± 9.44, $n = 6$); tail, ♂ 154–170 mm (160.0 ± 6.16, $n = 7$), ♀ 146–161 mm (153.6 ± 5.19, $n = 7$).
T. j. longicauda. E New Guinea (Mayr 1938): wing, ♂ 281–302 mm (290.6 ± 8.62, $n = 8$), ♀ 280–292 mm (286.7 ± 6.11, $n = 3$); tail, ♂ 175–204 mm (185.4 ± 9.18, $n = 8$), ♀ 164–188 mm (175.3 ± 12.06, $n = 3$).

CHICK: NE New Guinea, just hatched: wing 105 mm. Japen: wing 105 mm, tarsus 37 mm (Meyer 1874a).

WEIGHTS
T. j. jobiensis: ♀ 1360 g (Japen). *T. j. longicauda*: ♂ 1610 g (Humboldt Bay area); ♂ 1531, 1588, 1705 g (Sepik region).

CHICK: at hatching 110 and 125 g; at age of 6 weeks, 292 g (Ripley 1964).

GEOGRAPHICAL VARIATION
In size only; reports of darker body colour and sometimes more extensive red and black barring at base of neck in easternmost birds (e.g. Salvadori 1882; Meyer 1892; Rothschild and Hartert 1901) not substantiated later (Hartert 1930; Mayr 1938; this study). Nominate *jobiensis* from Japen distinctly smaller than *longicauda*, latter occurring in NE New Guinea between Astrolabe Bay and Milne Bay and perhaps south towards southern slopes of Central Mountain Range. However, boundary between races difficult to establish; according to Mayr (1938, 1941), nominate *jobiensis* occurs east to Humboldt Bay, *longicauda* eastwards from middle Sepik Basin. In birds examined, those from Sepik Basin indeed scarcely smaller than typical *longicauda* from Astrolabe Bay and Huon Peninsula, but neither are birds from northern slopes of Snow Mts; samples from Humboldt Bay area and lower Mamberamo R. very slightly smaller than Sepik birds, but much larger than Japen and therefore here included in *longicauda*. No birds from eastern shore of Geelvink Bay (opposite Japen) examined, but these probably near nominate *jobiensis*; boundary between the two races then formed by watershed west of Mamberamo Basin. Within *longicauda* as defined here, size probably increases gradually from lower Mamberamo (average wing 280 mm, tail 179, $n = 7$) to Milne Bay (average wing 300 mm, tail 193, $n = 2$) and also from lowland plains (e.g. Humboldt Bay area, average wing 282 mm, tail 173, $n = 5$) to mountain slopes (e.g. northern slope Snow Mts, average wing 287 mm, tail 178, $n = 2$), but wing and (especially) tail of nominate *jobiensis* appear distinctly shorter, not fitting within cline of decreasing size westward.

T. jobiensis has been collected on the southern slopes of the Central Range in Papua New Guinea at the Aroa R. in the Port Moresby district (Rothschild and Hartert 1901); an unidentified *Talegalla* is locally recorded elsewhere on the southern slopes (Lake Kutubu and Mt. Sisa in S Highland Province; Karimui area in Chimbu Province) (Diamond 1972; Dwyer 1981; Coates 1985), near the range of *T. fuscirostris*, presumed by Dwyer (1981) to be *T. jobiensis* on the basis of its voice; it may be that *T. jobiensis* replaces *T. fuscirostris* here on higher levels, but note that in the foothills of the Snow Mts it is *T. cuvieri* which occurs on southern slopes above altitudinal range of *T. fuscirostris*.

Range and status
Endemic to N New Guinea from Japen and Mamberamo Basin east to Milne Bay. Two specimens, adult and chick, known from the

Port Moresby area (Aroa R.), but status there, within range of *T. fuscirostris*, unknown. *T. j. jobiensis*: Japen and adjacent mainland (eastern and western limits on mainland unknown). *T. j. longicauda*: from the Mamberamo R. to the southeast point of New Guinea. Its occurrence in the Aroa R. area is presumably at a higher altitude than *T. fuscirostris*. (See also Chapter 3.)

Status: common in the Humboldt Bay area (Ripley 1964) and undoubtedly elsewhere.

Field characters

Length 53–61 cm. Very similar to *T. fuscirostris* but readily distinguished by its reddish to orange instead of pale yellow legs and feet, indistinct maroon-brown collar around the neck, elongated feathers on the crown, and dark reddish naked skin on the face and neck. Tibia completely feathered. As other *Talegalla* spp. (and in contrast with *Aepypodius* spp.), notoriously noisy, calling frequently day and night, the sound carrying for over a mile (J. M. Diamond, *in litt.*).

Voice

Vocal both day and night (Dwyer 1981). Series of two to four very loud braying notes, which resemble that of *T. fuscirostris* but are slower, the notes are fewer in number and longer in duration, given on a rising scale 'owagh-aagh' (2 sec), or 'agggh-owagh-aggh-ah' (6.5 sec) (Coates 1985). Also described as 'wankh-wankh' in an ascending scale (Ripley 1964).

Habitat and general habits

Frequents mostly well-drained, evergreen lowland forest up to at least 850 m. Fairly common in forest not usually occupied by other megapodes, for example on the northern watershed of the Sepik R., 190 river-miles from its mouth, the New Guinea Megapode *Megapodius decollatus* was common in the riverine and swamp forests bordering the river, while the Brown-collared Talegalla was found only in hilly forest far back from the edge of the river. Mounds in Saruwaged Mts (Huon) up to 1370 m (Mayr 1931); on Mt. Enassa on a sharp ridge under tall, thickly-mossed rainforest (Gilliard and LeCroy 1967*b*). A roosting place was located at about 7.6 m in a large tree (Ripley 1964).

Talegallas, not yet identified to species level (see 'Geographical variation') have been reported between 900 and 1200 m, northeast of Lake Kutubu and between 750 and 1500 m at Mt. Sisa, S Highland Province, and up to 1980 m in the Karimui area, Chimbu Province (Diamond 1972; Dwyer 1981; Coates 1985). Habits probably similar to *T. fuscirostris*.

Food

Unknown, but probably omnivorous as other megapodes (see also *T. fuscirostris*).

Displays and breeding behaviour

Undescribed. According to local information, this species builds a mound on flat ground beneath tall rainforest trees, which is constructed and owned by a single pair for a period of 3–5 years (Gilliard and LeCroy 1967*b*). This behaviour might imply monogamy as suggested for the Black-billed Talegalla.

Breeding

NEST: mound, which is huge, broad, and flat, constructed of humus, and measures up to 1.5 m in height and 0.9–3.6 m in diameter.

EGGS: similar in coloration and shape to the eggs of other *Talegalla* spp.; described as vinaceous russet to brown, showing white inner shell when outer layer flakes off during incubation. Similar in shape to the eggs of *Megapodius* spp., but slightly more tapered towards one end and less elongated. SIZE: *T. j. jobiensis*. 87–98.3 × 56–61.5 mm (94.7 × 60.0, $n = 8$, Schönwetter 1961); 95–98.3 × 60.4–61.5 mm ($n = 2$, Japen, Meyer 1890*b*). *T. j. longicauda*: 87–104 × 56–65.5 mm (96.0 × 61.8, $n = 20$, Schönwetter 1961); 97.3–100.4 × 62.3–62.4 mm ($n = 2$, Constantinehafen, Meyer 1890*b*). WEIGHTS: *T. j. longicauda*. 186–205 g ($n = 2$, Gilliard and LeCroy 1967*b*). Data from Mt. Sisa, S Highland

Province, most likely refer to this species: 232–235 g (233 ± 1.2, n = 3, Dwyer 1981).

BREEDING SEASON: according to local information, eggs can be found in any month of the year, totalling 15–17 eggs from one mound per year (Gilliard and LeCroy 1967b). This is contradicted by other information which suggests a short breeding season for this species. At Mt. Sisa (S Highland Province), at an altitude of c. 1450 m, eggs were for sale in Jan, while the birds were heard calling from Nov to Feb. Construction and maintenance of a mound was observed from Oct to Feb, while activity stopped around mid Mar. At lower altitudes (750–1000 m), the species was heard calling in late Mar and early Apr (Dwyer 1981). At Wapona, on the northern slope of the Maneao Range at 300 m, north-west of Mt. Simpson, E Papua New Guinea, eggs have been collected in Nov. A chick several weeks old was collected late July near Bodin on the upper Tor R., two others in Apr on Japen (Salvadori 1882).

Local names
Wayan (Japen), aro (Mt. Sisa, S Highland Province), nyonga (Adelbert Range), koreta (Mamberamo), waling (middle Sepik), óa, aloíya, wádi (Central Highlands, species not fully certain).

References
Meyer (1874a, 1890b, 1892), Salvadori (1877, 1882), Schlegel (1880a), Oustalet (1881), Ogilvie-Grant (1893, 1897), Rothschild and Hartert (1901), Hartert (1930), Mayr (1931, 1938, 1941), Rothschild et al. (1932), Rand (1942b), van Bemmel (1947), Ripley (1964), Gilliard and LeCroy (1966, 1967b), Rand and Gilliard (1967), Harrison and Frith (1970), Diamond (1972, 1985), Dwyer (1981), Coates (1985), Holmes (1989).

Genus *Leipoa* Gould

Leipoa Gould, 1840. *Birds of Australia*, 5, pl. 78 (= part 1).

One species, endemic to Australia.

A large aberrant member of Megapodiidae, characterized by contrasting colours of body, dense feathering on head and neck, short thin bill and short legs. In contrast to other large megapodes, skin bare only on lores, just round and behind eye, and immediately round ear; all remainder of head and neck densely covered with narrow lanceolate feathers, forming mane-like crest over crown and nape and a bunch of feathers down central throat. Feathers of forecrown hair-like, erect, up to 4 cm long. Wing fairly short and broad, tip rounded, but less broad and tip less bluntly rounded than in other large megapodes; length of wing relative to width of wing equalled only by a few members of *Megapodius*, and surpassed only by even longer and narrower wing of *Eulipoa*. Ten primaries: p5–p6 longest, p10 (outermost primary) 5.5–7 cm shorter than longest, p9 2–3.5 cm shorter, p8 0.5–2 cm shorter, p7 0–1 cm shorter, p4 0–2 cm shorter, p3 2–4 cm shorter, p2 4–6 cm shorter, p1 (innermost) 6–9 cm shorter. Tail long (together with *Alectura* relatively longest of all Megapodiidae), and combined with elongate and intricately patterned body and short legs give *Leipoa* a grouse-like appearance. Tail consists of 14–16 feathers; tail tip rounded. Upper tail-coverts long, reaching end of tail. Bill short, more slender and flattened dorso-ventrally than in other large megapodes, closely similar in shape and length to bill of (otherwise much smaller) Australian races of Orange-footed Megapode *Megapodius reinwardt*. Nostril narrow, covered by distinct operculum (flap) above. Unlike other Megapodiidae, feathers of body with bold pattern of contrasting bars, fringes, and shaft-streaks; bold bars otherwise found only in *Eulipoa*, contrasting fringes only in *Alectura* and (to a much lesser extent) in *Aepypodius*. Tarsus and toes short, relative to body size shortest of all Megapodiidae. Tibia feathered down to joint. Tarsus thick, covered by numerous hexagonal scales, widest in front. A small web between bases of middle and inner toe, reaching up to 15 mm from joint. Claws short but strong.

Malleefowl *Leipoa ocellata* Gould, 1840

Leipoa ocellata Gould, 1840. *Birds of Australia*, 5, pl. 78 (= part 1).

PLATE 3

Other name: Lowan.

Monotypic. Synonyms: *L. o. rosinae* Mathews, 1912, South Australia; *L. penicillata* Mathews, 1923, *nomen nudum*.

Description
PLUMAGES
ADULT: forehead, side of crown, side and rear of neck, and upper mantle medium bluish-grey or brown-grey, merging into plumbeous-black or sooty of crest on central crown and nape; feathers of crest narrowly tipped grey-brown when plumage fresh. Feathers of lower mantle grey on basal half, but this grey largely hidden; terminal half with dark rufous to chestnut bar (*c*. 6–18 mm broad), bordered by straight pale cream or pink-white bar (*c*. 3 mm wide) towards base, which is outlined in black at one or both sides, feathers bordered buff or buff-brown on tip. Pattern on scapulars and tertials basically similar, but cream or pink-white bar wider, more chevron-like, extending into pale shaft-streak pointing towards tip, dividing chestnut,

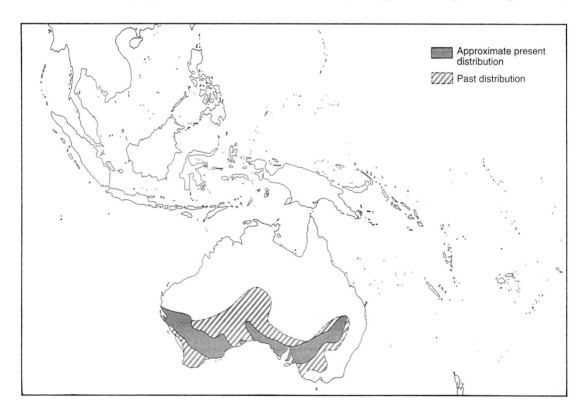

Approximate present distribution

Past distribution

Plates

Plate 1
Alectura and *Aepypodius*

1. **Australian Brush-turkey**
Alectura lathami p. 89
Large mound-building megapode from the forests of eastern Australia. Conspicuous long and broad tail. Sexually dimorphic. Male with bright red head and large inflatable yellow (nominate *lathami*) or purplish-white (ssp. *purpureicollis*) neck sac. Female less brightly coloured.
(a) *Alectura lathami purpureicollis*, adult male.
(b) *Alectura lathami lathami*, adult female.
(c) *Alectura lathami lathami*, adult male.
(d) *Alectura lathami lathami*, chick.

3. **Wattled Brush-turkey**
Aepypodius arfakianus p. 97
Medium-sized, mound-building megapode restricted to mountain forests of New Guinea, Japen, and Misol. Sexually dimorphic. Male with bright red and white (nominate *arfakianus*) or blue and white (ssp. *misoliensis*) wattles. Female less brightly coloured. Both sexes with rufous rump, which is characteristic for this genus.
(a) *Aepypodius arfakianus arfakianus*, adult male.
(b) *Aepypodius arfakianus misoliensis*, adult male.
(c) *Aepypodius arfakianus misoliensis*, adult female.
(d) *Aepypodius arfakianus misoliensis*, chick.

2. **Bruijn's Brush-turkey**
Aepypodius bruijnii p. 103
Large blackish-brown megapode, endemic to Waigeu, Irian Jaya, where it is probably restricted to mountain forests. Only known from 15 museum specimens. Sexually dimorphic. Male with one long wattle on the lower throat and a warted comb extending in two long red wattles on the back of the head. Female without these extreme ornaments. Both sexes with rufous rump, which is characteristic for this genus.
(a) *Aepypodius bruijnii*, adult male.

Plate 2
Talegalla

1. Red-billed Talegalla
Talegalla cuvieri p. 106
Large black mound-building megapode, endemic to the lowlands of northwest New Guinea, east to southern slopes of the Snow Mountains, where it overlaps range of Black-billed Talegalla. Unlike the Wattled Brush-turkey, with which it overlaps at the foot of mountains, it has no wattles and no rufous on rump. Bare skin of head and neck yellow or olive-green, iris yellow, bill red, and legs yellow or orange, contrasting with black feathering. Sexes similar.
(a) *Talegalla cuvieri*, adult.
(b) *Talegalla cuvieri*, chick, few weeks old.

3. Brown-collared Talegalla
Talegalla jobiensis p. 116
Large all-black mound-building megapode with short rough crest and indistinct chestnut markings at base of the neck. Endemic on Japen Island and in northern New Guinea between eastern shore of Geelvink Bay and eastern tip. May occur locally on southern slopes of central range of New Guinea, where it may then overlap with the Black-billed Talegalla. Bill, iris, and bare skin of head and neck red-brown or red. Legs red, which is obvious in bright sunlight but might be difficult to see in the shade of the forest. Sexes similar.
(a) *Talegalla jobiensis*, adult.

2. Black-billed Talegalla
Talegalla fuscirostris p. 110
Large all-black mound-building megapode, endemic to the lowlands of southern New Guinea as well as southern shore of Geelvink Bay and Aru Islands. Rather similar to Red-billed Talegalla, but bill and bare skin of head and neck dark slate-grey to blackish and iris brown, showing little contrast with black feathering. Legs yellow. Sexes similar.
(a) *Talegalla fuscirostris*, adult.
(b) *Talegalla fuscirostris*, chick.

Plate 3
Leipoa

1. Malleefowl
Leipoa ocellata p. 122
Large mound-building megapode, restricted to open mallee country of central and southern Australia. Feathering mainly grey, overlaid with a complicated pattern of black and chestnut marks. Unlike other large megapodes, it features no wattles or brightly coloured bare parts. Sexes similar.

(a) *Leipoa ocellata*, adult at mound.
(b) *Leipoa ocellata*, adult.
(c) *Leipoa ocellata*, chick.

Plate 4
Macrocephalon

1. Maleo
Macrocephalon maleo p. 130

Large burrow-nesting megapode, endemic to Sulawesi, Indonesia. It incubates its eggs at communal nesting grounds in the forest heated by volcanic activity or on sun-exposed beaches. Upperparts, neck, and conspicuous, flattened tail black, contrasting with salmon to cream-white belly. Distinct black casque on rear of head. Brightly coloured skin round eye. Sexes similar. Males have characteristic loud, rolling calls.

(a) *Macrocephalon maleo*, adult.
(b) *Macrocephalon maleo*, second immature, ± 8 months old.
(c) *Macrocephalon maleo*, first immature, ± 4 months old.
(d) *Macrocephalon maleo*, chick.

Plate 5
Eulipoa and *Megapodius*

1. Micronesian Megapode
Megapodius laperouse p. 152
Small olive-black megapode, restricted to the Mariana and Palau Islands, where it buries its eggs in mounds on coral islands or in burrows in volcanic soils. Differs from other small black megapodes in having a light grey head and neck with contrasting yellowish-red to bright red bare skin. Short crest conspicuous. Legs and bill bright yellow. Sexes similar.
(a) *Megapodius laperouse*, adult.
(b) *Megapodius laperouse*, chick.

3. Moluccan Megapode
Eulipoa wallacei p. 140
Small piedly coloured burrow-nesting megapode, endemic to the Moluccan islands and Misol (West Papuan islands). Plumage generally not conspicuous due to its secretive habits when feeding in deep forest and to its nocturnal life at the communal nesting grounds. White patches on underwing and on vent may have signalling function during the nocturnal flight to the nesting ground. Sexes similar.
(a) *Eulipoa wallacei*, adult.
(b) *Eulipoa wallacei*, chick.

2. Polynesian Megapode
Megapodius pritchardii p. 146
Small burrow-nesting megapode, restricted to tiny volcanic island of Niuafo'ou, Kingdom of Tonga. Entirely slate-grey with rufous or olive-brown back and wings. Short crest as in most other *Megapodius* species. Bare skin of head and neck red, bill yellow. Legs pale yellow to yellowish-red. A variable amount of white is often visible in the tail and at the primary bases. This loss of pigmentation occasionally occurs in small populations living on tiny islands. Sexes similar.
(a) *Megapodius pritchardii*, adult.
(b) *Megapodius pritchardii*, chick.

Plate 6
Megapodius

1. **Nicobar Megapode**
Megapodius nicobariensis p. 159
Medium-sized, brown megapode, restricted to the Nicobar Islands. It incubates its eggs in mounds, often built close to the beach. Bare patch on side of head pinkish-red, legs brownish. Sexes similar. Closely related to the Philippine Megapode which has the underparts, neck and head grey, and similar to the Sula Megapode which has brightly coloured red legs and which lacks the bare patch around the eye. Male and female not reported to perform duets.
(a) *Megapodius nicobariensis*, adult.
(b) *Megapodius nicobariensis*, chick.

3. **Philippine Megapode**
Megapodius cumingii p. 165
Medium-sized megapode from the Philippines, islands between the Philippines and North Borneo, and Sulawesi. Upperparts brown. Underparts, neck, and head grey. Bare patch around eye red, similar to Nicobar Megapode. Legs dark. Sexes similar. Mainly incubates its eggs in mounds or between decaying roots of trees. As for the Nicobar Megapode, duetting has not been reported for this species.
(a) *Megapodius cumingii tabon*, adult.
(b) *Megapodius cumingii gilbertii*, adult.
(c) *Megapodius cumingii*, chick.

2. **Sula Megapode**
Megapodius bernsteinii p. 175
Medium-sized, mound-building megapode restricted to the Sula and Banggai islands between Sulawesi and the Moluccas. Plumage entirely rufous-brown, legs red. Sexes similar. Duetting by male and female often heard at night.
(a) *Megapodius bernsteinii*, adult.

Plate 7
Megapodius

1. Dusky Megapode
Megapodius freycinet p. 181
All-black, medium-sized mound-building megapode from the northern Moluccas and West Papuan islands. Short but conspicuous pointed crest. Bare skin of head and neck mainly dull red. Legs dark. Sexes similar. Sympatric with the Orange-footed Megapode on Batanta and Sorong, which is mainly brownish-grey and has conspicuous orange legs. Male and female duet.
(a) *Megapodius freycinet*, adult.
(b) *Megapodius freycinet*, chick.

3. Melanesian Megapode
Megapodius eremita p. 198
Medium-sized, very dark megapode restricted to islands northeast of New Guinea. Naked skin of head and neck dull pinkish-red. Legs dark, often with greenish or olive tinge. No conspicuous crest. Sexes similar. Locally very common on volcanic sites, as on New Britain, where it buries its eggs in the ground. Male and female reported to duet.
(a) *Megapodius eremita*, adult.
(b) *Megapodius eremita*, chick.

2. Biak Megapode
Megapodius geelvinkianus p. 189
All-dark, mound-building megapode restricted to the Geelvink Bay islands, off Irian Jaya. Much smaller than the closely related Dusky Megapode from which it differs in having the legs partly or mostly reddish and bare skin of the head bright red. Sexes similar. Might be sympatric with the New Guinea Megapode on Japen, which is larger, brownish-grey, and which has dark instead of reddish legs. Male and female probably duet. Long considered a race of the Dusky Megapode.
(a) *Megapodius geelvinkianus*, adult.

4. Vanuatu Megapode
Megapodius layardi p. 205
Relatively large mound-building *Megapodius* species, restricted to Vanuatu. Generally blackish with conspicuous yellow legs. Bare parts of head and neck red. Sexes similar. Male and female are reported to perform duets.
(a) *Megapodius layardi*, adult.

Plate 8
Megapodius

1. Orange-footed Megapode
Megapodius reinwardt p. 213
Large mound-building *Megapodius* species, with size depending on race. Common throughout large parts of southeastern Indonesia, southern New Guinea, and Australia. Upperparts dark olive-brown, underparts grey. Legs and feet bright orange. Elongated pointed crest conspicuous. Sexes similar. Very vocal, often duetting.
(a) *Megapodius reinwardt*, adult.
(b) *Megapodius reinwardt*, chick.

3. Tanimbar Megapode
Megapodius tenimberensis p. 179
Large mound-building *Megapodius* species, restricted to Tanimbar Islands, Indonesia. Generally brownish-olive and grey with reddish legs. Bare skin of head reddish, crest short. Sexes similar. Long considered a race of the Orange-footed Megapode, but structurally rather different.
(a) *Megapodius tenimberensis*, adult.

2. New Guinea Megapode
Megapodius decollatus p. 208
Medium-sized, mound-building megapode from northern New Guinea and offshore islands, including Japen. Slightly smaller than the almost similar Orange-footed Megapode with which it is locally sympatric. Can best be distinguished by its dark instead of orange legs. Sexes similar. Male and female perform duets.
(a) *Megapodius decollatus*, adult.

4. Forsten's Megapode
Megapodius forstenii p. 193
Medium-sized, mound-building megapode restricted to the central Moluccan islands where it is sympatric with the Moluccan Megapode. Generally brownish-grey with short crest and olive or blackish legs. Bare skin of head and neck red, as in many other *Megapodius* species. Sexes similar. Male and female duet, often at night. Long considered a race of the Orange-footed Megapode.
(a) *Megapodius forstenii buruensis*, adult.
(b) *Megapodius forstenii*, chick.

dark rufous, or buff-brown of tip in two oval spots ('ocellae'), boldly outlined in black; pale fringes along feather tips broader, whiter (less buff), grey of feather bases sometimes partly barred black; on longer tertials, brown of feather tips and grey of bases with partial or complete black bars. Back and rump grey with slight buff tinge; feather tips with narrow dusky wavy bars, on lower rump with more regular black bars.

Upper tail-coverts and central pair of tail-feathers (t1) dark grey, tip cream-grey; marked with broad black bars, intervening grey space marked with narrow dusky irregular bars or marbling. T2 like t1, but darker grey on basal and middle portion and paler cream-buff on tip, all with wavy black bars; remaining tail-feathers mainly black, 2.5–3 cm of tip contrastingly cream or white, bases gradually more grey. Lower cheek, chin, and upper throat rufous-cinnamon with pale cream lines, cinnamon ending in sharp point downward. Lower throat to mid-chest black, ending in black point on mid-breast; black on lower throat with broad cream and rufous stripes, on mid-chest with cream or yellow-white stripes. Side of chest and breast medium bluish-grey, sometimes with slight brown or buff tinge; flank cream-yellow with bold black bars; remainder of underparts cream-yellow, almost white on vent, warmer cream-buff on under tail-coverts. Primaries sepia-brown, tip with 2–3 mm wide pale fringe, outer web with irregular sepia-black and off-white blotches, bars, or marbling; secondaries similar but with more regular broad cream-white and sepia-black bars on both webs, black widest subterminally; pattern on innermost secondaries more similar to that of tertials. Lesser upper wing-coverts medium grey with black subterminal bar and narrow off-white tip, latter tinged brown at border with black. Median and greater upper wing-coverts grey, barred black on larger coverts; tips with bold semicircular black dots broadly and contrastingly outlined by white. Under wing-coverts and axillaries greyish-cream with black subterminal bars. Much influence of bleaching and wear, dark spots on upperparts blackish-chestnut when plumage new, grading through paler chestnut to tawny when old; underparts (especially under tail-coverts) warm tawny-buff if new, isabel-white if old.

Sexes similar, or, according to Mathews (1910–11), undersurface of outer web of primaries more distinctly marbled and vermiculated in ♀♀ (♂♂ almost uniform).

CHICK: conspicuously spiny due to each feather of head and body ending in boldly patterned spiny tips. Top and side of head and side and rear of neck dark brown with narrow isabelline streaks; brown darkest on crown, isabelline more prevailing on lore and behind ear; narrow brown border encircles ear. Cheeks, chin, and throat cream or off-white. Spines of mantle streaked brown and cream, those of scapulars brown with cream or white bar on feather centre. Underparts soft, not spiny; yellowish-cream, tinged rufous on chest; side of breast and belly, chest, and flank faintly barred dark grey. Primaries black-brown, outer web and tip notched white; secondaries, tertials, and greater primary coverts black-brown with broad white tips and contrasting white bars or marbling along shafts.

IMMATURE: in first imm plumage, like ad, but under wing-coverts and axillaries closely marked with faint bars, less uniform and rufous than in older birds; undersurface of outer web of primaries distinctly vermiculated and mottled; as in other Megapodiidae, juv (first generation) outer primaries as well as some secondaries retained, short and strongly pointed, notched white on outer web and tip. Second imm plumage like ad, but second generation outer primaries slightly shorter and more pointed at tip than in ad, length of p10 to insertion in skin 165–175 mm ($n = 4$; in ad 173–185 mm, $n = 5$); undersurface of outer webs of primaries still distinctly marbled on older outer feathers, less so on newer inner feathers.

BARE PARTS
ADULT, IMMATURE: iris light hazel to brown. Bare skin of lore and round eye and ear clear blue, pale blue, or dusky grey, bluish-white just

below eye. Bill dark bluish-horn to black, tinged blue-grey at base. Leg and foot blue-grey, brownish-grey, dusky slate, or blackish-brown.

MOULTS
Primary moult serially descendant. Two to three series active on each wing: in seven birds of ten examined, first series just completed with regrowth of p10, second active or suspended at p4, p5, p6, or p7; in three birds, both with moult suspended, first series just reached p10, second to p8 or p7, third to p4. Only three of ten birds in active moult, but, as most of them undated, no indication about moulting season.

MEASUREMENTS
ADULT: S Australia; skins of 2 ♂♂, 1 ♀, and 10 birds unsexed (BMNH, RMNH, ZMB).

	n	mean	s.d.	range
wing	13	319.8	12.21	300–339
tail	12	215.7	10.27	198–235
tarsus	13	75.9	2.15	72.5–80.3
mid-toe	12	41.0	2.16	37.5–44.0
mid-claw	12	18.4	2.08	14.5–21.8
bill (skull)	12	35.2	3.00	30.8–41.8
bill (nostril)	9	15.8	1.30	14.2–17.5
bill depth	9	10.7	0.55	10.1–11.7

CHICK: wing 117 mm, tarsus 30.3 mm ($n = 1$, BMNH). Measurements of live specimens ($n = 10$): wing 112.1 ± 7.7 mm; tarsus 31.2 ± 1.0 mm; bill length 14.8 ± 0.7 mm; bill width 5.7 ± 0.4 mm; bill depth 5.9 ± 0.3 mm (Benshemesh 1992a).

WEIGHTS
ADULT: ♀ 1788–1901 g (1830, $n = 4$, Frith 1959b); ♀ 1520–2050 g (1768, $n = 5$, Booth 1987b); ♂ 2000–2500 g (D. Booth, in litt.); 1810–1925 g (Serventy and Whittell 1967). Both ♂♂ and ♀♀ reach a maximum weight in spring and a minimum weight during late summer to autumn (Benshemesh 1992a).

CHICK: at hatching 82–117 g (107.2 ± 8.42, $n = 29$, Priddel and Wheeler 1990a). Chicks may lose up to 10 g of their body weight in the week following hatching and do not gain much weight during the second week; from the beginning of the third week they grow slowly, but steadily; after 70 days body mass is approximately 300 g (Booth 1989a).

GEOGRAPHICAL VARIATION
None. *L. o. rosinae* described by Mathews (1912a) from S Australia; said to differ from nominate *ocellata* from Swan R., Western Australia, by its larger size and lighter coloration, but in view of strong influence of bleaching and abrasion on plumage and marked individual variation in size (depending on age), *rosinae* not recognized here, following Condon (1975).

Range and status
The Malleefowl originally occurred over large areas of southern, central, and western Australia, but is now restricted to patches of suitable habitat in southwest and central New South Wales, northwest Victoria, South Australia, and parts of southern Western Australia. Until the 1930s it occurred in the southern parts of the Northern Territory as well, with a single record from the Tanami Desert in the 1950s. Several pairs were introduced on Kangaroo I. (South Australia) between 1911 and 1936, where it is now extinct. For detailed distribution map see Blakers et al. (1984).

Habitat destruction and predation have reduced and fragmented the species' number in most parts of its range. In Victoria, the population is now estimated at less than 1000 pairs, with densities of up to 3.5 pairs per km^2. In 1985, the population in New South Wales was estimated at 750 breeding pairs. Between the 1950s and 1980, the population at Round Hill Nature Reserve, New South Wales, an area of 13 630 ha of marginal habitat, dropped from an estimated maximum of 0.6 active mounds per km^2 to 0.04 active mounds per km^2 (Brickhill 1985a); other reports mention a drop from 209 to three pairs for the same area (Priddel 1990). Furthermore, numbers have recently declined in Yalgogrin and at the Pulletop Nature Reserve. The population in South Australia is also considered small, although no detailed data are available from here nor from Western Australia.

Breeding densities have been found to be positively correlated with the amount of vegetation cover above 2 m, but show no relation to the abundance of either food-bearing shrubs or foxes (Benshemesh 1992a). The home range in an area of low rainfall mallee near Renmark in South Australia varied between 1.7 and 4.6 km^2, with considerable overlap of range in pairs using different mounds (Booth 1987b). Home ranges calculated by Booth are larger than suggested by Frith (1959b, 1962b), which is probably due to differences in technique; Frith's data are based on observations, whereas Booth used radio-tracking; estimates of home range derived from the latter would be expected to be both larger and more accurate. The home range of a mated, free-ranging, but remarkably tame ♀ observed for a period of 5 months in Wyperfeld National Park, Victoria, was estimated at between 49 and 75 ha, which is larger than suggested by Frith but smaller than measured by Booth (Benshemesh 1992a). Overall breeding density was 1.1 active mounds per km^2 in Booth's study, 2.5–5.5 active mounds per km^2 in the wetter areas near Griffith (New South Wales) studied by Frith, and 2 pairs per km^2 in the study by Benshemesh. The breeding range of one ♀ was more than twice the area held during the non-breeding season and included most of it (Booth 1987b).

Chicks disperse widely the first few days after hatching, with a mean rate of movement of 627 m/day (n = 19 chicks); three chicks averaged at least 2 km/day for one or more days (Benshemesh 1992a).

Field characters
Length ± 60 cm. Unmistakable. Large, heavily spotted or 'ocellated', short-legged galliform-like bird with small head and small bill of the semi-arid to arid scrubs and woodland, especially mallee, of southern Australia.

Voice
The ♂'s territorial call, most commonly produced near the mound, is the most characteristic of all calls. It consists of a loud booming 'uh-uh-uh-oome-oome-oome' (Bellchambers 1916) or 'ooee-ooee-ooee' with the emphasis on the last syllable of each note and audible up to a maximum of 800 m as reported by Frith (1959b); reports of audibility beyond 800 m not substantiated (D. Priddel, *in litt*). The sequence is often repeated three to five times. The rendering 'coo, coo-loo' given by Fleay (1937) is considered inaccurate (D. Priddel, *in litt.*). While booming, the ♂ takes a characteristic posture, bending his head and neck downwards against his breast while fluffing his neck-feathers and with his bill pointing backwards parallel to the ground (Frith 1959b; see Chapter 4, Fig. 4.4 for illustration, and Cooper (1966a) and Immelmann and Böhner (1984a) for photographs). The duet as described and illustrated by Böhner and Immelmann (1987)(see sonogram), in which the ♂'s booming territorial call is answered with a loud and long drawn-out note by the ♀, has not been reported as such during extensive field-work (D. Priddel, *in litt*). According to Bellchambers (1916), however, the ♀ replies with a long drawn-out and rising 'waugh' 'waugh', given at a much higher pitch than the call of the ♂. Frith (1959b) reported that the ♀ produces a high-pitched crow in answer to the ♂'s booming call, with her head pointing upwards (illustrated in Bellchambers 1916). This posture differs from the description and illustration in Böhner and Immelmann (1987), in which the ♀ stretches her body and neck slightly upwards and bends her head down until the tip of her bill nearly reaches the ground; this aberrant behaviour may have been due to the unnatural conditions (a tiny enclosure) in which the study pair was kept (D. Priddel, *in litt.*).

When working or feeding together, the ♂ and ♀ continuously utter low-pitched grunts (Frith 1959b, 1962a; Tarr 1965), audible only at very close range. Before and during laying, the ♀ croons continuously. This crooning,

126 Malleefowl *Leipoa ocellata*

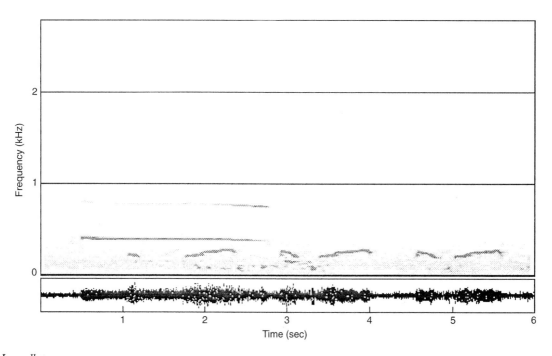

L. ocellata
J. Böhner, Adelaide Zoo, South Australia, 1982.

described by Tarr (1965) as 'ooma', and with emphasis on the first syllable and the second of lower pitch than the first (J. Benshemesh, *in litt.*), becomes increasingly loud and is repeated frequently shortly before the egg is laid, and the ♂ joins in at this stage. Bellchambers (1916) described this crooning call of the ♀ as 'whoo how', 'whoo how'. The alarm call varies between a soft, enquiring grunt (rendered 'ut ut ut' by Bellchambers 1916) and a loud sharp grunt (Frith 1959*b*, 1962*a*). However, birds often remain quiet when disturbed (Mattingley 1909).

Chicks are usually silent, but sometimes grunt when handled or threatened. At the age of one month, they sometimes utter deep grunting alarm calls very similar to that of the adults (D. Priddel, *in litt.*).

Habitat and general habits

The habitat of the Malleefowl differs from that of all other megapodes, for it lives in semi-arid areas in mallee and eucalypt woodland often with a dense shrub undergrowth. Highest densities occur in habitats with a nearly complete canopy, a rich shrub layer, and an open ground layer. In Victoria, it prefers areas with good overhead cover rather than a dense shrub layer, and shows no noticeable preference for an abundance of *Acacia*. In Western Australia, it occurs in low woodland of Bowgada *Acacia linophylla ramulosa*, Ti-trees *Melaleuca* spp., Casuarinas *Casuarina/Allocasuarina* spp., Hakeas *Hakea* spp., and Grevilleas *Grevillea* spp. At a location in South Australia, the habitat has been described as an association of Porcupine grass *Triodia irritans*, mallee (the major eucalypt species being *Eucalyptus socialis*, *E. incrassata*, *E. foecunda*, *E. gracilis*, and *E. cyanophylla*), and various shrubs such as *Acacia*, *Senna*, *Dodonaea*, *Eremophila*, and *Beyeria* (Booth 1987*b*).

Monogamous. Breeding starts at the age of 3 or 4 years. Pair bonds last up to 6 years or more, and probably remain intact until one of the partners dies (Frith 1959*b*). The record is held by a pair in the Little Desert, Victoria, which was known to have bred for 25 years before one of the birds disappeared (Benshemesh 1992*a*).

Polygyny has been reported only once: a single ♂ was mated with two ♀♀ which laid their eggs in separate mounds, both of which were maintained by the ♂ (Weathers et al. 1990).

Despite their long-lasting pair bond, ♂ and ♀ lead relatively solitary lives. The ♂ remains in the vicinity of the mound, both during the day and at night when he roosts in a nearby tree. As a daily routine, he leaves the roost before dawn and goes straight to the mound to start digging. Foraging is often postponed until 8 or 9 a.m. He spends most of the day in the vicinity of the mound, resting, sun-bathing, and digging. The ♀ roosts further away from the mound than the ♂, and usually visits it within the first hour after dawn, often at around sunrise. She forages during the morning and the evening and spends the remainder of the day resting at various locations. These observations of Frith (1959b, 1962a), confirmed by observations of D. Priddel (personal communication), differ from those of Immelmann and Böhner (1984a), who describe a much stricter pair bond between ♂ and ♀; their study was of a single pair of breeding birds in a tiny enclosure, though in their natural habitat. All activities took place with a high degree of synchronization between ♂ and ♀. Every morning when the pair met at the mound, they performed a 'greeting ceremony' of approximately 30 sec in which the birds walked around each other in opposite directions while preening their feathers and lowering their wings. Some of the elements of this ceremony, such as lowering of the wings and fluffing of the neck and head feathers, have also been observed in agonistic behaviour (Frith 1959b; see 'Displays and breeding behaviour').

Chick mortality, which is high initially, is related to the weight of the chick at hatching and thus to the weight of the egg. Chicks that survive the first 10 days are significantly heavier at hatching than those that died. The higher rate of survival of heavier chicks has been proved not to be related to body size; it is probably due to the fact that heavier chicks have larger food reserves (from the egg) stored in the abdominal cavity. Malleefowl chicks are usually active in the early morning and late afternoon, spending the greater part of the day resting in the shade of low bushes (Benshemesh 1992a).

Food

Frith (1962b) recorded fruits, buds (especially *Beyeria opaca*), and seeds of shrubs as accounting for 73% of the Malleefowl diet, herbs 10% and invertebrates 17%. Further vegetable food items included: *Acacia brachybotrya, A. buxifolia, A. hakeoides, A. rigens, A. stenophylla, Senna artemisioides, Eriostemon difformis, Santalum acuminata, Owenia acidula,* and *Pittosporum* sp. In contrast to the findings of Frith, who characterized the diet of the Malleefowl as primarily granivorous, Benshemesh (1992a) did not record any substantial consumption of seeds from shrubs. The diet of an adult ♀ studied over a five-month period consisted most frequently of herbs, in particular foliage (*Thysanotis baueri* and *Helichrysum leucoptera*) and flowers (*Lomandra effusa*). Flowering *Acacia rigens* shrubs were neglected. Surprisingly, lerps—the carbohydrate tests of psyllids (*Glycaspis* spp., Homoptera) which are rich in soluble sugars and other carbohydrates—comprised over a third of the diet. Also fungi, mostly *Mycena* spp., *Camponotus* ants, and termites were eaten when available. There was a distinct seasonal trend characterized by an increasing consumption of lerps and a decreasing consumption of lerps from Apr to Aug, while overall consumption more than doubled prior to egg laying. During this period the ♀ was also observed to peck at old bones, which might be an important source of calcium for egg production (Benshemesh 1992a).

In two adult birds, plant material accounted for 93.5 and 99% of the dry weight of crop and gizzard contents and consisted of seeds: *Cassytha melantha, Dodonaea bursariifolia, Enchylaena tomentosa,* and *Zygophyllum* sp., foliage: *Anthropodium strictum, Sclerolaena (Bassia)* sp., and *Zygophyllum* sp., and invertebrates: ants, bees, beetles, cockroaches, dragonflies, grasshoppers, spiders, and wasps (Booth 1986). Based on faecal analysis, Brickhill (1987b) identified fruits and seeds of mallee stranglevine *Cassytha* sp., wheat *Triticum* sp., Saloop *Einadia* sp., Saffron thistle *Carthamus* sp., and Wattle *Acacia* sp. as forming the main part of the diet, while green herbage was commonly eaten in spring and arthropods especially in autumn.

Displays and breeding behaviour

The complex process of mound building and mound maintenance has been described in detail for the Malleefowl in Chapter 6. At the onset of the new breeding season the pair selects a site, and the arduous task of construction begins all over again. The mound chosen may or may not be that of the previous year, and though a mound is sometimes used for 5 consecutive years or more, this is not necessarily by the same pair. In fact, there appears to be no pattern and this has led Frith (1959b) to suggest that mound selection is arbitrary.

During the breeding season ♂♂ are aggressive to intruders, although encounters near the mound are infrequent. Most disputes are vocal, but they escalate into threat display or fighting when the intruder comes too close to the mound. Frith (1959b) recognized three stages of aggression: threat display, intention to fight, and actual fighting. In full threat display, delivered when the opponent is visible, the crest and neck feathers are raised and the head is lowered until the bill touches the ground. The head is turned so that it is directed towards the intruder. The wings are opened and twisted forward and the feathers are spread. The tail is spread widely, displaying the prominent white terminal bar. On some occasions this display, often of a lower intensity, is directed to the ♀, who responds by giving the same display. If the intruder does not withdraw, the threat display of the mound owner is intensified by twisting the tail sideways, raising his breast with the feathers expanded, thus showing the intention to fight. The ♂ may now proceed towards the intruder, or walk in a tight circle, stamping frequently on the ground and displaying his open claws. During a fight, the ♂ jumps at his opponent, strikes with his bill and feet as well as with wings and chest. It normally results in some loss of feathers, but occasionally even in the death of one of the birds (Frith 1959b).

Copulation is simple, quick, and without noteworthy vocalizations (Frith 1959b), although Tarr (1965) reported vocalization resembling the booming territorial call preceding copulation. Copulation has been observed between Sept and Feb and occurs throughout the entire egg-laying period. Frith (1959b) never observed pre-copulatory display. The ♂ simply stops his labour at the mound and approaches the ♀ in a very low-intensity threat display (described above). She, in turn, crouches in a submissive posture with her wings partly spread and the ♂ walks on to her back, dismounting within a few seconds. After copulation, both ♂ and ♀ normally resume their activities.

When ready to lay, the ♀ approaches the mound, continuously uttering low-pitched crooning or clucking notes which trigger the ♂ to open up the mound. Both ♂ and ♀ participate in digging, with the amount of work varying between pairs. The ♀ ascends the mound regularly to observe the progress of the digging activities, taking over when the egg-chamber is opened up. She regularly probes the sand with her bill, apparently to test the temperature. She may now accept or reject the spot. In the case of the latter, the procedure starts all over again until a location with a suitable temperature has been found. Both ♂ and ♀ make clucking notes shortly before the egg is laid. When the ♀ finally lays her egg, she keeps her wings spread, while the ♂ stands on the rim of the mound, motionless. When the ♀ leaves the egg chamber, the ♂ walks in and starts to cover the egg (Frith 1959b).

Maintenance activities following egg-laying are mainly performed by the ♂ (Bellchambers 1916; Frith 1959b), and have been described in Chapter 6; Weathers *et al.* (1990) reported much more involvement in mound tending by the two mates of a polygynous ♂.

During the breeding season ♀♀ increase in weight by 200–500 g and become slightly heavier than their mates, who are on average 170 g heavier than the ♀♀ during the non-breeding season (Benshemesh 1992*a*).

Breeding

NEST: mound (see Fig. 5.2), varying in size, but most often 60–75 cm up to 1.5 m high and 2.7–4.5 m in diameter, with a circumference up to 13.5 m. The shape of the mound depends chiefly on weather conditions and the progression of the breeding season (see Chapter 6 for detailed account). Mounds are most commonly found on the lighter soils probably because of better drainage; soils of sandy clay loam and of heavier texture are often avoided (Frith 1959*b*). In contrast to *Megapodius* spp. which bury their eggs in separate excavations in different parts of the mound, the Malleefowl buries its eggs in the centre of the mound, with eggs laid progressively higher in the nest chamber as the season progresses (D. Priddel, *in litt.*).

EGGS: pinkish when fresh, but some turn buff with age. Moisture causes them to brown to a light tan. If moisture is excessive the coloured surface-layer flakes off, revealing the white underlying shell, as reported for eggs of other megapodes except *Aepypodius* and *Alectura*. SIZE: $91.9 \pm 2.6 \times 60.9 \pm 1.9$ mm ($n = 871$, Frith 1959*b*); $91.0 \pm 2.5 \times 58.5 \pm 1.2$ mm ($n = 43$, Vleck *et al.* 1984); $85.2–96.5 \times 54.6–62.2$ mm (91.6×58.9, $n = 56$, Schönwetter 1961); $88–102 \times 56–63$ mm (91×60, $n = 21$, Stirling Ranges and Hamelin Pool, Serventy and Whittell 1967); $84.7–95.3 \times 56.1–64.1$ mm ($90.4 \pm 3.85 \times 59.7 \pm 2.57$, $n = 10$, Shufeldt 1919). WEIGHTS: 92–202 g (168, $n = 281$, Booth 1987*b*); 117–275 g (187, $n = 844$, Frith 1959*b*); 148–195 g (173, $n = 43$, Vleck *et al.* 1984).

BREEDING SEASON: mound-construction usually June–Aug, sometimes as early as Mar, with some not finished until Nov. During a typical season, eggs are laid Oct–Dec and chicks hatch Dec–Feb. However, timing varies between years and between individuals, resulting in eggs being laid in Sept, and chicks hatching in Nov and Mar. During a dry spring, many nests are abandoned before egg laying begins while in wet summers many eggs are lost in saturated nests. Clutch-size varies between 2–34 eggs with a mean of 10.3–25.7 eggs per ♀ in different years depending on the availability of food, and an average of 5–10 days (mean 6.4 days) between eggs (Booth 1987*b*; Frith 1959*b*). In recent studies, mean clutch-size was recorded as 13.8 ± 4.1 eggs (Booth 1987*b*), 15.6 eggs ($n = 34$ nests) (Brickhill 1987*a*), and 19.8 eggs ($n = 43$ nests) (Benshemesh 1992*a*). The duration of the egg-laying season recorded in six breeding seasons (defined as the period between the first and last egg laid in a single mound) varied between 71 and 138 days (average 103 days) (Frith 1959*b*). The incubation period depends on the mound temperature, and at 34°C is 62–64 days. Bellchambers (1916) reported an incubation period of 55–77 days under captive conditions. In the wild, eggs have not been recorded hatching in less than 60 days (D. Priddel, *in litt.*). Mean hatching success varies between 49.5 and 79.2%, and the mean number of hatchlings per nest varies between 7.8 and 10.9 chicks (Booth 1987*b*; Brickhill 1987*a*, *b*; Frith 1959*b*; Priddel and Wheeler 1990*b*). The hatching success of 676 eggs monitored during three breeding seasons at Wyperfeld, Victoria, was 85.4%; of the remaining eggs, 8.0% were addled, 2.4% predated, and 4.2% broken (Benshemesh 1992*a*).

Local names
Lowan (Victoria). Also gnow, ngow, ngow-o, ngow-o-ou, nganamara, ngamara, marrakko or marra-ko.

References
Schlegel (1880*a*), Oustalet (1881), Ogilvie-Grant (1893, 1897), Mattingley (1909), Mathews (1910–1911, 1912*a*), Bellchambers (1916), Shufeldt (1919), Ashby (1922, 1929), Lea and Gray (1935), Fleay (1937), Griffiths (1954), Frith (1955, 1956*a*, 1957, 1959*b*,

1962*a*, *b*), Schönwetter (1961), Tarr (1965), Cooper (1966*a*), Serventy and Whittell (1967), Abbott (1974), Condon (1975), Blakers *et al.* (1984), Immelmann and Böhner (1984*a*, *b*), Vleck *et al.* (1984), Brickhill (1985*a*, 1987*a*, *b*), Kimber (1985), Booth (1986, 1987*b*, 1989*a*), Böhner and Immelmann (1987), Priddel (1990), Priddel and Wheeler (1990*a*, *b*), Robinson *et al.* (1990), Weathers *et al.* (1990), Benshemesh (1992*a*).

Genus *Macrocephalon* S. Müller

Macrocephalon S. Müller, 1846. *Archiv für Naturgeschichte*, **12** (1), p. 116.

Synonym: *Megacephalon* Temminck, 1846.

One species, endemic to Sulawesi.

A large black megapode with large black knob-like casque and contrasting salmon belly. Head bare, area round eye contrastingly coloured. Small rounded tubercle behind nostril. Flattened bare crown runs smoothly into large smooth casque, which consists of a bony rounded knob-like shield at rear connected to nape by somewhat narrower stem, together with head and bill resembling a tinker's hammer. Neck down from lower cheek covered with scanty short feathers, somewhat denser at hindneck. Bill about equal in length to head (without casque), relatively longest of all Megapodiidae; deep at base, laterally compressed, and with relatively long sharp tip. Nostril large, almost round. Wing short, broad, tip bluntly rounded; ten primaries, p5–p6 longest, outermost feather (p10) 6.5–8 cm shorter, p9 2.5–4 cm shorter, p8 1–2 cm shorter, p7 and p4 0–1 cm shorter, p3 0.5–1.5 cm shorter, p2 1–2 cm shorter, innermost (p1) 2–3.5 cm shorter. Tail moderately long (but, together with *Aepypodius*, relatively shortest of all large megapodes), consisting of 18 feathers, of which third pair from centre is longest and central and outer pairs distinctly shorter; tail compressed laterally, each tail half fan-shaped as seen from the side. Upper tail-coverts short. Tarsus and toes rather short (compared with body size, shortest of all Megapodiidae, except for *Leipoa*; but about equal to *Alectura*); claws relatively shortest of all megapodes. Tibia feathered almost down to tibio-tarsal joint. Tarsus entirely covered with small hexagonal scales. Basal phalanges of all front toes connected by distinct webs, extending up to 2.5 cm from tarsus/toe joints (no webs in any other Megapodiidae, except for small ones found between inner and middle toe in *Alectura* and *Leipoa*).

Maleo *Macrocephalon maleo* S. Müller, 1846

Macrocephalon maleo S. Müller, 1846. *Archiv für Naturgeschichte*, **12** (1), p. 116.   PLATE 4

Other names: Maleofowl or Gray's Brush-turkey.

Monotypic.

Description
PLUMAGES
ADULT MALE: head and casque bare, neck down from lower cheek, chin, ear, and lower base of casque thinly covered with reduced black feathers. Upperparts, tail, upper and under wing-coverts, and flight feathers deep black, feathers of upperparts and upperwing sometimes with slightly paler sooty grey fringes, flight feathers sometimes slightly browner or greyer, depending on abrasion. Upper chest, side of breast, lower flank, feathering of thigh,

Maleo *Macrocephalon maleo* 131

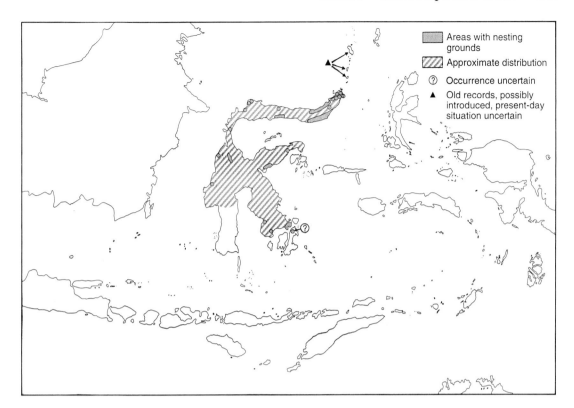

and under tail-coverts black, remainder of underparts white with strong salmon-pink tinge, white ending into point on mid-chest.

ADULT FEMALE: like ♂, but slightly smaller (see Measurements), underparts less bright, and area round eye sometimes slightly duller (see 'Bare parts'). Best distinguishing measurements are: (1) length of bill, head, and casque, measured from tip of bill to rear of casque: ♀ 82–92 mm (86.6 ± 3.21, $n = 15$), ♂ 90–99 mm (94.9 ± 3.14, $n = 16$); (2) maximum width of casque: ♀ 25–34 mm (30.4 ± 2.87, $n = 16$), ♂ 32–38 mm (34.6 ± 2.12, $n = 16$); (3) maximum depth of casque from flattened surface at rear of crown to bottom at upper neck: ♀ 21–26 mm (23.2 ± 2.39, $n = 16$), ♂ 26–33 mm (29.1 ± 1.97, $n = 17$); (4) bill depth at middle of nostril: see Measurements. Note that imm with underdeveloped casque may strongly resemble ad ♀.

CHICK: central forehead, whole crown, nape, and side of neck fuscous-olive-brown, finely streaked buff on side of crown and on nape and side of neck, sometimes on central crown also. Upperparts fuscous-brown, tinged olive from upper mantle to back, darker fuscous on outer scapulars, tertials, rump, and upper tail-coverts; back and rump with broad ill-defined paler streaks. Small patch round eye bare. Side of head behind eye yellowish-olive. Lore, cheek, chin, and throat yellow-buff. Chest, side of breast, and under tail-coverts dark fuscous-olive-brown, lower chest faintly streaked buff; remainder of underparts yellowish-buff or pale cinnamon-buff. Flight feathers dark fuscous-brown with slight olive tinge, secondaries with white fringe along tip of outer web; upper wing-coverts like scapulars (shorter more olive, longer more fuscous), longer ones sometimes with off-white fringe along tip.

IMMATURE: in first imm plumage, forehead, crown, and nape sooty greyish-black, lateral feathers all or partly white, forming mottled white stripe along side of forehead and crown, bordered by broad black line from above eye to nape; feathers of hindcrown and nape slightly elongated, forming short crest up to 2 cm long; no casque. Lore and side of head round eye bare. A broad black stripe from base of lower mandible backwards, extending above and below ear opening to upper hindneck. Chin and throat contrastingly white. Lower neck sooty grey, merging into black of upperparts and black of chest. Body as in ad, including salmon tinge of belly; black sometimes slightly browner than ad, depending on wear. Outer primaries and some secondaries and upper wing-coverts still juv, brown, relatively short; outer primary (p10) sharply pointed, length to insertion in skin 114–117 mm ($n = 2$), in ad, 145–159 mm (150.6 ± 3.72, $n = 15$); tail short, 122–127 mm (124, $n = 3$). In second imm plumage, rather like ad, but casque only slightly developed; head bare, except for a strip of downy grey feathers from forehead to top of casque and for short white feathers on chin and upper throat; neck covered with short grey feathers. Innermost primaries are those of third generation, outer primaries of second generation; tip of second generation p10 more pointed than in ad, length of p10 134–146 mm (139.4 ± 5.22, $n = 5$); tail 136–146 mm (139.7 ± 3.61, $n = 6$).

BARE PARTS
ADULT: iris dark brown, red-brown, or dark red. Bare skin of lore, top of head, casque, base of bill, and hindneck black; bare ring round eye orange-yellow to orange-red (♂) or yellow to orange-yellow (♀); remainder of side of head yellow to bluish-yellow (♂) or pale olive to greenish-yellow (♀). Partly bare skin of throat and foreneck bluish-yellow or olive-yellow. Upper mandible reddish-orange, sides pale grey-blue, tip and cutting edges light yellow-horn to orange-yellow; lower mandible grey-blue with yellow-horn tip. Mouth orange. Leg and foot light blue, grey-blue, or slate-blue (once 'black' in southeast peninsula of Sulawesi); soles yellowish or ochre; claws pale horn or light yellow-horn.

CHICK: iris brown or dark brown. Skin round eye flesh-coloured. Bill brown-red or dark purple-red with light cutting edges and tip. Leg and foot flesh-red or dark purple-red.

IMMATURE: in first plumage, iris red-brown; ring round eye yellow, remaining bare skin on side of head blue; bill black with red-orange tip; leg and foot grey-blue, claws yellow. In second plumage, like ad ♀.

MOULTS
Primary moult serially descendant. In ad, two ($n = 21$), three ($n = 8$), or four ($n = 1$) moulting series in each wing, wing thus showing several generations of feathers; distance between series 3–7 feathers (5.2 ± 1.04, $n = 25$) in birds with two series in each wing, 2–6 feathers (4.1 ± 1.03, $n = 13$) in birds with three or four series in each wing. Of all 52 series of 29 ad, moult active in 24, suspended in 28; alas, only 12 of these 29 ad dated, and hence not clear whether a clear moulting season exists in population: in Jan–Mar, moult in four birds active, in two suspended; in June–Aug, moult in four birds active, in one suspended; in Oct, in one active; in undated birds, in seven active, in ten suspended. In imm, moult of second generation of primaries starts with innermost feather (p1) before first generation outermost (juv) feathers are fully grown; when moult series of second generation reaches p6–p7 (occasionally p5 or p8) and outer primaries thus still juv, third generation starts with p1, this in turn followed by a fourth series when third generation reaches p6–p7. Of 12 imm examined, nine in active moult, three with primary moult suspended (these latter older imm with moult of third generation advanced and fourth generation just started or probably about to start); as in ad, only a few birds dated, and hence no information on start of moult. In both ad and imm, moult most often seems to start with p1 in Jan–Feb, followed by another series in May–June; moulting series may suspend with almost any primary: up to p1 new (five birds), p2 (five), p4 (one), p5 (five), p6 (seven), p7 (three), p8 (three), and p9 (three).

MEASUREMENTS
ADULT: N, C, and SE Sulawesi, skins (BMNH, RMNH, SMTD, ZMA, ZMB).

MALES	n	mean	s.d.	range
wing	23	301.5	6.24	287–310
tail	13	148.1	5.40	140–155
tarsus	23	90.9	2.53	87.0–96.3
mid-toe	17	51.4	2.38	47.0–55.3
mid-claw	17	18.2	1.32	15.4–20.9
bill (skull)	17	46.9	2.36	43.6–50.8
bill (nostril)	15	28.4	1.08	26.9–30.4
bill depth	14	20.1	0.81	18.9–21.4

FEMALES	n	mean	s.d.	range
wing	17	291.4	5.99	278–299
tail	12	145.9	4.96	137–152
tarsus	18	87.0	2.81	81.3–89.7
mid-toe	16	50.0	2.18	47.0–53.8
mid-claw	14	16.9	0.92	15.4–18.5
bill (skull)	15	43.8	1.70	41.2–46.2
bill (nostril)	15	26.1	1.50	23.9–28.5
bill depth	11	18.3	0.78	17.2–19.8

Live birds, Dumoga-Bone National Park, N Sulawesi: wing, ♂ ($n = 5$) 302–310 mm (305.6 ± 3.29), ♀ ($n = 2$) 296, 301 mm (Argeloo 1992c).

CHICK: just after hatching, wing 132, 136, 140, 142 mm; tarsus 36.5, 38.0, 38.5 mm (RMNH). On day of hatching: wing 126–142 mm (133.5 ± 4.51, $n = 37$, Dumoga-Bone N.P., N Sulawesi); wing-length increases on average c. 1 mm/day during the first week (R. Dekker, personal observation).

WEIGHTS
♂ 1389–1588 g (1513.2 ± 86.3, $n = 4$), ♀ 1503–1758 g (1673 ± 147.2, $n = 3$)(AMNH, Guillemard 1886); ♂ 1365, 1512, 1574, 1622 g, ♀ 1430 g; a ♀ of 1599 g carried fully developed egg, a specimen of 1724 g could also be ♀ with egg (Dumoga-Bone N.P., N Sulawesi, Argeloo 1992c).

CHICK: on day of hatching 125–173 g (145.1 ± 11.05, $n = 40$, Dumoga-Bone N. P., N Sulawesi); weight decreases especially during the first 3–5 days (R. Dekker, unpublished data), sometimes 11–14 days (B. Winn, *in litt.*). At hatching, when still in the ground, chicks weigh 60.7–69.1% (65.2 ± 2.23, $n = 24$) of the initial egg-weight; when emerging from the ground weight reduced to 55.6–65.2% (59.8 ± 3.18, $n = 15$)(R. Dekker, unpublished data).

GEOGRAPHICAL VARIATION
None.

Range and status
Endemic to Sulawesi but absent from the largely deforested southwestern province. Possibly still occurs on the neighbouring islands of Lembeh and Bangka although no records exist for the last few decades. A recent report suggests its presence on Butung (Buton) (Pramono 1991). Formerly reported from Siao, Great Sangi and Tahulandang (Meyer 1879, 1890a), north of Sulawesi, but possibly introduced. According to local information, still present on the southwest point of Great Sangi in 1991, breeding on a coastal nesting-ground (M. Argeloo, personal communication).

MacKinnon (1978, 1981) roughly estimated the population to be 3000 adults for N Sulawesi and 5000–10 000 for the entire island, but these figures lack any reliability as they are based not on counts but on wrong assumptions of the number of eggs laid by a ♀ in the course of the year, and therefore of the number per nesting ground. The status of the species, widely regarded as vulnerable, can better be estimated by the number of nesting grounds. These data show that the situation gives cause for concern, as most sites are severely threatened by over-exploitation of eggs and habitat destruction (see Dekker 1990a for distribution map). So far, 85 nesting grounds have been reported, of which at least 22 have been abandoned due to human interference (19 coastal and three inland) (Kukila 1990; Baltzer 1990; Dekker 1990a; Pramono 1991; M. Argeloo, in preparation; S. van Balen, personal communication). Of the remaining 63 nesting grounds, no data are available for 12 sites, while the other 51 sites are still in use. Only one or two of these are totally undisturbed and considered 'not threatened'. All others are threatened (some severely) and are visited by only a small number of birds; without conservation management most of these will be abandoned in the near future. In addition, there are 17 potential sites, mainly beaches, which need to be verified as being used by Maleos for egg-laying (M. Argeloo, in preparation). In 1985 the populations at the Tambun and Tumokang nesting grounds in the Dumoga-Bone National Park were estimated at between 150 and 200 pairs each, with maximum productions of 10–15 eggs per day; during the 1985–86 egg-laying season Sept–June, the total production for the two sites taken together was approximately 3500 eggs (Dekker and Wattel 1987).

Field characters
Length *c.* 55 cm. Unmistakable. Medium-sized to large blackish megapode with salmon underparts and black casque (or cephalon). Easily located by its characteristic voice. Very different from the Philippine Megapode *Megapodius cumingii gilbertii* from Sulawesi, which is much smaller, generally brown and grey in colour, seemingly tail-less, and lacks a cephalon on the head. Imm Maleos look like small adults, except for the cephalon which is still lacking, and can thus easily be separated from *M. cumingii*. The bicoloured Maleo chicks have dark fuscous-olive-brown upper-parts and wings and yellowish-buff underparts, which makes them distinctly different from the much smaller, entirely brownish *M. cumingii* chicks.

Voice
Six different calls are known. The most characteristic is a rolling call, produced only by the ♂, a loud, warm, vibrating 'kee-ourrrrrrrrrrrr' or 'coo-ourrrrrrrrrrrrr', lasting 2–2.5 sec (see sonogram). Described as 'grrrrrrr' by Meyer (1879). The ♂ stretches its neck upwards while giving the first note, then retracts it, bringing the head against its breast while producing the second part of its call during which the upper and lower mandible are clapped against one another (R. Dekker, personal observation).

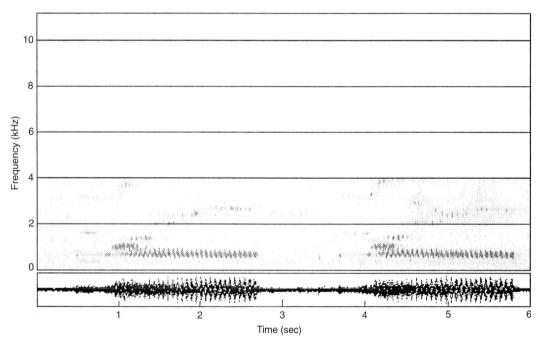

M. maleo
M. Argeloo, Dumoga-Bone N.P., N Sulawesi, Indonesia, 1991.

Watling (1983b) described the posture while calling slightly differently, the bird standing with its head down instead of upwards, its neck outstretched and its throat 'wobbling madly' as it produces the call. Commonly heard at the nesting ground as well as away from it. It may have a territorial function, but it is also produced when the pair are separated due to disturbance. The ♀ responds to this call with a duck-like 'kuk' until she rejoins her mate. When attacked by or chasing other Maleos at the nesting ground, the ♂ or both sexes produce goose-like 'gak-gak-gak' calls, rather similar to the call of the ♀ described above. While digging or walking on the nesting ground, and only audible at close range, a constant moaning irregular 'mm-mm, mm-mm, mm-mm' (in pairs of notes) is uttered by both sexes. A similar soft, indeterminate sound from the throat (but continuous, not in pairs of notes), uttered at a rate of approximately three per second, is regularly heard from birds sitting in trees. Finally, after copulation (and only observed in captivity), a high shrill 'whaaaa-ee' is given (not known by which sex), initially rising in pitch but with the last note slurred downwards.

Once a chick of 12 days was heard regularly producing what sounded almost like the rolling call of the ♂, though much softer. None of the other chicks of hundreds handled has ever been heard doing this. The only other chick-sound is 'kuk', similar to that of the ♀ described above, and uttered when taken in the hand.

Habitat and general habits

Tropical forests of Sulawesi, from sea level up to at least 1200 m. Also in secondary vegetation, coconut plantations, and other man-made habitats when forced to cross such areas to reach a nesting ground.

Monogamous, pair bond probably for life. ♂ and ♀ not separated for more than a few metres while foraging, egg laying, or sleeping in trees on large horizontal branches. Shy and wary when on the ground, but less shy and therefore much easier to approach when sitting in a tree, where they seek safety if disturbed. When the pair is separated, for example after disturbance or while approaching the nesting ground, the ♂ utters its characteristic rolling call which allows the responding ♀ to relocate him. When restless or offended, and possibly also sometimes when approached by the ♀, the ♂ raises his head, stretches his neck, and lowers his wings. He raises his tail to its full extent with the central tail feathers pointing vertically upwards while fluffing his pink breast feathers, making him appear much bigger. In this posture, he bobs heavily with his entire body, bends up and down, and flicks his tail (R. Dekker, personal observation).

Aggressive behaviour against other pairs at the nesting ground occurs when pairs are digging close to one another or trying to steal one another's burrow. Behavioural observations of Maleos away from the nesting ground, either singly or in pairs, are scanty and lacking in detail, and refer mainly to birds walking or foraging on the forest floor.

Food

Omnivorous; diet consists of fruits, seeds, and invertebrates such as beetles, ants, termites, and snails. A stomach collected in the Dumoga-Bone N.P. consisted of eight or nine specimens of land snail (six species), ten specimens of one species of freshwater snail as well as a whipscorpion (R. Dekker, personal observation). The discovery of both land snails and freshwater snails in this stomach might indicate that Maleos feed not only on the forest floor, but also along river beds.

Fruits from a stomach of a bird collected by Von Rosenberg (1878) were identified as those of *Pangium edule* (Flacourtiaceae), which is surprising as the large seeds of this tree contain prussic acid and are therefore highly poisonous.

Displays and breeding behaviour

This section is based on personal observation by R. Dekker.

This burrow-nesting species lays its eggs early in the morning, or sometimes in the afternoon, in holes in geothermally heated soils in the forest or on sun-exposed beaches. In most cases the pair reaches the vicinity of the communal nesting ground the evening before egg laying and spends the night in a tree close to the site. Early in the morning the birds descend to the nesting ground and start looking for a suitable place. If not disturbed by other egg-laying pairs or by external sources, they may choose the same burrow as they have used on an earlier occasion, which might thus be considered as their 'nest'. Both ♂ and ♀ participate in digging activities and work alternately, with one bird keeping watch and chasing intruders while the other is scratching. As in other megapodes, the Maleo scratches with one leg several times in succession before shifting to the other leg. If the birds are chased from their burrow or the burrow does not satisfy the birds for some reason, they start elsewhere. The whole procedure of digging, egg laying, and burying the egg takes several hours, the time varying depending on soil temperature (and thus egg depth), soil structure, and the amount of disturbance. While digging, both birds regularly take a bill-full of sand, presumably to measure soil temperature, as recorded for other species. References in the literature to ♀♀ fainting after laying the large egg have never been substantiated. The depth at which the egg is buried depends on the soil temperature and thus the soil structure, the distance from the heat source, and the weather conditions in the days or weeks prior to egg laying. They are not laid 'as deep as possible' as suggested by MacKinnon (1978). Egg depth varies between 10–15 cm and 80–100 cm. Eggs are buried most frequently at a depth of 30–50 cm. Of 556 eggs, 82% (465) were buried in a vertical position, 14% (77) horizontal, and 4% (24) diagonal.

Copulation has never been observed at the nesting ground, and evidently takes place elsewhere in the forest. Under captive conditions, copulation is preceded by vigorous and audible scratching as the ♂ throws sand and herbs up into the air. This activity is interrupted by short breaks during which the ♂ quickly turns round in small circles at approximately the same spot, after which soil scratching continues. After taking several steps forward he steps back while scratching heavily again. Subsequently, the tip of the wing of the ♂ facing the ♀ is pointed upwards, his tail slightly raised and his breast feathers fluffed. The ♀ ignores the ♂ when he walks past her, but then starts vigorously scratching the soil herself for a few seconds, and this is followed by the ♂ doing the same. He then approaches the ♀, who lowers her abdomen and tail to the ground. The ♂ mounts and copulation takes place, 4 min after the ♂ initiated the ceremony with his scratching behaviour. When mounted, the ♀ lowers her head to the ground with the head of the ♂ just above hers and following her movements, while the ♂ or ♀ produces a high shrill 'whaaaa—ee' (see 'Voice'). Copulation lasts only a few seconds, then the ♂ rolls over the ♀, after which the ♀ scratches a few strokes and the ♂ walks away, fluffing and preening his feathers while producing the soft 'mm-mm, mm-mm, mm-mm' sound described under 'Voice'. On one occasion, 5 min after the ♂ started scratching and just before copulation took place, the ♀ lowered one of her wings while the ♂ fluffed his breast feathers, raised his wing tip, and pressed his breast to the ground. After copulation (though in a different context) the ♂ picked up food items from the ground and fed them to the ♀ (R. Dekker, personal observation), as has also been observed in the Nicobar Megapode *M. nicobariensis* and the Polynesian Megapode *M. pritchardii*.

The number of eggs laid by a ♀ in the course of a year or season is unknown, but was estimated to be one every 12 or 13 days or approximately 30 eggs per year by MacKinnon (1978, 1981), and one every 7–9 days or 8–12 eggs during the individual egg-laying season of 2–3 months by Dekker (1990a). Guillemard (1886) estimated the number of eggs to be about 16 or 18, based on inspection of the ovary. Coomans de Ruiter (1930) mentioned a period of 14 days between eggs, and a total of six to eight eggs per season, based on information from local egg collectors.

Breeding

NEST: burrow at volcanic soils and sun-exposed beaches, lake shores, river banks, and even dirt roads along the coast. Depending upon the substrate and soil temperature, the burrows vary considerably in size and shape from a shallow crater in loose, dry sand to broad, deep or irregular excavations in more solid soil. The bottom of the burrow consists of a layer of loose soil, due to the regular digging activities of the birds. Individual burrows can be used regularly during one season by one or more pairs and thus may contain a number of eggs. Of the 85 nesting grounds mentioned above, 48 are coastal and heated by the sun, and 37 are located inland and rely mainly on geothermal heat (Dekker 1990*a*; M. Argeloo, in preparation). The coastal nesting ground at Bakiriang, C Sulawesi (illustrated in Watling 1983*b*) was the most spectacular known; although now under heavy human pressure, birds were still present there in 1991 (M. Indrawan, personal communication). The highest recorded geothermal nesting ground, heated by hot springs, is at 1200 m altitude in the Lore Lindu National Park in C Sulawesi (Watling 1983*a*).

EGGS: similar to eggs of *Megapodius* spp. and of the Moluccan Megapode *Eulipoa wallacei*, being elliptical, without clear pointed or blunt side, pale reddish-buff to pinkish-brown. Outer layer flakes off during incubation showing white shell underneath. Slightly more elongated than eggs of *Megapodius* and *Eulipoa*, length to width ratio on average 1.70–1.73 (Dekker and Brom 1990). SIZE: 92.1–112.6 × 57.6–65.5 mm (105.4 ± 2.84 × 61.9 ± 1.42, n = 233, Dekker and Brom 1990). Various published data from smaller samples fall within this range. WEIGHTS: 178–267 g (231.5 ± 13.23, n = 233, Dekker and Brom 1990). Various published data from smaller samples fall within this range. Average weight loss of nine eggs during 59.1 ± 5.99 days of incubation (or 6.4 ± 2.70 days prior to hatching) is 0.44 g or 0.19% of the fresh egg weight (R. Dekker, unpublished data).

BREEDING SEASON: all year round though with clear peak season from Oct to May or June at volcanically heated inland nesting grounds in N Sulawesi, with only a few pairs laying July–Sept (Dekker 1990*a*). Egg laying at coastal nesting grounds, where the sun provides the incubation heat, occurs only or mainly during the dry season. Egg-laying season on south coast of N Sulawesi from Sept to Mar, but on north coast of N Sulawesi (less than 50 km away) from Mar to Sept. However, eggs reported from Panua nesting beach along the south coast of N Sulawesi every month of the year (Uno 1949), though with a distinct peak from Jan to May. During July and Aug birds were absent from the Kamarora and Saluki geothermally-heated nesting grounds in Lore Lindu, C Sulawesi, indicating seasonal breeding here as reported for similar sites in N Sulawesi. Egg laying during the dry season from Sept to Mar on sun-exposed river banks in Morowali, C Sulawesi (Simonson 1987), and in May–July and Nov–Jan in SE Sulawesi, though details about type of nesting ground lacking (White and Bruce 1986). Incubation period dependent on soil temperature, ranging from 62 to 85 days for 104 chicks in a hatchery at the Tambun nesting ground, N Sulawesi (Dekker 1988*d*).

Local names
All N Sulawesi: senkawor, sengkawur or songkel (Minahassa), suangke (Bintauna), tuanggoi (Bolaang Mongondow), tuangoho (Bolang Itang), bagoho (Suwawa), mumungo (Gorontalo), panua (general), molo or moleo (Mengkoko Mts).

References
Quoy and Gaimard (1830), Müller (1846), Schlegel (1862, 1880*a*), Von Rosenberg (1878), Meyer (1879, 1890*a*), Oustalet (1881), Guillemard (1885, 1886), Ogilvie-Grant (1893, 1897), Blasius (1896, 1897), Vorderman (1898*a*), Shufeldt (1919), Riley (1924), Coomans de Ruiter (1930), Stresemann (1941), Uno (1949), Lint (1967), Howes (1969), Lucas

and Stettenheim (1972), MacKinnon (1978, 1981), Crowe and Withers (1979), Watling (1983a,b), Van den Berg and Bosman (1986), White and Bruce (1986), Dekker and Wattel (1987), Simonson (1987), Dekker (1988d, 1990a), Starck (1988), Baltzer (1990), Dekker and Brom (1990), Kukila (1990), Pramono (1991), Argeloo (1992a, c).

Note on structure and possible function of the Maleo's helmet

Macrocephalon, meaning 'big head', refers to the large swollen hindpart of the skull of the adults, the casque or cephalon. This structure, which is absent in other megapodes, has often been linked with the ability of the birds to measure the ground temperature. Observations of the breeding behaviour of several megapode species, including the Maleo, seem to have indicated that this possibility can be ruled out, however, since megapodes regularly take a bill-full of sand while digging a hole to bury their egg. Possibly thermosensitive nerve endings in their tongue or elsewhere inside the bill play a role in measuring the temperature.

Two aspects of the functional anatomy of the head of the Maleo should be mentioned. (For an explanation of scientific terms, see the drawing of a Maleo skull.)

The casque could, in combination with an intensive vascularization by extensive retia (networks) in the dermal part of the skin, function as a heat exchange system comparable to the situation described for Helmeted Guineafowl *Numida meleagris* (Crowe and Withers 1979). It consists of a leathery keratin, supported by a bony swelling, which is formed as an extension of the strongly pneumatized spongeous bone of the parietal bone only (see illustration). Radiograms made by C. Niemitz, Berlin, show that the frontal and supraoccipital bones, though also pneumatized, do not contribute at all to the formation of the bony casque. They also show that the casque is an additional structure not affecting intracranial architecture (for example, the elevation angle of the brain).

The skin covering the casque is topped by a thick epidermal layer of stratified epithelium. It is keratinized for about half its width (10–12 cell layers in thickness, which colour vivid red in Azan staining). Its black leathery look is caused by black pigmented cells or melanocytes which occur only in the epidermis, from the horny layer (*stratum corneum*) down to the basal layer (*stratum basale*). In the Azan staining procedure the deeper part of the epidermis, the germinative layer (the *stratum germinativum*, consisting of the *stratum transitivum*, *stratum intermedium* and *stratum basale*, adjacent to the *lamina basalis*) is not stained except for the brownish-black melanocytes. The corium or dermis is about double the thickness of the overlying epidermis. It is rich in fibres of connective tissue (collagen, staining brilliant blue) and extraordinarily rich in small blood vessels or capillaries, easily identified because of their red blood cell (erythrocyte) contents. Similar retia with attributed thermal regulation are described in Helmeted Guineafowl (Crowe and Withers 1979). The black colour of the

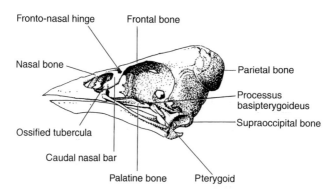

skin, as is known from radiation physics theory, is also an adaptation to maximal heat exchange.

Structural adaptations of the skull may also act as shock absorbers while cracking hard-shelled nuts by forceful hammering, as suggested by Starck (1988). Various adaptations in the skull seem to reflect these shock absorption mechanisms:

(a) the strong median and two lateral spine-like caudal extensions of the nasal bone extend the area of contact where the nasal bone articulates with the frontal bone by way of the fronto-nasal hinge (see illustration). Ontogenetic evidence is needed to confirm that this is indeed the true separation of these bones rather than a more caudal situated intrafrontal hinge. In addition, this interlocking area seems to be connected by ligamentous tissue to the pronounced anterior (rostral) part of the frontal bone,

(b) a short and stout pterygoid, articulating via an enlarged and wide processus basipterygoideus forms a firm, bilateral supporting strut limiting caudal excursions of the strong and horizontal palatine bones, and

(c) extra stability is given for the tensile forces working in the region of the caudal nasal bar which is strengthened by a rostrally pointing drop-like bony swelling of the caudal nasal bar, a structure unique to the Maleo. These ossified tubercula are covered by the same black keratin that covers the casque. As in other birds in which the head and neck are used in digging actions (e.g. guineafowl), the bare skin of the head permits easy removal of attached sand.

Genus *Eulipoa* Ogilvie-Grant

Eulipoa Ogilvie-Grant, 1893. *Catalogue of Birds in the British Museum*, **22**, p. 445 (key), p. 462 (diagnosis).

One species, restricted to the Moluccan islands and Misol (off western New Guinea).

A small boldly coloured megapode, close to *Megapodius* in some characters, but some peculiar features necessitate recognition of a distinct genus for the single species; its range is entirely contained within that of the Dusky Megapode *Megapodius freycinet* and Forsten's Megapode *M. forstenii*. Head and neck feathering fairly dense, except for ring round eye, which is almost bare or covered with tiny feathers only. Lower mantle, scapulars, and median upper wing-coverts barred deep maroon and blue-grey; greater coverts and outer webs of tertials maroon; distal part of outer web of outer primaries contrastingly paler than remainder of feathers. Whitish vent and white patch on underwing contrast strongly with remainder of plumage; may have a signalling function in this apparently nocturnal species. Remaining plumage delicate brown, olive, and grey, as in *Megapodius*. Bill short and slender; nostril a narrow slit, partly covered by membrane above (less oval and open than in *Megapodius*). Wing more narrow and pointed at tip than in any other megapode, probably indicating better flying capability; 10 primaries, p7 longest, outermost (p10) 2.5–3.5 cm shorter, p9 0.5–1 cm shorter, p8 and p6 0–0.5 cm shorter, p5 *c.* 0.5 cm shorter, p4 1–2 cm shorter, p3 2.5–3.5 cm shorter, p2 3.5–4.5 cm shorter, p1 4.5–5.5 cm shorter (in *Megapodius* spp. with about similar wing length, such as the Philippine Megapode *M. cumingii gilbertii* and *M. f. forstenii*, p6 usually longest and p1 3–4 cm shorter). Tail rather longer than in *Megapodius* when compared with wing-length (but Orange-footed Megapode *M. reinwardt* almost equal); however, seemingly long tail feathers mainly due to long pointed tip of each feather, an aberrant feature among Megapodiidae; tail consists of 12 feathers, outermost *c.* 1.5 cm shorter than longest (central) pair. Tarsus relatively and absolutely shorter than any *Megapodius* (for example, absolutely shorter than in smaller-bodied Micronesian Megapode *M. laperouse* and Polynesian Megapode *M. pritchardii*), but toes and (especially) claws long for the bird's size; difference from *Megapodius* in tarsus/toe and toe/claw proportions probably point to difference in locomotion or digging. Front of tarsus covered with a single row of transverse scutes, as in *Megapodius*.

Moluccan Megapode *Eulipoa wallacei* (G. R. Gray, 1860)

Megapodius wallacei G. R. Gray, 1860. *Proceedings of the Zoological Society of London*, 1860, p. 362, pl. 171. (East Gilolo = Halmahera.)

PLATE 5

Other names: Moluccan Scrubfowl, Painted Megapode, Wallace's Scrubfowl.

Monotypic.

Description
PLUMAGES
ADULT: forehead medium grey with slight brown or olive tinge, grading into olive-brown on forecrown and this in turn to deep rufous-brown or black-brown on hindcrown and on short broad crest on nape; side of crown and crest more olive-brown; in worn plumage, hindcrown and crest fade to dark olive-brown. Lore, side of head to ear, chin, throat, and upper front and side of neck partly bare, partly covered with short pale brown-grey feathers; skin almost fully

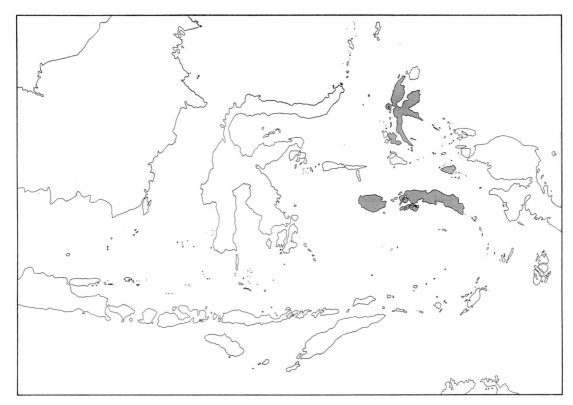

bare round and behind eye, relatively closely feathered on lore, lower cheek, and chin. Hindneck and lower front and side of neck covered with grey-brown feathers. Upper mantle and lesser upper wing-coverts warm olive-brown; feathers of lower mantle, scapulars, and median upper wing-coverts deep reddish-maroon with contrasting light blue-grey tip 3–5 mm wide; hidden feather bases contrastingly olive-brown or light-grey, fringes of some grey tips on mantle or scapulars washed olive-brown; grey tips of median coverts sometimes bordered paler grey or white subterminally; some outer lesser coverts partly washed maroon. Greater upper wing-coverts maroon, outer washed dark olive-brown. Back, rump, and upper tail-coverts medium bluish-grey, upper tail-coverts partly suffused olive-brown. Underparts mainly dark bluish-grey or slate-grey; mid-belly slightly paler grey, chest, side of breast, and flank sometimes slightly washed olive-brown; vent contrastingly white with slight grey, yellow, or isabelline tinge. Tail dark sepia-brown to greyish-black. Greater upper wing-coverts and primaries brownish-black, broad outer fringe of outermost primary (p10) contrastingly pale olive-brown (if fresh) to isabelline (when worn); pale outer fringe gradually shorter towards inner primaries, restricted to small pale dot on tip of outer web on (p3–)p4–p5; p9 and (often) p8 with isabelline notch near base of outer web. Inner web of secondaries greyish-black, outer webs dark olive-brown on base, grading to pale olive-brown or pale greyish-olive on tip; innermost secondaries and tertials pale olive-drab, tertials with broad maroon outer border. Under wing-coverts and axillaries dark plumbeous-grey, greater under primary coverts contrastingly white. Sexes perhaps different, but number of trustworthy sexed birds examined small: nape of ♂ appears to be more black-brown, of ♀ more rufous; maroon and grey bands on wing of ♂ extend over all greater and median upper wing-coverts except outer two to three, in ♀ bands more restricted, distinct only on inner greater and median coverts, reduced to ill-defined chestnut and grey blotches on middle coverts.

CHICK: crown and hindneck dark rufous-brown; forehead and side of head and neck rich buff-cinnamon. Upperside of body deep fuscous-brown, grading to rufous-brown on scapulars and upper wing; back to tail-coverts rufous-chestnut; feathers of lower mantle, scapulars, and inner median and greater upper wing-coverts contrastingly barred black and rufous-pink (if fresh) or pale buff (if worn) subterminally. Chin buff, merging into grey-brown on throat, chest, and flank, this in turn merging into buff-brown of belly and vent. Primaries greyish-black, tips sometimes partly fringed pale dirty buff; secondaries and tertials blackish with narrow rufous or buff fringes on tips, tertials and (sometimes) secondaries with rufous-pink or buff notches or traces of bars on outer webs. Longer under wing-coverts conspicuously white. Some individual variation, some more fuscous (less rufous) on head and body than bird described above, and some variation in extent and colour of pale barring, but in all examined at least inner wing-coverts, scapulars, and tertials contrastingly barred rufous or buff and black, unlike *Megapodius freycinet* and *M. forstenii*, and rump more rufous-chestnut than in these species. Also, tarsus relatively shorter and claws relatively longer than in *Megapodius*, independent of stage of development of chick; in *E. wallacei* ratio of tarsus to middle claw in chick 2.45–3.05 (2.77 ± 0.20, $n = 6$), in *M. freycinet* (from Halmahera, Bacan, and Misol) and *M. forstenii* (from Buru and Ceram) 2.88–4.04 (3.36 ± 0.36, $n = 18$). However, best character for identification are white longer under wing-coverts (light or dark brown-grey in *M. freycinet* and *M. forstenii*).

IMMATURE: closely similar to ad. In first imm plumage, side of head and all neck closely feathered, only area just round eye bare (except for tiny feathers); outermost primaries still juv, p10 shorter and more sharply pointed at tip than ad; length of p10 from tip to insertion in skin below *c.* 105 mm (in older birds, 110–125 mm, average 114.7 ± 4.39, $n = 13$). In second imm plumage, closely similar to ad;

second generation of outer primaries apparently not much different in shape from those of ad; birds with length of p10 (to insertion in skin) less than 115 mm and with total wing length below *c.* 196 mm and tail below *c.* 70 mm probably imm, but no proof.

BARE PARTS
ADULT, IMMATURE: iris light to dark brown. Bare skin of side of head and neck and of throat light pink to dark flesh colour. Base of bill green, olive-grey, or dark olive, tip light yellow to greenish-yellow; in some, dark red patch between nostrils (Vorderman 1898*b*). Tarsus light yellowish-grey, dark grey-green, dark olive, or front blackish-olive and rear yellowish-olive; toes dark slate to blackish-olive.

CHICK: iris hazel-brown. Eyelids grey. Bill black, cutting edges narrowly grey. Leg and foot brown-black (data from slide by M. Argeloo).

MOULTS
Primary moult serially descendant. Of 16 birds examined for primary moult, three had a single moult centre on each wing only, these probably younger imm. Remainder, probably older imm and ad, had either two or three moult centres on wing, separated from each other by 3–6 (4.3 ± 0.85, $n = 12$) feathers. No information on moulting season, as most birds examined were undated; moult active in singles from May and Nov, suspended in singles from Jan, Apr, and June; of undated birds, moult active in five, suspended in six. Egg-laying ♀ had moult suspended in Nov.

MEASUREMENTS
ADULT: whole geographical range, sexes combined; skins of 5 ♂♂, 4 ♀♀, and 7 unsexed birds (BMNH, RMNH, SMTD, ZMA, ZMB); wing includes data of 1 ♀ from Vorderman (1898*b*), 1 ♂ and 3 unsexed birds from Stresemann (1914*a*), and 5 ♀♀ from Siebers (1930).

	n	mean	s.d.	range
wing	26	203.1	6.35	193–214
tail	16	71.8	5.29	65–80
tarsus	16	56.6	3.09	52.4–62.3
mid-toe	16	34.6	1.98	31.5–38.9
mid-claw	15	20.5	1.47	17.6–23.5
bill (skull)	15	25.7	1.92	22.6–27.8
bill (nostril)	13	12.0	0.86	10.6–13.1
bill depth	12	7.8	0.70	6.8–9.1

No sexual differences found, but samples small; e.g. wing, ♂ 194–214 mm (203.5 ± 8.09, $n = 6$), ♀ 195–211 mm (203.7 ± 5.46, $n = 10$). No marked difference between northern and southern Moluccas: e.g. wing, Ternate, Halmahera, and Bacan 193–211 mm (201.7 ± 6.38, $n = 17$); Buru and Ceram, 194–214 mm (205.8 ± 5.70, $n = 9$). In live birds: wing, ♀ 204, 206, and 208 mm; tarsus 64 mm (Haruku, M. Argeloo, *in litt.*).

CHICK: at hatching or shortly thereafter: wing 86–96 mm (92.2 ± 3.82, $n = 6$); tarsus 25.5–28.1 mm (26.6 ± 1.15, $n = 6$).

WEIGHTS
ADULT MALE: 510 g (Ceram, Stresemann 1914*a*); ♀ 490–510 g (498.8 ± 10.31, $n = 4$, Haruku, M. Argeloo, *in litt.*; R. Dekker, *in litt.*).

CHICK: 68 and 78 g at hatching ($n = 2$, Haruku, M. Argeloo, personal communication).

GEOGRAPHICAL VARIATION
None found; see 'Measurements'.

Range and status
Moluccan islands of Buru, Ceram, Ambon, Haruku, Bacan, Halmahera, Ternate; may occur on Morotai, Kasiratu, Obi group, or Saparua, but no proof. Collected on Misol by Ripley (1960). Reported recently from Wafawel (on Buru), though not as common as *M. forstenii* (Jones and Banjaransari 1989), and from S Ceram, Haruku, and Pulau Pombo in Oct–Nov 1991 (R. Dekker, personal observation).

Natural occurrence on some of the smaller islands, such as Ternate, Ambon, and Misol, is doubted by some. Ternate and Ambon were important trading posts of bird skins in the past, and specimens from there might have been trade skins from elsewhere. However, well-labelled skins from trustworthy collectors exist for these islands. Von Rosenberg (1878) found *E. wallacei* rare on Ternate in about 1870, restricted to the thinly-populated western part of the island; birds were collected by him and by Bernstein (May 1862) and van Musschenbroek (Jan 1875); it may now be extinct there. Martin (1894) considered the species to be very common on the west coast of Haruku, a few miles from Ambon. Observations in 1991 show that the communal nesting ground of Kailolo, NW Haruku, still attracts thousands of birds yearly (see 'Breeding'). If data from local people are correct (40 000–50 000 eggs per year), the population using Kailolo might be 4000–5000 pairs. These birds originate not only from the island itself, but also from S Ceram. Specimens reported from Ambon, such as a chick collected by Müller (Apr 1828) and a specimen collected by Platen (Sept 1881) almost certainly originated from this nesting area also. The forested Salahutu Mt., NE Ambon, still seemed suitable for the species in 1991 (R. Dekker, personal observation).

On Misol, very rare according to some local hunters (Ripley 1960), but they had a name for the species and thus apparently were familiar with it. On Ceram, it was already considered to be rare by Stresemann (1914*a*). On Buru, Siebers (1930) found only a single colony on 50 km of beach investigated, while some other colonies reported to him either had strongly declined or were completely lost. Not present on Saparua or Nusa Laut, E of Haruku (Martin 1894). Being a communal nester on or close to beaches, *E. wallacei* is probably under heavy human pressure.

Field characters
Length 29–31 cm. Easily distinguished from *Megapodius* spp. with which it occurs sympatrically (*M. forstenii* on the southern Moluccan islands, and *M. freycinet* on the northern Moluccan islands and Misol), by combination of overall olive coloration and coarsely maroon-banded back.

Voice
The following descriptions are based on observations and tape-recordings made at night at the nesting ground, and include sounds likely to be produced during agonistic interactions over nesting burrows (R. Dekker, personal observation): irregular, rapid series of sharp, nasal 'kèp' 'kèp' or 'kèw' 'kèw'. Also 'ki-ouw kouw', 'kou - kouw - kouw - kouw' or 'kùk - kuk-uk (uk)' in which the first note lasts longer than the remaining three; sometimes the initial, longer note is omitted. Duetting (as in *Megapodius*) has not been heard at the nesting ground; not known if any occurs.

Voice at the nesting ground also described as 'noisy crowing' (de Wiljes-Hissink 1953) and 'loud wailing cries' (Wallace 1869).

Habitat and general habits
Away from the nesting ground it inhabits hill and mountain forest between 750 and 2000 m altitude, though sometimes lower. On Misol two ♂♂ were collected at 100 m and 300 m (Ripley 1964). On Bacan it has been observed at altitudes ranging from 750 to 1650 m in dense evergreen rainforest and subtropical wet moss forest in Oct (Ripley 1960). Also recorded at the forest edge in disturbed forest at 230 m and in coastal scrub on Ceram (Bowler and Taylor 1989; R. Dekker, personal observation).

Shy. Nocturnal when at the nesting grounds, where it arrives after dusk for egg-laying. Away from the nesting ground probably mainly diurnal. Occurs singly or in small groups when away from the nesting ground. No data available on pair-bond or behavioural aspects; probably monogamous. Chicks react to unfamiliar sounds or sights (in captivity) by freezing (West *et al.* 1981) and are especially active at night (Siebers 1930; de Wiljes-Hissink 1953).

Food
Not reported, but probably omnivorous.

Displays and breeding behaviour

This burrow-nesting species is the only megapode known to lay its eggs at night. The nesting sites are visited only for egg laying. Some or most birds reach the site on the wing in the evening, often flying in over areas of open water such as the strait between S Ceram and Haruku. Heinrich (1956) collected a female at 1500 m altitude on Bacan carrying a fully developed egg, which supports the conclusion that the journey to the nesting ground is undertaken shortly before or on the night of egg laying. While digging and egg laying, two individuals remain close together in clusters of many birds, suggesting that ♂ and ♀ visit the site together. However, Siebers (1930) collected only ♀♀ at a nesting beach, which might suggest that ♂♂ do not accompany ♀♀ to the beach (compare burrow-nesting Polynesian Megapode *M. pritchardii* and Melanesian Megapode *M. eremita*). Aggressive behaviour between pairs is commonly observed when the distance between them or between their burrows is too small (M. Argeloo and R. Dekker, personal observation). It has been suggested that this species occasionally lays eggs in mounds of *Megapodius* spp. (Rand and Gilliard 1967). This idea originates from Siebers (1930), who claimed to have found an egg of *E. wallacei* in a mound of *M. forstenii* on Buru; thus the observation is based on the size, shape, and coloration of a single egg. However, eggs of the two species cannot in fact be separated, because of the great similarity in their shape and coloration and the overlap in size.

Breeding

NEST: burrow at sun-exposed beaches, but maybe sometimes at volcanic soils. New burrows almost like rabbit holes, 50–100 cm deep. Nesting grounds known from Galela and Gamkonora on Halmahera, the island of Meti, near the mouth of the Wa'Kasi and Wa'Tina Rivers on Buru, Kailolo on Haruku, Pulau Pombo, and near the mouth of the Tala R. and at Amahai on S Ceram. The site at Amahai is situated on the bulldozed margins of an airstrip 6 m above the high tide line, with burrows widely scattered in small open areas of sand and stones among secondary vegetation and even in the steep wall bordering the beach (de Wiljes-Hissink 1953; R. Dekker, personal observation). The nesting ground at Kailolo, Haruku, is the largest currently known and measures approximately 1.5 ha. It is separated into four open spots of fine white sand surrounded by vegetation, 50–150 m from the beach. This location is probably mainly used by birds from S Ceram, only a few kilometres away, where the muddy and rocky coastline is unsuitable for egg laying. The nesting ground on the northern tip of the small sandy island of Pombo, just opposite Kailolo, has never been large and is nowadays used by a few pairs which must be part of the Kailolo breeding population (R. Dekker, personal observation).

EGGS: elongate-oval, brownish buff to reddish-brown without gloss; similar in coloration and shape to *Megapodius* and *Macrocephalon* eggs. Outer layer partially flakes off during incubation showing white shell underneath. SIZE: 75.4–83.0 × 46.9–50.5 mm (80.0 ± 1.99 × 49.0 ± 1.04, $n = 19$, Haruku, R. Dekker, personal observation); 75–80 × 44–50 mm (77.6 ± 1.60 × 47.6 ± 1.61, $n = 20$, Pulau Pombo, near Ambon, West *et al.* 1981—note that eggs probably originate from Haruku); 78–86 × 50–57 mm ($n = 10$, Ambon, Blasius and Nehrkorn 1883); 75–82 × 46.5–51 mm (77.7 ± 2.11 × 49.1 ± 1.54, $n = 10$, Buru, Siebers 1930); 76.5–81.2 × 47.6–51.0 mm (79.1 ± 1.71 × 49.3 ± 1.18, $n = 10$, Galela (on Halmahera), coll. RMNH); 72.8–80.3 × 46.0–51.0 mm (76.6 ± 2.85 × 49.0 ± 1.87, $n = 5$, Elpah-putih, S Ceram, coll. RMNH); 73.5–83 × 45.2–51.8 mm (78.7 × 48.5, $n = 15$, Schönwetter 1961). WEIGHTS: 92–116 g (106.4 ± 5.98, $n = 42$, Haruku, R. Dekker, personal observation); 82–101 g (94.1 ± 5.09, $n = 20$, Pulau Pombo, near Ambon, West *et al.* 1981—note that eggs probably originate from

Haruku). Relative egg weight has been measured on three occasions from females carrying eggs: egg of 93 g from ♀ of 505 g = 18.4% of her body weight (R. Dekker, personal observation), egg of 97 g from ♀ of 490 g = 19.8%, and egg of 101 g from ♀ of 490 g = 20.6% (M. Argeloo, *in litt.*).

BREEDING SEASON: on Haruku said to be throughout year but with distinct peak during the dry season (Oct–Apr/May), similar to reproductive pattern of the Maleo *Macrocephalon maleo*. Hundreds of birds at the Kailolo nesting ground, Haruku, in Oct–Nov 1991 (R. Dekker, personal observation) and Jan–Feb 1992 (M. Argeloo, personal communication). However, strangely enough reported absent here late Nov 1992 (A. Lewis, *in litt.*). Eggs collected near Amahai, S Ceram, in Apr (de Wiljes-Hissink 1953), fresh digging here in Nov (R. Dekker, personal observation). Eggs from Buru in Dec, from Bacan in June. Chicks collected in Jan ($n = 2$, Pulau Pombo and Haruku), Feb ($n = 1$, Haruku), Apr ($n = 1$, Ambon), and Aug ($n = 2$, Halmahera). Incubation period as in other megapodes largely dependent on incubation temperature; 97–101 days ($n = 4$) under artificial conditions in the Jakarta Zoo (West *et al.* 1981).

Local names

Mamoa (Ternate), man'lato (coastal Buru), man'titin (inland Buru), holeiko (Bacan), mulëhu (Ambon), maleo or meleo (Haruku and Ceram).

References

Gray (1860), Wallace (1863, 1869), Schlegel (1866, 1880a), Von Rosenberg (1878), Oustalet (1881), Salvadori (1882), Blasius and Nehrkorn (1883), Ogilvie-Grant (1893, 1897), Martin (1894), Vorderman (1898b), Stresemann (1914a, b), Toxopeus (1922), Siebers (1930), de Wiljes-Hissink (1953), Heinrich (1956), Ripley (1960, 1964), Schönwetter (1961), Rand and Gilliard (1967), West *et al.* (1981), White and Bruce (1986), Bowler and Taylor (1989), Holmes (1989), Jones and Banjaransari (1989).

Genus *Megapodius* Gaimard

Megapodius Gaimard, 1823. *Bulletin Général et Universel et des Annonces et de Nouvelles Scientifiques*, **2**, p. 450.

Includes *Alecthelia* Lesson, 1826; *Amelous* Gloger, 1841; *Megathelia* Mathews, 1914.

Thirteen species, here considered to form allospecies of a single superspecies, together ranging from the Nicobar Islands to western Polynesia.

Small to medium-sized megapodes, clad in sober colours, generally various shades of olive, rufous, or brown above, and usually paler or darker grey below, but some species virtually uniform black. Feathering on head variable, only restricted patch on side of head bare in some species (e.g. *M. reinwardt*), virtually entire head and neck (except for crown and nape) in others (e.g. *M. layardi*). A crest on hindcrown, short and broad in most species, longer and more pointed in others (especially *M. reinwardt*), but generally not exceeding 3–4 cm in length. Bill short and slender; nostril oval. Wing rather short, tip broadly rounded; 10 primaries, p6 longest, p7 equal or very slightly shorter, p5 0–0.5 cm shorter, p10 (outermost) and p1 (innermost) 2–4 cm shorter in smaller species, 3.5–5.5 cm shorter in largest ones, 2.5–4.5 cm shorter in intermediate ones. Tail relatively very short (*M. nicobariensis*, *M. pritchardii*, *M. cumingii*, and *M. bernsteinii*), relatively rather long (*M. reinwardt*), or rather short (all other species); rounded, composed of 12 feathers. Upper tail-coverts do not reach to the tip of the tail feathers. Tarsus rather short (relatively shortest in *M. reinwardt*), but relatively somewhat longer in *M. forstenii*, *M. eremita*, *M. pritchardii*, and *M. geelvinkianus*. Toes rather long, claws relatively longest of all Megapodiidae (except *Eulipoa*). Front of tarsus covered with single row of large broad scutes.

Polynesian Megapode *Megapodius pritchardii* G. R. Gray, 1864

Megapodius Pritchardii G. R. Gray, 1864. *The Annals and Magazine of Natural History*, (3), **14**, p. 378. (Niuafo'ou.)

PLATE 5

Other names: Polynesian Scrubfowl, Niuafo'ou Scrubfowl, Tongan Megapode, Malau.

Monotypic. Synonym: *huttoni* Buller, 1870, Niupo [=Niuafo'ou].

NOTE ON NOMENCLATURE: *Megapodius stairi* G. R. Gray, 1861, and *M. burnabyi* G. R. Gray, 1861, are often considered synonyms of *M. pritchardii*. Both names were described on the basis of single eggs from Samoa and Tonga prior to the description of *M. pritchardii*. A study by Steadman (1991) of both holotypical eggs which are kept in the collection of the BMNH revealed that neither *M. stairi* nor *M. burnabyi* is unequivocally synonymous with *M. pritchardii*. Based on size and colour, and because *Megapodius* eggs in general lack species-specific characters, both eggs could well belong to other extant or extinct *Megapodius* spp. According to Steadman (1991), both names are best regarded as *nomina dubia* (names of doubtful application).

Description
PLUMAGES
ADULT: Sexes similar. Forehead and crown dark ash-grey with slight brown tinge. Feathers of nape slightly elongated, dark ash-grey, forming indistinct short, broad crest. Area on lore and round eye and ear virtually bare; light grey feathers surrounding bare patch and on neck short, reduced, leaving variable amount of skin of head and neck bare; those from above eye to nape forming pale grey streak, contrasting strongly with darker cap, those on chin and upper throat almost white. Light grey of lower neck gradually shades into plumbeous-grey or dark grey on mantle, chest, side of breast, and upper flank. Upper scapulars and shorter upper wing-coverts (along bend of wing) dark brown with dark ash-grey fringes along tips; remainder of scapulars and upper wing-coverts as well as tertials brown, tinged olive in some birds,

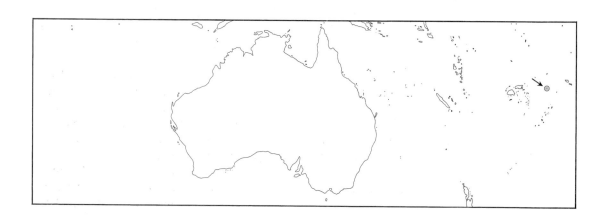

more cinnamon in others; some grey of feather-centres sometimes visible. Back and rump like remainder of upperparts, but slightly paler; upper tail-coverts white with variable amount of grey suffusion on tips. Belly, lower flank, vent, and under tail-coverts pale brown-grey with olive suffusion on feather tips, appearing somewhat mottled. Central pair of tail feathers (t1) brown-grey, other tail feathers white with variable amount of brown-grey suffusion along shafts and on tips, especially on outer web of outermost feather. Primaries dark grey, bases of webs suffused white, especially those of outer webs of middle primaries (which show much white). Secondaries, upper primary coverts, axillaries, and under wing-coverts dark fuscous-brown.

CHICK: forehead, lore, cheek, chin, and throat ochre; no bare skin visible at side of head. Crown and hindneck fuscous-brown. Mantle and shorter upper wing-coverts fuscous-brown with some isabelline bars; scapulars, tertials, back and longer upper wing-coverts contrastingly marked with broad deep black bars and narrower pale buff ones. Rump and upper tail-coverts rufous-brown. Chest, side of breast, and flank pale brown-grey, some light rufous wash towards rear; remaining underparts ochre or isabelline-white. Flight-feathers and primary coverts greyish-black, outer webs of secondaries marbled black and pale buff, tips of outer webs of primaries faintly marked buff.

IMMATURE: in first imm plumage, similar to ad, but less white in tail and flight feathers and some early-grown second generation body feathers with rufous fringes on upperparts or buff on underparts. Some secondaries, outer primaries, and tail still juv, shorter and narrower than in ad; flight feathers with traces of buff marks, primaries contrasting in length and width with second generation innermost; juv outermost primary (p10) and tail much shorter than in ad, p10 more pointed at tip. In second imm plumage, outermost primaries of second generation, somewhat shorter than third (and eventually fourth) generation of middle and inner primaries; p10 somewhat pointed, length from tip to insertion in skin 102–104 mm (102.5 ± 0.87, $n = 3$).

Some individuals vary in extent of albinism in primaries, tail, and tail-coverts. Loss of pigmentation occasionally occurs in small populations living on small islands; see, for example the Pacific reed-warblers *Acrocephalus* spp., some Pacific *Aplonis* starlings, and the Truk Monarch *Metabolus rugensis*.

BARE PARTS

ADULT, IMMATURE: iris brown. Bare skin of head and neck vermilion or dark red. Bill bright yellow or orange-yellow, sometimes with dusky base. Leg and foot pale yellow, orange-yellow, yellowish-red, or light red; in ♂ more orange-yellow, in ♀ light yellow (Finsch 1877), discernible only when a pair is seen together in good light (Todd 1983). Sexual difference in coloration not recognized by D. Rinke (personal communication).

CHICK: iris brown. Bill red-brown or brown, fading to orange-yellow on cutting edges. Leg and foot red-brown or brownish-orange with darker brown front of tarsus and upper surface of toes.

MOULTS

Primary moult serially descendant. Only three in moult examined (one from Nov, others undated); all had primary moult suspended, outer moult series apparently just completed with regrowth of p10, inner series suspended with up to p4 (in one) or p6 (in two birds) new.

MEASUREMENTS

ADULT: Niuafo'ou; skins of 3 unsexed birds (RMNH, ZMB), and data of 4 birds from Finsch and Hartlaub (1867), Oustalet (1880), and Ogilvie-Grant (1893), as well as wing of 4 ♂♂, 2 ♀♀, and 1 unsexed bird from Amadon (1942); perhaps some imm included.

	n	mean	s.d.	range
wing	14	187.7	5.13	179–196
tail	6	54.5	2.53	50–58
tarsus	6	58.1	1.00	57.0–59.5
mid-toe	5	32.6	1.06	31.1–34.0
mid-claw	4	17.1	0.37	16.5–17.3
bill (skull)	5	24.9	1.53	22.6–26.8
bill (nostril)	5	10.9	0.67	10.0–11.8
bill depth	3	7.3	0.15	7.2–7.5

In 'a large number', wing 185 mm (178–192), tarsus 58 mm (56–60), middle toe 30 mm, middle claw 18 mm, bill 24 mm (22–26) (Friedmann 1931). In live birds, wing 194.0 ± 2.76 mm (190–198, $n = 6$), bill 12.3 ± 0.37 mm (12.0–12.9, $n = 6$) (D. Rinke, *in litt.*).

CHICK: newly hatched: wing 80–89 mm (84.0 ± 3.00, $n = 18$); tarsus 25–30 mm (27.7 ± 1.24, $n = 18$, Friedmann 1931; Todd 1978; Rinke 1986*b*; BMNH); wing 81–90 mm (85.4, $n = 8$, Rinke 1991).

WEIGHTS
ADULT: 295–365 g (332.2 ± 21.67, $n = 9$, D. Rinke, *in litt.*); ♀ 362–454 g (415.2 ± 36.33, $n = 6$, D. Todd, *in litt.*).

CHICK: newly hatched: 38.5–58.0 g (48.4 ± 6.7, $n = 9$, Todd 1978); 40–51 g (46.2 ± 4.65, $n = 4$, Rinke 1986*b*); 43–51 g (48.5, $n = 8$, Rinke 1991).

GEOGRAPHICAL VARIATION
None.

Range and status
Endemic to the small, volcanic island of Niuafo'ou, Kingdom of Tonga, Central Polynesia (15°36′S, 175°38′W), where they are concentrated around the inner slopes of the caldera and on islands in the central lake (Weir 1973). Natural occurrence elsewhere is difficult to prove as live birds and eggs of this species have been transported to other Pacific islands. However, bone fragments of this species have recently been found on 'Eua, Tonga, approximately 700 km south of Niuafo'ou (Steadman 1991, see also Chapter 3). Live birds obtained on Savaii I., Samoa (Finsch and Hartlaub 1867); eggs obtained in Samoa and in the Ha'apai group, central Tonga (Gray 1861) possibly from Niuafo'ou; said to occur on Niuatoputapu, northern Tonga, east of Niuafo'ou (Gräffe 1870), though confusion of names possible as both islands are commonly referred to as Niua (D. Todd, *in litt.*). Reports of occurrence on Viti Levu and Kadavu (Fiji) unconfirmed and might relate to the Bar-winged Rail *Nesoclopeus poecilopterus* (D. Todd, *in litt.*). Introduced on Tafahi (N Tonga, near Niuatoputapu) in 1968, but apparently without success (Rinke 1986*b*). Several chicks have been released and various dozens of eggs buried on Late and Fonualei in 1992 and 1993 (Curio 1992*b*; Rinke 1994). A fully grown Polynesian Megapode which was seen on Fonualei in 1994 by D. Rinke must originate from one of the seven chicks or one of the 37 eggs transferred to the island in June 1993 (Rinke 1994). If a megapode population manages to survive on Fonualei it will mean a range extension for the species.

Population: H. Bregulla (in Todd 1983) estimated the population (in 1968) at 100 birds for the entire island, a figure which was revised to *c*. 2000 birds by Weir (1973). Todd (1978) estimated the population at 200–400 (or more) birds, later recalculated as *c*. 820 adults based on density figures (Todd 1983). In 1992, the population was estimated at *c*. 200 pairs (A. Göth and U. Vogel, *in litt.*). The carrying capacity for the entire island is estimated to be 2500 birds (Todd 1983; Rinke 1991); however, this figure is only theoretical since only 500 of the 1500 ha of seemingly suitable habitat is occupied by the species (D. Rinke, *in litt.*).

Note: highly isolated occurrence in Pacific is artefact of human interference, with related species on nearby islands exterminated by man (see Chapter 3). Nearest extant megapode is the Vanuatu Megapode *M. layardi*, which is sometimes considered nearest relative. However, head pattern of *M. pritchardii* (paler grey side of head contrasting with darker cap)

Polynesian Megapode *Megapodius pritchardii*

similar to the Micronesian Megapode *M. laperouse* and the Nicobar Megapode *M. nicobariensis*, while in relative proportions of tail, leg, and bill *M. pritchardii* is also remarkably similar to *M. nicobariensis*, both showing, for instance, relatively shortest tail of all *Megapodius*, combined with a rather long tarsus; *M. layardi* is quite different in these respects. *M. pritchardii*, *M. laperouse*, and *M. nicobariensis* are probably relics of an early distribution wave of old *Megapodius* stock from Australasia (see Chapter 3). The most recent and widespread species in Australasia is apparently the Orange-footed Megapode *M. reinwardt*, which is at the opposite end of the *Megapodius* variation spectrum, showing the relatively longest tail and shortest tarsus of all (see Roselaar 1994).

Field characters

Very small; length *c.* 28 cm. The only megapode on Niuafo'ou. Slate grey with rufous or olive-brown back and wings; crest short and slate grey; feet and bill pale yellow to orange-yellow. Differs from Spotless Crake *Porzana tabuensis* (which has dark brown upperparts, greyish-black underparts, and red or orange legs and eyes) by its larger size, paler overall plumage, and yellow legs and bill.

Voice

Calls heard mostly in the early morning and occasionally during the rest of the day and night. Loud call of three whistled notes sometimes given as duet (see sonogram), 'kwey kwee kwrrr', produced by the ♂, with the ♀ joining in on the third note (Rinke 1991; D. Rinke, personal communication), or only the first two notes produced by the ♂, with the ♀ producing the third note (D. Todd, *in litt.*). If one pair calls, neighbouring pairs often call in response. Monosyllabic alarm call is a loud 'kreek'. During agonistic behaviour a crooned trill, which is answered by 'kwe kwe kwe .....' or ' kway kway kway' (Todd 1983; D. Todd, *in litt.*). At the nesting hole the ♀ produces a quiet, tuneful, bubbling sound, only just audible a few metres away. A soft bubbling trill has also been heard from ♀♀ moving to

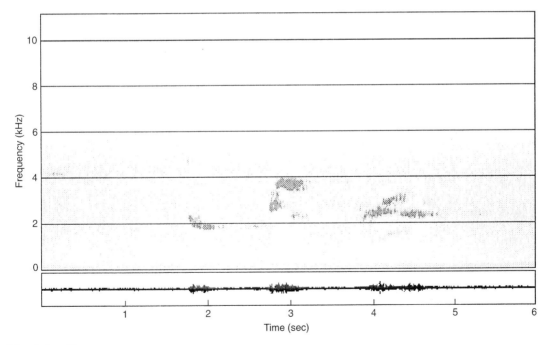

M. pritchardii
D. Rinke, Niuafo'ou, Kingdom of Tonga.

and from the nesting area; this call might act as an appeasement signal to owners of territories through which the ♀♀ are moving (Todd 1978, 1983).

Habitat and general habits

The information in this and the following sections comes mainly from Todd (1983) and is in part quoted from him. Niuafo'ou is the peak of an active, basaltic, shield volcano (which had its latest eruption in 1946) with a summit caldera, formed by the collapse of a composite cone. Two large lakes separated by a line of volcanic ash hills cover most of the caldera floor, leaving the island with a land area of about 35 km². The climate varies little during the year. The monthly mean daily temperature ranges from about 25 °C in Aug to about 28 °C in Jan. The annual rainfall averages about 2700 mm, most of which tends to fall during the hotter part of the year (Todd 1983). Mostly confined to the forests covering the inner slopes of the crater of Niuafo'ou and the islands within the central lake, rarely encountered in forested areas elsewhere, hence their range is extremely small (about 5–10 km²). Over 90% of the megapodes seen away from the laying sites were in areas of forest or thicket. Few were seen, or heard, in areas with a dense shrub layer. None were found in the more open habitats of the lava fields and areas of grassland.

Frequently found feeding in pairs, which suggests monogamy. Pairs usually remain within two or three metres of each other. The ♂ has been observed to offer food to the ♀. Having uncovered a food item, the ♂ calls to the ♀ with repeated soft, piping notes. If the food item is slow-moving, the ♂ simply points to it with his bill with his head inclined away from the ♀, leaving her to pick it up. However, if the item is more active, the ♂ picks it up himself and holds it in his bill with his head lowered for the ♀ to take. When the ♀ is very close, the ♂ often does not bother to call and the posture he assumes is enough to attract her attention. In the course of half an hour's feeding, a ♂ may offer food to the ♀ more than 20 times. The fact that this behaviour was observed so often (in over 60% of the occasions when pairs were observed feeding for periods of 5 min or more) suggests that food supply is critical for the pair (Todd 1983). This behaviour might also suggest that pair bond is for life.

Although there is no conclusive evidence, there are indications that pairs establish and defend feeding territories which may be at a distance from the nearest nesting ground. Since the ♀ is normally recorded at the nesting site alone, the ♂ probably stays behind in their territory to guard and defend it against intruders. When disturbed, the birds usually run rather than fly away, but if they do take to the wing they sometimes find safety by perching in trees. Fights are characterized by wing beating, kicking, pecking, and shoving. ♂♂ take the initiative in territorial conflicts between pairs, which normally end in the intruders being chased from the territory. Following each encounter, both pairs call continuously for 5–10 min. Territory size is estimated to be about 1.1–1.6 ha (Todd 1983).

Main predators on Niuafo'ou are feral cats (which selectively take laying ♀♀), Barn Owls *Tyto alba* (Weir 1973), and man (see also Chapter 9).

Food

In the study by Todd (1983), insects represented 53% of the larger items that could be recognized, land snails 25%, centipedes 13%, and worms 9%. The fallen fruit of *Syzygium* spp. made up the remaining 4%, although it was not clear whether the birds were eating the fruits or collecting insects from within them. According to Finsch (1877), F. Hübner recorded snail-shells, small crabs, centipedes, and in a few cases seeds in the stomachs of the birds he collected. Captive birds have been recorded as eating cockroaches, termites, ants, worms, and coconut (Weir 1973).

Displays and breeding behaviour

Polynesian Megapodes do not build mounds but lay their eggs at communal, geothermally heated nesting grounds. They arrive at the nesting sites between dawn and 10.00 a.m. A few birds appear later in the day. The ♀ alone is responsible for all the work at the burrow. Only on one occasion was a ♂ seen to accompany a ♀ at the nesting ground. When the birds reach the site, they often walk from burrow to burrow looking for a suitable place to bury their egg. While digging, the ♀ regularly leaves the burrow and looks around. Digging is resumed by removing sand from the entrance of the burrow and working her way back inside. The feet are used alternately for about six powerful, back kicks at a rate of about 60 kicks per minute. After egg laying, she reappears and kicks sand back down the hole. After it has been filled, digging becomes more random so that freshly turned soil is scattered over a wide area. The whole process involved in egg laying, recorded on several occasions, lasts between 2 h 10 min and 3 h 40 min (Todd 1983).

Breeding

NEST: burrow at geothermally heated sites. Whether heat from solar radiation or decaying roots plays an additional role in the incubation of the eggs has still to be established. Eggs are reported to be laid in burrows between roots of trees (Curio 1992b), but it is unclear whether these burrows are dug especially here to give structure and stability to the burrow or because of heat production in the roots. Burrows are 15–20 cm in diameter and 90–150 cm deep (Weir 1973). Eggs were found at depths ranging from 0.2–1.7 m inside the burrow. A total of 11–14 nesting grounds have been reported for the island, most of which are divided into several sub sites, totalling 26–29 different locations (Todd 1983; D. M. Todd, unpublished report). All sites, except one on the southwest coast and three unconfirmed ones on the north coast, are within the crater. None is known from the steep west and north shores of the crater. About one-third of the nesting grounds listed by Todd are now abandoned (Rinke 1991). It is still uncertain whether there are nesting areas near the seashore as reported by Weir (1973) and shown by Todd. For breeding sites see Todd (1983, Fig. 4, p. 75).

EGGS: elongate-oval; mat, without gloss; brownish-buff to reddish-brown when newly laid, changing to dull buff or ochre-brown during incubation when outer layer also partially flakes off showing spots of white shell underneath. Similar in shape and coloration to eggs of *Macrocephalon* and *Eulipoa*. SIZE: 70–80 × 39–47 mm (75.6 ± 2.8 × 44.8 ± 1.8, $n = 30$, Todd 1978); 74.6 ± 3.0 × 44.7 ± 1.8 mm ($n = 100$, Todd 1983); 73.2–76.4 × 41.1–44.5 mm (75.1 × 43.1, $n = 6$, Rinke 1986b); 71.4–79.7 × 40.4–49.0 mm (74.8 × 44.4, $n = 70$, Steadman 1991). WEIGHTS: 65–82 g (75.9 ± 5.2, $n = 30$, Todd 1978); 75.0 ± 5.1 g ($n = 100$, Todd 1983); 71–82 g (78, $n = 6$, Rinke 1986b).

BREEDING SEASON: according to local people, eggs can be found every month of the year, apparently without a clear peak season. Curio (1992b), however, reported low digging and laying activities in Jan–Feb, while Weir (1973) suggested a peak during Apr–May. No significant variation in number of eggs was recorded between May and Sept 1979 (Todd 1983). Incubation period 47–51 days ($n = 4$, Todd 1983) or up to 67 days (Curio 1992b).

Local name
Malau.

References
Gray (1861, 1864a), Finsch and Hartlaub (1867), Buller (1870), Gräffe (1870), Finsch (1877), Oustalet (1880, 1881), Schlegel (1880a), Ogilvie-Grant (1893, 1897), Friedländer (1899), Friedmann (1931), Amadon (1942), Preston (1969), Weir (1973), DuPont (1976), Todd (1978, 1983), Rinke (1986a, b, 1991, 1994), White and Bruce (1986), Pratt et al. (1987), Steadman (1991), Curio (1992a, b).

Micronesian Megapode   *Megapodius laperouse* Gaimard, 1823

Megapodius La Pérouse Gaimard, 1823. *Bulletin Général et Universel des Annonces et de Nouvelles Scientifiques*, **2**, p. 451. (Tinian, Mariana Is.)

PLATE 5

Other name: Micronesian Scrubfowl.

Polytypic. Two subspecies. *Megapodius laperouse laperouse* Gaimard, 1823, Mariana Is; *Megapodius laperouse senex* Hartlaub, 1867, Palau Is.

Description
PLUMAGES

ADULT: *M. l. laperouse*. Sexes similar. Forehead and crown dark ash-grey; short rough crest on nape dark or medium ash-grey, slightly pointed at rear. Area round eye and ear virtually bare, remaining side of head and neck covered with short sparse light ash-grey feathers, variable amount of bare skin (40–90%) visible; narrow strip at rear of neck fully feathered. Upperparts and upper wing black, feathers faintly fringed dark grey on mantle and scapulars, and tinged olive-brown on scapulars, back, median and greater upper-wing coverts, and tertials; rather similar to the Biak Megapode *M. geelvinkianus*. Rump and upper tail-coverts dull sooty black or black-brown with distinct olive tinge and fringes; often contrastingly more olive than remainder of upperparts. Chest and flank sooty black with dark plumbeous-grey feather fringes, belly sooty black, feathers with olive-grey to fuscous-olive fringes, vent medium grey with broad ill-defined pale olive-grey or fuscous-brown fringes; lower flank brown-grey with

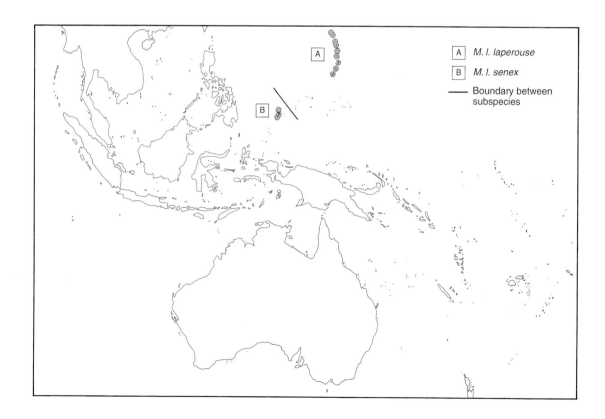

broad ill-defined ochre feather tips. Flight-feathers and tail greyish-black; outer webs of secondaries tinged olive.

CHICK: *M. l. laperouse*. Upperparts dark brown with olive tinge, latter most pronounced on mantle and scapulars. Lower mantle, scapulars, back, tertials, and outer webs of inner secondaries barred rufous. Rump and upper tail-coverts rufous-brown. Lores, side of head, chin, and throat buff, remainder of underparts light brown, fading to buff on mid-belly and vent. Feathers of nape slightly elongated.

IMMATURE: *M. l. laperouse*. In first imm plumage, like ad, but upperparts dark brown, head and throat tinged pale brown, underparts mottled light and dark brown; upper wing-coverts and secondaries fringed rufous, crest less pronounced. Grey head of ad obtained before brown tinge of upperparts lost. Some secondaries, outer primaries, and tail still juv, much shorter and narrower than in ad, secondaries partly notched isabelline on outer web, juv outermost primary (p10) pointed at tip, length of p10 from tip to insertion in skin 95 mm ($n = 1$; in ad, 101–114 mm, average 107.0 ± 6.06, $n = 4$), length of juvenile tail 52 mm ($n = 1$). In second imm plumage, like ad, but side of head (except area round eye) and side of neck more densely covered with reduced feathers; upperparts slightly more dark olive-brown, less sooty black with restricted olive tinge; outer primaries and tail of second generation, still somewhat shorter than ad, outer primaries relatively shorter and more pointed at tip than neighbouring third generation of middle or inner primaries; length of p10 97–107 mm (average 103.7 ± 4.13, $n = 5$).

BARE PARTS
ADULT: iris orange-brown to dark brown in nominate *laperouse*, golden-yellow to tan in *senex*. Bare skin of head and neck in both races yellowish-red, bright orange-red, or bright red, sometimes red with more orange or flesh-pink neck. Bill yellow, buff-yellow, orange-yellow, or orange-red, upper mandible clove-brown to black at base (latter apparently more extensively so in nominate *laperouse* than in *senex*). Leg and foot yellow, buff-yellow, yellow-orange, reddish-yellow, or orange, joints of toes or all upper surface of toes dark grey or black, dark tinge sometimes partly extending onto tarsus.

CHICK: iris brown, bill horn-brown to slate-black, leg pink-flesh, front of tarsus and toes slate-black, soles dull yellow.

IMMATURE: no information on iris. Skin of head, leg, and foot red-brown.

MOULTS
Primary moult serially descendant. Most birds examined undated. An imm had p1–p3 (innermost primaries) of second generation, p4–p5 growing, outer feathers (p6–p10) juv (first generation); five older imm (one from Aug, Saipan, one from Feb, Pagan, one from Jan, Palau, remainder undated) had inner three to seven primaries new, third generation, next one growing, outer primaries of second generation; two still older ones had p1 and p2 of fourth generation, p3 and p4 growing, p6 to p7 or p8 third generation, p8 or p9 growing, p10 or p9–p10 old, second generation. Four ad had moult suspended, all with an outer moult series just completed with regrowth of p10 and inner series suspended with p5 or p6. In nominate *laperouse* on Guguan, no active primary moult in four ♀♀ trapped at a nesting ground, late May and early June; of two unsexed birds, one in active moult, other without moult; in mid-Sept, two ♂♂ and two unsexed birds in moult, three unsexed birds without active moult (Glass 1988). In *senex*, birds from Aug, Sept, and Nov in body moult, in Dec in wing moult (Baker 1951).

MEASUREMENTS
ADULT: *M. l. laperouse*. Mariana Is, skins of 1 ♂ and 2 ♀♀ (BMNH, RMNH, ZMB); wing, tail, and tarsus include data of 2 ♂♂ and 3 ♀♀ from Baker (1951).

	n	mean	s.d.	range
wing	7	182.4	1.89	178–185
tail	8	61.0	2.78	55–63
tarsus	8	54.9	0.77	54–56
mid-toe	3	30.3	0.61	29.8–31.0
mid-claw	3	16.2	1.39	15.4–17.8
bill (skull)	3	24.5	0.81	23.6–25.1
bill (nostril)	3	10.9	1.08	10.0–12.1
bill depth	3	7.2	0.48	6.7–7.7

Wing, ♂ 155–169 mm, ♀ 158–170 mm; tail, ♂ 54–62 mm, ♀ 56–65 mm; tarsus, ♂ 51–54 mm, ♀ 50–55 mm; bill, ♂ 22.5–24 mm, ♀ 23–25 mm (Taka-Tsukasa 1932). Data supplied by Glass (1988) perhaps included some imm: wing of 2 ♂♂, 3 ♀♀, and 7 unsexed birds 155–189 mm (168.5, $n = 12$), tarsus 48–58 mm (50.5, $n = 11$), culmen 12–20 mm (13.7, $n = 11$). Wing 169–189 mm (Oustalet 1891).

M. l. senex: Palau Is, skins of 1 ♀ and 8 unsexed birds (BMNH, RMNH, ZMB).

	n	mean	s.d.	range
wing	9	186.6	3.90	182–193
tail	8	61.5	1.58	59–64
tarsus	9	59.7	2.48	56.0–63.2
mid-toe	7	33.3	1.52	31.4–35.5
mid-claw	9	18.3	0.85	17.4–19.7
bill (skull)	9	25.0	1.03	23.6–25.3
bill (nostril)	9	11.8	0.53	11.0–12.4
bill depth	8	7.6	0.46	7.0–8.3

In 3 ♂♂ and 7 ♀♀, wing 171–189 mm (mean 182.2), tail 46–68 mm (58.2), tarsus 45–60 mm (55.3), bill 22.7–30 mm (26.1) (Baker 1951). Wing, ♂ 176–181 mm, ♀ 177–187 mm; tail, ♂ 59–67 mm, ♀ 62–68 mm; tarsus, ♂ 58–61 mm, ♀ 55–58 mm; bill, ♂ 25.5–26 mm, ♀ 24–26 mm (Taka-Tsukasa 1932). See also Oustalet (1881) and Amadon (1942).

CHICK: *M. l. laperouse*. Wing 81 mm, tarsus 26 mm (Oustalet 1881).

WEIGHTS
ADULT: *M. l. laperouse*. ♂ 275–291 g ($n = 2$, Guguan), ♀ (with egg) 400–455 g ($n = 4$, Guguan), sex unknown 295–372 g (325.8 ± 27.13, $n = 6$, Guguan), sex unknown 230–330 g ($n = 2$, Saipan) (Glass 1988). *M. l. senex*: no data.

CHICK: 34.2 g (age c. 7 days, Saipan, Nov 1990, D. Stinson, *in litt.*).

GEOGRAPHICAL VARIATION
M. l. senex from Palau Is differs from *M. l. laperouse* from Marianas in light ash-grey forehead, crown, and nape, somewhat paler still towards side and rear, similar to colour of remainder of head and neck, not forming contrastingly darker cap; body and upper wing dull black with plumbeous tinge, darker and more plumbeous than in nominate *laperouse*; plumbeous-black of mantle gradually turns into dusky olive-brown of scapulars, tertials, and upper wing-coverts; rear body on average less pale than in nominate *laperouse* (but in both races, imm browner and with more olive tinge than in ad, and head in both brownish; chicks similar); upper mandible of *senex* apparently less extensively black at base than in nominate *laperouse*; length of wing, tail, and bill rather similar, but tarsus, toes, and claws of *senex* distinctly longer than in nominate *laperouse*. Not known whether any variation in colour or size occurs within nominate *laperouse* on Mariana Is; variation in wing length large in some samples, but this probably mainly due to inclusion of not fully grown imm rather than mixing of different-sized populations from several islands.

Range and status
On Marianas, nominate *laperouse* recorded (from north to south) from Farallon de Pajaros (Uracas), Maug, Asuncion, Agrihan, Pagan, Alamagan, Guguan, Sarigan, Anatahan, Saipan, Tinian, Agiguan (Aguijan), Rota, and Guam, i.e. from all islands except Farallon de Medinilla, although one was believed to be heard calling from near the northeast shore of Medinilla in the early 1980s (Reichel and Glass 1991).

Status: nominate *laperouse*. Uracas: rare, only one report of nesting burrow; most recent record is of one ad seen summer 1971 and 1972. Maug: common, nesting ground on

North island estimated at 150–300 birds in May 1987 (Reichel *et al.* 1987). Asuncion: rare, no definite breeding records but reported for the island as recently as 1987 (Reichel *et al.* 1987). Agrihan: rare, nesting area and large-scale egg-collecting reported in 1978. Pagan: rare, four nesting areas known to residents; reported absent from large areas of habitat. One geothermal nesting ground was buried under volcanic ash during the eruption of Mt. Pagan in 1981 (D. Stinson, personal communication). Alamagan: rare. Guguan: abundant, active and prolific in nesting area in volcanic cinder fields; estimate of 1500–2200 adults in Sept 1986; adult to chick ratio given as 1:1. Sarigan: common, chicks and adults seen in Sept 1983; estimate of 50–100 pairs in Sept 1983, but as many as 300–400 in Sept 1990 (D. Stinson, *in litt.*). Anatahan: uncommon, juveniles seen in 1971–72. Medinilla: absent. Saipan: rare, rediscovered here in 1978; juveniles seen in 1986, estimate of about 50 birds for the island (Reichel and Glass 1991); formerly reported as 'common' (Oustalet 1896). Tinian: rare or absent, no definite breeding records; not found in 1982, but one bird was seen in 1985; island almost certainly does not have a viable breeding population (Wiles *et al.* 1987; G. Wiles and D. Stinson, *in litt.*). Agiguan: uncommon, estimate of 11 birds in 1982 and at least 10 birds, including two pairs, in 1987; also recent report of a mound and juvenile. Rota: already reported absent in 1819, but collected in 1887; not reported here in 1982 (Glass 1988). Report of juvenile found in 1985 doubtful (D. Stinson, *in litt.*). Historic presence shown by Steadman (1992). Guam: already reported absent in 1819; no definite (breeding) records since, although specimen in AMNH collected here.

In Palau (or Belau) Is, *senex* recorded from volcanic islands of Babelthuap (or Babeldoab) and Koror (Oreor), and from coralline islands of Kayangel (Ngcheangel) group, Arakabesan (Ngerekebesang), Urukthapel (Ngeruktabel), Aulong (Ulong or Auror), Eil Malk (Mecherchar), Garakayo, Ngabad (Ngebad), Ngesebus (Ngedbus), Ngerukeuid and Kmekumer islands, Gayangas, Arumidin, Peleliu (Beliliou), and Angaur (Ngeaur), as well as many small offshore islets.

Status: rare or extinct already in late nineteenth century on larger inhabited islands of Babelthuap, Arakabesan, and Koror (Finsch 1875); in 1945, mainly restricted to isolated undisturbed offshore islets, with few birds elsewhere, for example 20–30 Garakayo, 5–10 Ngabad, 10–20 birds on Peleliu and fewer than ten on Angaur (Baker 1951). At present, common south of Koror, but still rare or uncommon Babelthuap, Arakabesan, and Koror. Highest densities on a small, flat, sandy, uninhabited island in the atoll of Kayangel in the far north of the archipelago, with mounds not more than 50 m apart (Pratt *et al.* 1980). Common in the Ngerukewid Is Wildlife Reserve (Seventy islands) of SW Palau, where found on nine islands each larger than two hectares. Conservative estimates on small islands ranged from 0.76 to 1.60 birds per ha, or 69 to 103 birds per km^2, for Ngerukewid and Kmekumer in 1988, with an average of 1 mound per 10–15 birds (Wiles and Conry 1990). Also common in southern Angaur in 1991 (G. Wiles, *in litt.*).

Field characters
Small. Length *c*. 28 cm. The only megapode on the Palau and Mariana Is. Generally blackish with short pale grey crest and bare skin of head and neck red. Legs and feet yellowish.

Voice
As in other *Megapodius* spp. the voices of ♂ and ♀ differ. The ♂'s (territorial) call is the most frequent call. It consists of a loud, raucous introductory 'keek' (described as 'skeek' by Pratt and Bruner 1978), followed by a pause and two further 'keek' notes of lower volume and slightly descending pitch. It is heard at all times of the year and is given most frequently early and late in the day; it is often answered by other birds in different locations. During 70 hours of observation, the ♂ territorial call was heard only once at the nesting ground, when a digging ♀ was raped (Glass 1988).

156 Micronesian Megapode *Megapodius laperouse*

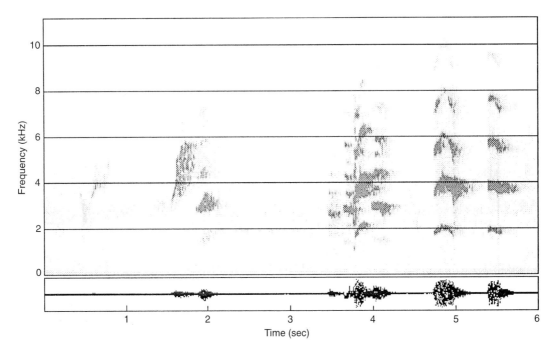

M. l. laperouse
P. Glass, CNMI Division of Fish and Wildlife, Guguan, Mariana Is, Mar 1988.

The ♀'s call consists of an introductory 'kek' (described as 'kuk' by Pratt and Bruner 1978), followed by a series of five to nine short 'kek' notes rapidly ascending in staccato fashion, followed by a short pause and a louder, sharper 'keek'. It is frequently given after successfully defeating and chasing off a rival digger at the nesting area. Also given in conjunction with the ♂'s call during duetting (see below) (Glass 1988).

According to Pratt *et al.* (1987), the duet call is initiated by one bird with 'keek-keek-keek- ...', gradually building in speed and pitch to a loud 'keek-keeer-kew' (on Palau) or 'keek-keeer-keet' (on the Marianas), and answered by the other bird with a rapid, chuckling cackle rising in pitch and slowing near the end. According to Glass (1988), the duet (see sonogram) is normally initiated by the ♀ with a loud introductory 'kek', followed immediately by the ♂ call and the remainder of the ♀ call simultaneously. So far the duet has been heard only during Nov and Dec, possibly an indication of the breeding season (see 'Breeding'). It comprised 14% of the total calls heard (*n* = 59) during regular call count surveys (Glass 1988).

Curiosity and alarm calls are made by both ♂ and ♀ and consist of a series of relatively low 'kek' notes at regular intervals of approximately the same pitch and volume, sometimes preceded by a 'keek' note of higher pitch and volume. The alarm call is louder and more intense than the curiosity call. However, both sounds are of much lower volume and pitch than the ♂'s territorial call and thus cannot be heard over a great distance (Glass 1988).

The calls of the two subspecies differ somewhat but are similar in overall pattern and quality. Contact note used by foraging pairs of *M. l. senex* is a single loud 'keek'. Also a loud crow 'keek-keer-kew' with the notes dropping in pitch; the homologous call of nominate *laperouse* does not drop at the end. *M. l. senex* also produces a lower pitched 'cuk-cuk-cuk' than nominate *laperouse* (Pratt and Bruner

1978; Pratt *et al.* 1980). The vocalization of *senex* has also been described as two kinds of loud notes, either separately by the same individual or as a duet; one bird gives six staccato notes on the same pitch 'cuk-cuk-cuk-cuk-cuk-cuk' while the other joins in with three long cries, each with a downward inflection and lower pitch than the preceding 'keer, keer, keer' (Marshall 1949).

Habitat and general habits
On Saipan, the species was most often observed in limestone forest remnants, with only 28% of sightings during surveys in 1984 in 'tangantangan' *Leucaena leucocephala*. Megapodes no longer used 'tangantangan' habitat after a typhoon knocked down all the *Leucaena* trees and a dense impenetrable viny tangle resulted. On Guguan the birds inhabit virtually all parts of the island, except the most barren volcanic slopes far from any line of vegetation and the most dense swordgrass thickets. Even unforested areas covered by beach morning glory and tussock grass have birds although the densities are not as high as in the forest. On Tinian, one bird was seen in limestone forest, 6–10 m tall, with dominant species such as *Cynometra ramiflora*, *Guamia mariannae*, *Ficus tinctoria*, *Premna obtusifolia*, *Pisonia grandis*, and the vine *Mikania scandens*.

Normally seen alone or in small groups; the chicks appear to be solitary. Foraging usually consists of vigorous digging under ferns, branches, and piles of leaf litter in much the same manner as nest excavation. Small chicks dig in a similar way. Foraging has also been observed on trees, usually in and around a large 'bird's nest' fern. Pair bond maintained throughout the year, from year to year, in relatively small home ranges; one individual, probably a ♂, resighted six times in a period of 9 months within 70 m of ringing site, and others within 150 m of ringing-site. This ♂ allowed both adults and immatures into its territory to forage, but all except its mate were chased 5–10 m when they approached closer than 3–4 m. Strict pair bonding and rigid territorial advertisement observed on Saipan. Home range estimated from call-counts data approximately 1 ha (Glass 1988).

On Palau, *senex* is rarely found in luxurious stands of forest on the rich deep soil of the gently sloping mountains of Babelthuap. However, it is abundant on the smaller limestone islets adjacent to Babelthuap, in forest on steep rocky ridges which are practically without soil and are dissected by fissures, potholes, narrow ravines and caves (Marshall 1949). On Peleliu, occurs in relatively undisturbed rainforest areas as well as in viny and brushy secondary vegetation, with preference for 'jungle' with heavy undergrowth and ground litter. Nine out of ten mounds in Ngerukewid Island Wildlife Reserve in strand forest behind sandy beaches. Regularly reported flying distances of several kilometres between small islands. Normally encountered as individuals or pairs. Duetting suggests monogamy also in this subspecies.

Food
M. l. laperouse (stomach contents from two specimens from Guguan and Sarigan; percentages based on visual estimate, no volumes or weights taken). In first specimen: seeds (possibly *Colubrina asiatica*) 50%, grit 15%, Coleoptera (Tenebrionidae) 10%, unknown insect larvae 6%, Hymenoptera (Formicidae, 15 small) 6%, Lepidoptera (scales) 3%, Lepidoptera larvae 2%, Coleoptera larvae 2%, bark and plant fragments 2%, Diptera (Phoridae) 1%, spiders (small), Mallophaga, unknown material 3%. In second specimen: grit 20%, Hymenoptera 10%, Coleoptera 10%, unknown insect 10%, unknown arthropod 10%, plant fragments 5%, Lepidoptera? 5%, unknown 30% (D. Stinson, *in litt.*). The literature further mentions small fruits, snails, ants, and ant larvae. *M. l. senex*: seeds, vegetable matter, insects (for example, Blattidae), and crabs.

Displays and breeding behaviour
M. l. laperouse (Guguan): the ♀ is responsible for digging a burrow at the nesting ground, not

or only occasionally accompanied by the ♂. As in other burrow nesters except the Moluccan Megapode *Eulipoa wallacei*, ♀♀ visit the nesting ground at all hours of the day but not during the night. Aggressive behaviour occurs when ♀♀ approach each other too closely, with the intruder being chased off; very wary and silent around the nesting ground. The entire process of searching, digging, and egg laying takes 1.5–3 h. Digging as in other megapodes, with several strokes (three to nine) with one foot before changing to the other (Glass 1988).

M. l. senex: both sexes participate in mound building. Favoured sites for mound building are sandy beaches surrounded by steep forested slopes, with the mounds not placed in direct sunlight. Individual birds rather than pairs have been observed visiting mounds on three occasions (Pratt *et al.* 1980).

Breeding

NEST: *M. l. laperouse*. Burrows in sun-heated or geothermally heated soil on volcanic islands; reported to build mounds on non-volcanic, coral islands in the Mariana island chain such as Aguiguan, located at the base of a large fallen tree, measuring 2 m in diameter and only 0.2 m high (Stinson 1989), but probably larger when material is available (D. Stinson, *in litt.*). Generally, the burrow is similar to that of the Polynesian Megapode *M. pritchardii*, with a diameter of approximately 20 cm in moist soil, but is shaped as an irregular crater in dry soil, dug at an angle of 45°. Depth ranges from 48 to 94 cm (average 73 cm, $n = 15$). A major nesting ground on Guguan is located at the southwest edge of a volcanic cinder field on a slope of 30°, measuring 230 × 35 m, with 230 burrows and a maximum of 15 birds simultaneously at the site in 1986. A smaller nesting ground is located along cliff-edges at the lower slopes of the volcanic peaks with *c.* 30 burrows (Glass 1988). *M. l. senex*: mound, constructed of sand with small amounts of organic litter when situated close to the beach, but entirely of leaf litter and detritus when located further inland. Generally 1–2 m high and 2–6 m in diameter. On E Peleliu, mounds constructed of rough coralline rubble (Wiles and Conry 1990).

EGGS: similar in shape and coloration to eggs of *Macrocephalon maleo*, *Eulipoa wallacei*, and other *Megapodius* species (see under *M. pritchardii*). SIZE: *M. l. laperouse*. 65.5–71 × 41.5–45 mm (68.6 ± 1.62 × 43.3 ± 1.25, $n = 9$, Yamashina 1932), 69.1–72.3 × 42.0–44.2 mm (70.5 ± 1.63 × 43.4 ± 1.19, $n = 3$, Guguan, Glass 1988), 70.9–74.9 × 45.0–46.7 mm (73.0 × 45.9, $n = 5$, Schönwetter 1961). *M. l. senex*: 71–81 × 47–51 mm (76.5 ± 2.90 × 48.3 ± 1.34, $n = 13$, Yamashina 1932), 70–79.5 × 44.5–51 mm (75.8 × 47.8, $n = 10$, Schönwetter 1961), 72.4–83.4 × 44.5–47.8 mm (77.7 × 46.4, $n = 7$, Steadman 1991). WEIGHTS: 75–80 g (77.0 ± 2.65, $n = 3$, Guguan, Glass 1988).

BREEDING SEASON: *M. l. laperouse*. Nesting observed in May, June, and Sept. On Saipan and Guguan, breeding activity increases from Sept to Feb, the middle to latter part of the rainy season. On Saipan copulation in Jan. On Pagan eggs in Feb and May, on Agrihan in June, a chick in July. On Guguan a marked difference is noticed in calling frequency between Sept, when calling is more or less continuous throughout the day and night, and May or June, when it is much less frequent (Glass 1988). In 1989 calling was much more frequent in June than in Aug on Aguiguan. *M. l. senex*: nesting probably all year round, with peak in egg production during the southeast monsoon in Apr–Nov, especially May and June. Two chicks on Garakayo in Sept from eggs laid June–July. ♀ with egg in body cavity collected on Peleliu in Sept. However, eggs collected on Auror, Ngesebus, and Peleliu in Jan, while six out of ten mounds in Ngerukewid Island Wildlife Reserve were active in Jan (Wiles and Conry 1990).

Local names

Sasangal, sasangat, sasengay, sasségniat, sassenat, sesengi (Marianas), apagaj, bagai, bakai, betai (Palau).

References
Gaimard (1823), Hartlaub and Finsch (1872), Finsch (1875), Schlegel (1880a), Oustalet (1881, 1891, 1896), Ogilvie-Grant (1893, 1897), Bolau (1898), Hartert (1898b), Safford (1904), Taka-Tsukasa (1932), Yamashina (1932), Amadon (1942), Baker (1948, 1951), Marshall (1949), Schönwetter (1961), Greenway (1967), Falanruw (1975), Pratt and Bruner (1978), Pratt *et al.* (1980, 1987), Lemke (1983a, b), Engbring *et al.* (1986), Glass and Villagomez (1986), White and Bruce (1986), Reichel *et al.* (1987), Wiles *et al.* (1987), Glass (1988), Stinson (1989), Wiles and Conry (1990), Reichel and Glass (1991), Steadman (1991, 1992).

Nicobar Megapode *Megapodius nicobariensis* Blyth, 1846

Megapodius nicobariensis Blyth, 1846. *Journal of the Asiatic Society of Bengal,* **15,** p. 52. (Middle group of the Nicobar Is.)

PLATE 6

Other name: Nicobar Scrubfowl.

Polytypic. Two subspecies. *Megapodius nicobariensis nicobariensis* Blyth, 1846, Middle Nicobar Is from Teressa and Tillanchong to Katchall and Nancowry; perhaps this race on Batti Malv (Northern Nicobar Is) and formerly on the Coco I. group (Burma), north of the Andaman Is. (Synonym: *trinkutensis* Sharpe, 1874, Trinkut I.). *Megapodius nicobariensis abbotti* Oberholser, 1919, Little and Great Nicobar in southern group of Nicobar Is.

Description
PLUMAGES
ADULT MALE: *M. n. nicobariensis.* Forehead and crown olive-brown, sometimes tinged grey; crown in most birds rather contrastingly bordered below by pale ash-grey supercilium, extending backwards across nape, ending in short broad crest. Lore and side of head backwards to ear and down to upper cheek virtually bare. Short feathers of chin and throat whitish-grey; neck medium grey. Upperparts, including upper wing and back to upper tail-coverts, uniform saturated buffish olive-brown, sometimes slightly tinged grey on upper mantle, often more cinnamon on upper wing. Chest and side of breast olive-brown, feathers with ash-grey fringes; remainder of underparts extensively medium ash-grey, some olive-brown of feather bases sometimes partly visible. Feathering on thigh olive-brown or ash-grey. Flight feathers dark brown-grey, outer webs with light grey borders. Under wing-coverts and axillaries medium grey. In fresh plumage, feathers of crown and upperparts more distinctly fringed or tinged rufous-olive or slightly cinnamon. Pronounced individual variation; three morphs recognized here, though no true morphs, as intermediates occur. Grey morph, described above, has underparts extensively medium ash-grey; in intermediate morph, chest, flank, and thigh are extensively olive-brown, only belly mainly grey; in olive morph, entire underparts olive-brown, only foreneck and sometimes restricted patches on mid-belly showing grey. In extreme form, olive morph is extensively tinged cinnamon-buff, also on pale grey above eye and across nape; upperparts appear saturated olive-cinnamon, belly greyish olive-cinnamon.

ADULT FEMALE: like ad ♂. Already Hume (1874) suggested that ♀♀ may be more often in grey morph, ♂♂ more often in olive morph, but later on thought that pronounced olive tinge on underparts was due to immaturity. However, of ten ad ♂♂ examined, 30% were

Nicobar Megapode *Megapodius nicobariensis*

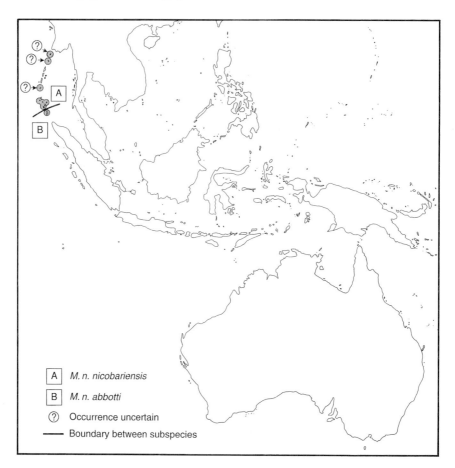

grey morph, 30% olive morph, remainder intermediate; in 12 ad ♀♀, 75% were grey morph, 25% intermediate, none olive. Of 10 ♂♂ in second imm plumage, 20% grey, 50% olive, 30% intermediate, of three imm ♀♀, 33% olive, remainder intermediate. Thus, colour apparently dependent on both age and sex, olive morph being more common in ♂♂ and imm, and least common in ad ♀♀, though variation considered to be independent of age and sex by Oberholser (1919).

CHICK: *M. n. nicobariensis*. Crown, upperparts, and upper wing rufous fuscous-brown, sometimes slightly tinged olive on crown and mantle; scapulars with rather faint rufous arcs or bars. Forehead and side of head from just above eye down to chin and throat buff-ochre, not sharply defined from fuscous-brown of upperparts. Underparts dull cinnamon-brown, sometimes with slight grey tinge. Flight feathers grey-brown, secondaries with more rufous-brown outer web, freckled black. No bare patch round eye. Feathers of crown, nape, and hindneck slightly elongated. *M. n. abbotti*: crown, upperparts, and upper wing blackish-brown. Forehead and side of head slightly olive with greyish tinge, extending on nape. Scapulars with rather faint rufous arcs or bars. Underparts blackish-brown on breast becoming fuscous-brown with grey tinge on belly. Flight feathers blackish-brown with grey tinge, vaguely freckled greyish-brown and black; tertials and secondaries more rufous-brown on outer web, freckled black. Lower back inconspicuously rufous and black.

IMMATURE: in first imm plumage, like ad, but crown and upperparts generally tinged buffish-cinnamon, less olive; underparts strongly olive-cinnamon or olive-brown, grey restricted to chin, throat, and sometimes chest (thus, like ad olive or intermediate morph, but general tinge more cinnamon). Grey of chin and throat pale, almost white; feathers short, but bare skin not visible. Lore and area round eye bare, except for short sparse feathers. Outer primaries and often tail still juv, much shorter than in ad, new inner primaries contrastingly longer, broader and more rounded at tip; tip of juv p10 pointed, often marked or freckled buff, length of p10 from tip to insertion in skin 90–105 mm (n = 2; in ad, 121–142 mm, mean 127.9 ± 6.30, n = 6), tail 57–59 mm (n = 2). In second imm plumage, similar to ad (for colour morphs, see ad ♀), but outer primaries still second generation (late-grown first imm plumage), middle ones in third, inner in third or fourth; second generation p10 often more pointed at tip than ad (but less than juv), length to insertion in skin 112–124 mm (117.6 ± 3.91, n = 7).

BARE PARTS

ADULT: iris light brown, hazel-brown, deep brown, or red-brown. Skin of head and neck including eyelids mauve-pink, bright light red, cherry-red, bright brick-red, or vermilion-red. Bill light-horn or brown-horn at base, sometimes tinged red or olive, tip gradually paler, light greenish- or yellowish-horn, cutting edges pale yellow-horn. Front of tarsus dark or light horn-brown, sometimes with green tinge, scutes marked lighter horn; tibio-tarsal joint and side and rear of tarsus light horn, horn-pink, light red, dull brick-red, or dark reddish-horn; upper surface of toes dark horn-brown to black-brown, darker towards tips of toes, soles pink-horn, flesh-red, dull red-brown, or pale yellow.

CHICK: iris greyish-brown. Bill brownish-grey or horn. Leg and feet mottled dark reddish- or purplish-brown, toes slightly darker.

IMMATURE: like ad, but rather restricted amount of bare skin on head pink to light red, tarsus light horn with pale pink-horn rear, toes dark horn.

Among both ad and imm, a few birds develop a large black bare callosity on crown, somewhat resembling bare black skin on top of head of Maleo *Macrocephalon maleo*. In most pronounced form, the entire skin of the crown is bare and somewhat swollen, with some additional black warts on lore and above eye (for example, in ad ♀ from Nancowry and ad ♂ from Katchall); in others, only part of skin of crown is bare and black, the remainder red and feathered (for example, in ad ♂ and imm ♀ from Camorta). This aberration perhaps caused by a disease (Ogilvie-Grant 1893).

MOULTS

Primary moult serially descendent. Virtually all birds examined for moult are from Jan to Mar, and hence no indication of moult season or duration of primary moult; of 16 ad examined from this period, 88% had moult suspended, only 12% active; two birds examined from early Apr both in active moult. In all birds, two or three series of primary moult in each wing, but often difficult to judge as differences in colour and wear of oldest feathers of earlier series and neighbouring newest feather of next series often slight. In imm, second generation of inner primaries starts to grow approximately June–Sept, before juv outer primaries full-grown; in eight birds in second imm plumage from Jan to Mar, second generation had just completed with regrowth of p10, third series active (in five birds) or suspended (in three birds) with up to p3–p8 new; in more advanced birds, also a fourth series had started with innermost primaries; unknown whether these birds are 1 or 2 years old.

MEASUREMENTS

ADULT: *M. n. nicobariensis*. Middle Nicobar Is (Teressa, Bompoka, Tillanchong, Katchall, Camorta, Trinkat, and Nancowry); skins of

19 ♂♂, 16 ♀♀, and 7 unsexed birds (BMNH, RMNH).

	n	mean	s.d.	range
wing	42	234.6	4.57	227–242
tail	35	69.3	3.64	62–75
tarsus	37	71.3	2.55	66.5–77.3
mid-toe	35	39.8	1.66	36.8–42.7
mid-claw	39	22.3	1.13	20.2–25.3
bill (skull)	38	29.4	1.50	26.5–32.6
bill (nostril)	37	13.7	0.71	12.2–15.6
bill depth	18	8.8	0.42	8.1–9.5

No statistically significant differences between sexes, e.g. wing, ♂ 228–242 mm (235.2 ± 4.36, $n = 19$), ♀ 227–242 mm (233.9 ± 5.19, $n = 16$); tarsus, ♂ 67.6–77.3 mm (71.0 ± 2.70, $n = 16$), ♀ 66.5–75.6 mm (71.3 ± 2.59, $n = 15$). No variation between islands, e.g. wing, Teressa, Bompoka, and Tillanchong 227–241 mm (235.8 ± 5.04, $n = 6$); Katchall 228–242 mm (232.6 ± 6.31, $n = 5$); Camorta 228–239 mm (233.6 ± 3.83, $n = 10$); Trinkat 228–238 mm (233.4 ± 4.12, $n = 7$); Nancowry 228–242 mm (235.4 ± 5.19, $n = 7$). In another sample of 19 birds from the same islands, measured with somewhat different technique, wing 206–233 mm (224.6 ± 5.50); tail 64–74 mm (69.4 ± 2.93); tarsus 63.5–69.5 mm (67.1 ± 1.53); bill to nostril 13–14.5 mm (13.7 ± 0.50) (Oberholser 1919). According to Ali and Ripley (1969), wing occasionally up to 250 mm.

ADULT: *M. n. abbotti*. Great and Little Nicobar; skins of 3 ♂♂ and 3 ♀♀ (Oberholser 1919; measured with slightly different technique: compare with Oberholser's sample above).

	n	mean	s.d.	range
wing	6	219.5	5.96	215–230
tail	6	67.2	2.77	64.0–71.5
tarsus	6	68.2	2.42	65.5–71.5
bill (nostril)	6	13.8	0.52	13.0–14.5

In Oberholser's samples, wing of one of six *abbotti* over 223 mm, in five of 19 nominate *nicobariensis* below 224 mm.

CHICK: at hatching or up to 6 days later: wing 99, 102.5, 104 mm; tarsus 26, 28.6, 30.4 mm (Hume 1874; BMNH). *M. n. abbotti*: live bird, 1 day old: wing, 101 mm; tarsus 29.5 mm; bill to nostril 6.0 mm (Dekker 1992).

WEIGHTS
ADULT: *M. n. nicobariensis*. Range: ♂ 595–907 g (756.5 ± 130.9, $n = 4$), ♀ 850–1021 g (926 ± 87.1, $n = 3$), unsexed, 964 g (Shufeldt 1919; Baker 1928; BMNH).

CHICK: *M. n. abbotti*. 1 day old: 55 g (Dekker 1992).

GEOGRAPHICAL VARIATION
Slight. Wing of *abbotti* on average slightly shorter than in nominate *nicobariensis* (see Measurements). In plumage, *abbotti* darker above and below, especially primaries being darker on outer webs, contrasting less with inner webs: outer web cinnamon-brown, inner web sepia (in nominate *nicobariensis*, outer web described as pale cinnamon-brown, inner web light sepia: Oberholser 1919); such differences may also exist between specimens of different age or at different stage of plumage wear, and further study on validity of *abbotti* needed.

Range and status
At the end of the nineteenth century, the Nicobar Megapode occurred in the southern Nicobars (Great and Little Nicobar: race *abbotti*), middle Nicobars (Teressa, Bompoka, Tillanchong, Camorta, Trinkat, Nancowry, and Katchall: nominate *nicobariensis*; no records for Chaura or Coral Bank), and Northern Nicobars (Batti Malv: race unknown; no record for Car Nicobar). Earlier, this or a related species may have occurred throughout the Andaman Is also, becoming extinct there due to human or other predators, except for at least one tiny offshore island in the north: Hume (1874) found disused megapode mounds on Table I. (just north of Great Coco I., near 14°N, politically belonging to Burma), but did not see the owners, which were described by the local lighthousekeeper as 'brown hens with large feet'. Reports of occurrence in the Coco Is (Seymour Sewell 1922; Abdulali 1964) probably refer to

Hume's observations on Table I. Seymour Sewell (1922) reported that he had seen a mound on Little Andaman. In the BMNH, a 6 day old chick is present, bought 'with many others' near Port Blair, South Andaman, in May 1898; on the label, it is stated to have come from 'one of the southern Nicobars, probably (if read well) from Preparis', which island is, however, at the northern limit of the Andaman chain, near Table I. Whatever its origin is, this chick differs from the available chicks of nominate *nicobariensis* and *abbotti* in more pronounced sooty grey and rufous bars on upper wing, tertials, lower mantle, and back, and by broad pale cinnamon fringes and black submarginal marks on secondaries.

Status: a letter from the Indian National Trust for Art and Cultural Heritage (9 Feb 1988) stated that the species had become extinct on all islands except Great Nicobar, where no more than 400 birds survived. However, a survey conducted in 1992 revealed that race *abbotti* is still common on Great Nicobar, with an average of 1.6 active mounds per kilometre of coastline in Mar and an estimated 780 breeding pairs in the coastal strip (Dekker 1992). The other islands are predominantly covered with primary forest, and it is expected that the species will still occur there as well.

Field characters

Length 37–40 cm. Endemic to the Nicobar Is. A terrestrial brown or reddish-brown bird with pinkish-red bare patch around the eye and greyish crown, normally seen in pairs in forest close to the beach. Characteristic voice. Confusion with any other terrestrial bird impossible.

Voice

M. n. abbotti: ♂ territorial call 'kyouououou-kyou-kou-koukoukoukoukou(kou)', rising in pitch on the first note, gradually descending on the staccato series; duration *c.* 3.5 sec (see sonogram). Similar to the Philippine Megapode *M. cumingii*, in that the ♂ call is not answered in duet by the ♀, but unlike other *Megapodius* spp., such as the Sula Megapode *M. bernsteinii*,

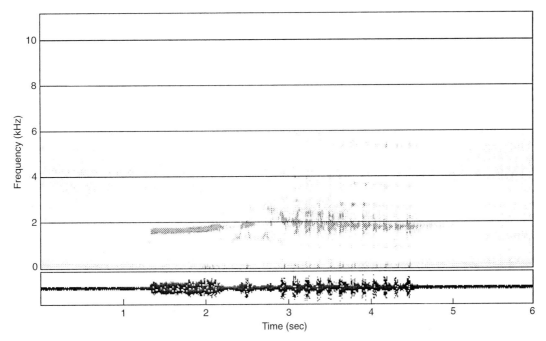

M. n. abbotti
R. Dekker, Great Nicobar, India, Mar 1992.

Dusky Megapode *M. freycinet*, and Melanesian Megapode *M. eremita*, which do have a duet. Contact-call 'wèk, wèk' and a throat-wobbling 'woubwoubwoub', probably produced by both sexes. A rapid series of soft suppressed 'uhk' 'uhk' 'uhk' 'uhk' while working at the mound (R. Dekker, personal observation).

Contact call also described as a loud cackling 'kuk-a-kuk-kuk', repeated quickly several times; the 'peculiar guttural crow reminiscent of the croaking of bull frogs: ... kiouk, kiouk, kok, kok, kok (etc.)' (Ali and Ripley 1969), is probably the territorial call described above.

Habitat and general habits

Dry deciduous forests fringing the coast as well as evergreen forest in the central parts of the islands. Also found in patches of secondary vegetation and stands of *Areca*, *Casuarina*, and *Pandanus*. Mounds often close to the beach as in many other *Megapodius* spp., though also located further inland.

Keeps in pairs or (probably outside the breeding season) in groups of 6–30 birds of both adults and immatures. ♂ and ♀ spend the day looking for food on the forest floor, with the ♀ regularly eating food items which the ♂ picks up and gives to her, either holding the item in his bill or dropping it in front of her (R. Dekker, personal observation), as has also been described for the Polynesian Megapode *M. pritchardii*. Regularly calls at night, though probably not active in any other way after sunset (Ali and Ripley 1969; R. Dekker, personal observation). When disturbed at the mound either walks, runs or flies away.

Food

Land snails, seeds, vegetable matter, insects, and other small invertebrates from the forest floor. Stomach contents of specimen from Tillanchong contained a beetle *Scarabus plicatus* and a snail *Helicina zelebori* (Ali and Ripley 1969).

Displays and breeding behaviour

Monogamous. ♂ and ♀ visit the mound during the breeding season on the day of egg laying only and not on intermediate days. No mound maintenance recorded during the egg-laying period in Mar. Digging activities do not necessarily start early in the morning, but may commence around noon or even early in the afternoon. ♂ and ♀ share equally in excavating and filling in the hole in which the egg is laid. They often stand in line, with the ♀ raking leaf litter from the base of the mound towards the ♂, who rakes it up to the top and into the hole in which the egg has been laid. When disturbed while digging, the birds leave the mound and the ♂ regularly produces his territorial call until he and his mate return.

Eggs can be laid at all hours of the day. Eggs buried 40–50 cm deep ($n = 3$). Digging activities involved in egg laying can take up to 7 h. Interval between two eggs from same pair 9 days ($n = 1$). Individual mounds are used by one to three pairs (average two pairs, six mounds observed) and for several or many years in succession depending upon mound type (see 'Breeding'). (Dekker 1992). Sixteen eggs, some fresh and some partly incubated, were taken from a single mound on Great Nicobar in Mar 1992 (S. Bashkar, personal communication).

Breeding

NEST: three different types of mound could be recognized on Great Nicobar: (1) true mounds, regular in shape, often higher than 1 m and built on an open spot and not against large (buttressed) trees; (2) mounds irregular in shape, often (much) lower than 1 m and built against the buttresses or trunk of a large living tree; (3) mounds similar in shape to type 2, but built against, around, or under a dead, rotting tree stump. A typical mound of type 1 measured 1.15–1.80 m in height and 7.22–8.50 m in diameter; type 3 mounds are usually not higher than 0.30–0.50 m. All mound types consist of sand, leaves, twigs, branches, other vegetable matter, or coral debris, depending upon location and distance from the beach; in general, mounds of type 1 contain more vegetable matter than the sandy mounds of types 2 and 3. Mounds are often built within the forest, a few metres from the edge of the sandy shore above the high-tide mark (Dekker 1992). The largest mound reported from Trinkat measured 2.4 m in height and 18 m in circumference, and contained three eggs. Those built away from the shore are reported to be smaller, 0.9 m high and 3.6–4.2 m in diameter.

EGGS: similar in shape and coloration to eggs of *Macrocephalon maleo*, *Eulipoa wallacei*, and other *Megapodius* species (see under *M. pritchardii*). SIZE: *M. n. nicobariensis*. 76.4–85.5 × 46.2–57.1 mm (82.6 × 52.3, n = 84, Baker 1935), 76.4–88.7 × 49.0–55.5 mm (82.1 × 52.4, n = 77, Steadman 1991); *M. n. abbotti*: 83.1 × 50.8 mm (n = 8, Baker 1928), 81–85.2 × 47.5–52 mm (83.3 × 49.9, n = 4, Baker 1935), 81.2–85.3 × 49.7–51.6 mm (83.4 × 50.3, n = 4, Steadman 1991), 78.6–84.7 × 50.9–51.9 mm (81.8 ± 3.06 × 51.2 ± 0.58, n = 3, Great Nicobar, Dekker 1992). WEIGHTS: *M. n. abbotti*. 114–124 g (118.7 ± 5.03, n = 3, Great Nicobar, Dekker 1992).

BREEDING SEASON: according to local information restricted to the dry season (east monsoon) from Nov or Dec to Apr. However, eggs of nominate *nicobariensis* have been collected not only in Apr, but also in May and June, those of *abbotti* in Mar and Apr. Chicks in Mar (1), May (1) and Aug (1).

Local names
Kuchav (Nicobarese language), kongah.

References
Hume (1874), Sharpe (1874), Oustalet (1881), Ogilvie-Grant (1893, 1897), Oberholser (1919), Shufeldt (1919), Seymour Sewell (1922), Baker (1928, 1935), Abdulali (1964, 1967, 1978), Ali and Ripley (1969), Steadman (1991), Dekker (1992).

Philippine Megapode *Megapodius cumingii* Dillwyn, 1853

Megapodius Cumingii Dillwyn, 1853. *Proceedings of the Zoological Society of London*, 1851 (1853), p. 119, pl. 39. (Labuan I.)

PLATE 6

Other names: Philippine Scrubfowl, Tabon Scrubfowl, Cumings Megapode.

Polytypic. Seven subspecies. *Megapodius cumingii gilbertii* G. R. Gray, 1861, Sulawesi and neighbouring small islands of Talisei, Tendila, Lembeh, and Togian group; *Megapodius cumingii cumingii* Dillwyn, 1853, small islands off Sabah (N Borneo) from Omadel (in east) via Banggi (in north) to Labuan (in west) and probably local on mainland, as well as Balabac and Palawan (SW Philippines) with small neighbouring islands (synonym: *lowii* Sharpe, 1875, Labuan I.); *Megapodius cumingii dillwyni* Tweeddale, 1877, Mindoro, Marinduque, Luzon, and small islands north of Luzon, north to Batan (20°26′N); *Megapodius cumingii pusillus* Tweeddale, 1877, central Philippines (Masbate, Cebu, and probably this race on others), W Mindanao, and Basilan, perhaps grading into nominate *cumingii* on the Sulu islands and on Maratua I. (east of Borneo) (possible synonyms: *balukensis* Oberholser, 1924, Balukbaluk I., and *tolutilis* Bangs and Peters, 1927, Maratua I.); *Megapodius cumingii tabon* Hachisuka, 1931, E Mindanao; *Megapodius cumingii talautensis* Roselaar, 1994, Talaud Is; *Megapodius cumingii sanghirensis* Schlegel, 1880, Sangihe Is.

Description
PLUMAGES
ADULT: *M. c. gilbertii*. Sexes similar. Crown dusky olive-brown, grading to dark brownish-grey on forehead, side of crown, nape, and hindneck; feathers of nape and hindneck slightly elongated, forming short broad crest on nape. Rather restricted area round eye and ear virtually naked, lore, cheek, and upper side

Philippine Megapode *Megapodius cumingii*

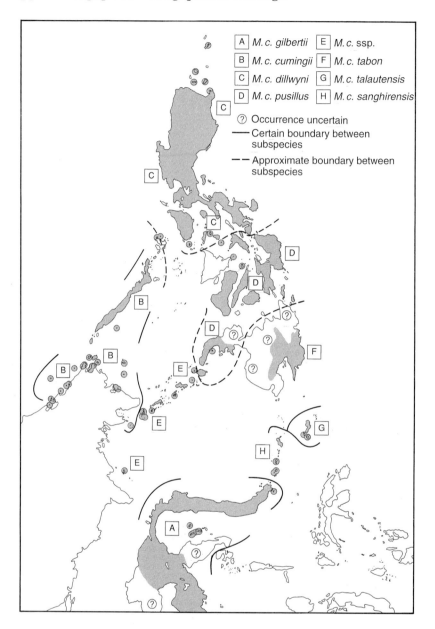

of neck rather sparsely covered with reduced dark brown-grey feathers, chin, throat, and remainder of neck more fully covered with dark brown-grey or dark plumbeous-grey feathers. Entire upperparts from upper mantle to upper wing-coverts, tertials, and upper tail-coverts dark olive-brown, rather like upperparts of Forsten's Megapode *M. forstenii*, but *M. c. gilbertii* has upperparts slightly browner and with a slight cinnamon cast (also, upper mantle of *M. forstenii* contrastingly plumbeous-grey, and rump to upper tail-coverts darker and browner); some dark grey of feather-bases often visible on back, rump, and upper tail-coverts, dark olive-brown feather tips occasionally much reduced: in some birds, rump similar in colour to upperparts, in others, contrastingly darker and greyer. Entire underparts dark to medium plumbeous-grey, often with slight olive-brown wash, mainly on belly, occasionally extending faintly to chest (but much less olive-brown than belly and vent of *M. forstenii*); thighs, under tail-coverts, axilliaries, and under wing-coverts dark grey to blackish-grey. Tail, flight feathers, and upper primary-coverts blackish-grey (if fresh) or dark grey-brown (if worn), outer webs extensively tinged dark olive-brown.

CHICK: *M. c. gilbertii*. Sulawesi: all upperparts including rump dark brown, rather faintly barred sooty and dark rufous-brown on lower mantle, lower scapulars, and back (if worn and bleached, bars darker and paler brown); forehead, crown, and upper mantle slightly tinged olive-grey, side of head and entire underparts light brown with olive-grey suffusion, more buffish on chin and cheek, darker olive-fuscous on chest and flank, more rufous towards vent; isabelline feather bases partly visible. Flight feathers and tertials uniform sooty-black. Chicks from Sabah and Palawan (nominate *cumingii*) similar, but general tinge slightly paler and greyer; barring on upperparts faint, as in Sulawesi, but faint rusty-red marbling present on edges of secondaries. Philippine birds (*dillwyni* and *tabon*) similar to nominate *cumingii*, but upperparts distinctly barred dull black and rusty-red on all scapulars, upper wing-coverts, and (sometimes) mantle, secondaries partly marbled and freckled rusty-red.

IMMATURE: like ad, but outer primaries, some secondaries, and tail still from first (juv) generation, shorter and narrower than in ad; outermost primary (p10) pointed, length from tip to insertion in skin in small-sized race *gilbertii* 93–101 mm (97.1 ± 3.37, $n = 4$), in intermediate-sized races nominate *cumingii*, *dillwyni*, and *sanghirensis* 89–105 mm (98.0 ± 8.18, $n = 3$), in large-sized *pusillus*, *tabon*, and *talautensis* 121 mm ($n = 1$); in ad, 102–127 mm (117.8 ± 6.37, $n = 8$) in *gilbertii*, 121–140 mm (129.3 ± 5.25, $n = 14$) in nominate *cumingii*, *dillwyni*, and *sanghirensis*, and 130–150 mm (138.4 ± 7.31, $n = 14$) in *pusillus*, *tabon*, and *talautensis*; juv tail 48–57 mm (*gilbertii*, $n = 2$), 62 mm (*tabon*, $n = 1$), or 53–63 mm (58.4 ± 3.97, $n = 5$) (intermediate-sized races). In second imm plumage, outer primaries from second generation, still somewhat shorter than in ad, p10 slightly more pointed, sometimes with pale fringe along outer web; length of p10 100–114 mm (108.9 ± 4.37, $n = 8$) in *gilbertii*, 109–132 mm (123.8 ± 5.44, $n = 13$) in intermediate-sized races, or 128–130 mm ($n = 2$) in large-sized races.

BARE PARTS

ADULT: iris light brown, warm brown, light reddish-brown, burnt sienna, or dark chocolate-brown; occasionally, in some birds orange-yellow (Sulawesi) or golden-brown (Mindanao). Bare skin of head and neck bright red, vermilion-red, dull red, dark grey with red tinge, or almost black (Sulawesi, Sabah, Palawan), in ♂♂ often brighter than in ♀♀; dirty crimson, dull salmon, and dark purple-brown recorded on Mindanao. Bill dark horn-brown to black on base and pale yellow to horn-yellow on tip, or uniform olive-green, horn-yellow, yellow-brown, dark horn-brown, or black-brown (independent of locality; variation perhaps related to age or season). Leg and foot uniform brown, dark brown, dark grey, or slate-black (Sulawesi); light brown, dark olive-brown, dark grey, or black-brown on front of tarsus and upper surface of toes, with remainder of foot paler brown, yellow-grey, or more or less strongly tinged red (Sabah, Palawan), or dark brown to dull red on front of tarsus and on toes with remainder (including tibio-tarsal joint) extensively dull red to red-brown (Luzon to Mindanao).

CHICK: iris brown-grey to dark hazel. Bill dark brown. Leg and foot dark horn-brown to almost black.

IMMATURE: iris brown or dark brown. Bare skin of head yellow-brown to salmon-pink. Bill uniform dark horn-brown. Leg and foot uniform dirty brown to dark slate-grey.

168 Philippine Megapode *Megapodius cumingii*

MOULTS

Primary moult serially descendant. In chick, first moulting series (second generation) starts with shedding of innermost primary (p1) when outermost feathers (p9–p10) of juv (first generation) have just started to grow. When first moulting series reaches p4–p8 (on average 6.5, n = 15) and juv p9–p10 fully grown, a second moulting series starts with p1. In older imm and ad, two (in 46 birds examined) or three series (in eight) in each wing; also, three birds show several irregularly scattered growing feathers in wing and these have moult between wings asymmetrical. In birds with two series in each wing, newest feather of each series separated by four to eight feathers (on average, 4.6 ± 0.74, n = 14). Of 27 imm examined, 85% in active primary moult, remainder had moult suspended; of 60 ad, 82% in active moult. Though moult unlikely to be synchronous between all populations in huge range of *M. cumingii*, virtually all birds with suspended moult occur in Dec–Apr, when 30–50% of the dated birds examined for each month show suspension of a primary moult series (in total, 41% of 22 birds suspended moult), while only two birds (Aug, Nov) show suspension of 21 examined May–Nov. In nominate *cumingii*, ad birds which have just started a new moult series (p1, p2, or p3 growing) occur Jan (1), Feb (1), May (1), Sept (1), and Dec (2), in *dillwyni*, *pusillus*, and *tabon* Mar (3), Aug (1), and Sept (1), in *gilbertii*, *sanghirensis*, and *talautensis* Jan (2), Mar (1), Apr (1), June (1), July (3), Sept (1), Oct (1), and Dec (2). No dated imm examined which had just started moult of second generation primaries; those with moult more advanced (second generation p5, p6, or p7 growing) occur Feb (2), Mar (1), Apr (3), Oct (1), and Nov (2).

MEASUREMENTS

ADULT: *M. c. gilbertii*. Sulawesi, Lembeh, and Togian Is, skins of 10 ♂♂, 10 ♀♀, and 6 unsexed birds (BMNH, RMNH, SMTD, ZMA, ZMB).

	n	mean	s.d.	range
wing	24	205.1	5.04	196–213
tail	23	64.9	2.65	60–69
tarsus	24	61.2	2.92	55.5–66.4
mid-toe	19	34.1	1.56	31.0–36.5
mid-claw	22	17.6	1.15	15.3–19.8
bill (skull)	23	27.2	1.25	25.0–29.5
bill (nostril)	24	12.1	0.84	10.9–13.5
bill depth	14	7.8	0.48	7.0–8.5

Sexes closely similar, e.g. wing, ♂ 197–212 mm (205.7 ± 4.56, n = 10), ♀ 196–210 mm (202.9 ± 5.19, n = 8). Data on wing from literature: Sulawesi, ♂ 201–220 mm (mean 210.0, n = 5), ♀ 189–218 mm (mean 203.9, n = 8) (Stresemann 1941); ♂ 205, 207, 211 mm, ♀ 196 mm (Blasius 1897; Vorderman 1898a; Ripley 1941). Wing of birds from Togian Is: 197, 199, 206 mm (BMNH, SMTD, ZMB).

M. c. cumingii: small islands and coast of Sabah from Labuan I. to Sandakan, and Banggi Is; skins of 1 ♂, 2 ♀♀, and 14 unsexed birds (BMNH, RMNH, SMTD).

	n	mean	s.d.	range
wing	16	224.4	4.01	218–232
tail	13	71.8	2.42	68–76
tarsus	16	64.0	2.96	60.7–68.8
mid-toe	13	37.5	1.58	34.8–39.8
mid-claw	14	18.8	0.82	17.5–20.4
bill (skull)	15	27.6	1.39	25.4–29.5
bill (nostril)	15	12.0	0.78	10.5–13.1
bill depth	7	8.1	0.51	7.3–8.8

Palawan, skins of 6 ♂♂, 7 ♀♀, and 1 unsexed bird (BMNH, RMNH, SMTD, ZFMK, ZMB).

	n	mean	s.d.	range
wing	13	228.7	4.58	222–236
tail	14	73.9	2.10	70–78
tarsus	14	67.4	2.51	63.7–71.5
mid-toe	13	36.9	1.94	34.3–40.4
mid-claw	13	19.7	1.06	18.3–21.4
bill (skull)	14	28.3	1.05	25.7–29.4
bill (nostril)	14	12.7	0.67	11.3–13.7
bill depth	12	8.0	0.43	7.4–8.7

Of this latter sample, wing of ♂ 222–230 mm (225.5 ± 2.68, n = 6), ♀ 225–236 mm (230.9 ± 3.94, n = 6).

M. c. dillwyni: Luzon, Marinduque, and small islands north of Luzon; skins of 2 ♂♂, 2 ♀♀, and 6 unsexed birds (BMNH, ZMB); wing, tail, and middle toe include data of 2 ♂♂ and 1 ♀ from Fuga I., north of Luzon (Riley 1924).

	n	mean	s.d.	range
wing	12	237.7	6.23	230–248
tail	12	75.2	5.21	61.5–83
tarsus	9	68.5	3.22	65–73.2
mid-toe	11	39.1	1.70	35.5–41.4
mid-claw	9	20.8	1.83	18.5–23.4
bill (skull)	9	29.2	2.16	27.0–31.9
bill (nostril)	9	13.1	0.90	12.0–14.8
bill depth	9	8.1	0.50	7.3–9.0

M. c. pusillus: Masbate, Zamboanga peninsula (W Mindanao), and Basilan; skins of 5 ♂♂ and 3 ♀♀ (BMNH, SMTD).

	n	mean	s.d.	range
wing	8	245.4	5.48	239–254
tail	8	76.4	3.10	73–82
tarsus	8	70.7	2.04	67.6–73.3
mid-toe	8	39.7	1.38	37.5–41.0
mid-claw	8	21.6	0.98	20.5–23.1
bill (skull)	8	28.8	1.47	27.6–30.7
bill (nostril)	7	12.7	0.57	11.6–13.7
bill depth	6	8.2	0.43	7.7–8.7

Balukbaluk (off Basilan), ♀ (probably imm), wing 211 mm, tail 73.5 mm, tarsus 61.5 mm (Oberholser 1924). Sulu islands, ♀, wing 217 mm, tail 77 mm, tarsus 62 mm (Blasius 1897). Maratua (off E Borneo), ♀, wing 230 mm, tail 76 mm, tarsus 64 mm (Bangs and Peters 1927).

M. c. tabon: E Mindanao; skins of 3 ♂♂ and 1 ♀ (BMNH, RMNH, ZMB).

	n	mean	s.d.	range
wing	4	257.2	2.63	255–261
tail	4	84.2	2.90	81–87
tarsus	4	73.8	2.21	71.0–76.4
mid-toe	4	42.7	3.06	38.4–45.4
mid-claw	4	22.6	1.27	21.3–24.1
bill (skull)	4	31.0	1.83	28.6–32.5
bill (nostril)	4	13.6	0.79	12.6–14.4
bill depth	4	8.6	0.20	8.5–8.9

M. c. talautensis: Talaud Is; skins of 1 ♂, 2 ♀♀, and 7 unsexed birds (BMNH, SMTD, ZMB).

	n	mean	s.d.	range
wing	10	244.3	3.97	240–254
tail	10	75.1	2.33	72–79
tarsus	9	72.7	2.77	69.6–78.3
mid-toe	10	40.4	1.83	37.8–44.3
mid-claw	10	23.5	1.40	21.7–25.2
bill (skull)	10	29.0	1.52	27.0–31.4
bill (nostril)	10	13.4	0.98	12.2–14.8
bill depth	9	9.0	0.20	8.7–9.3

Wing of population on Karakelong 240–247 mm (243.4 ± 2.28, n = 7), on Kaburuang 241–254 mm (246.3 ± 6.81, n = 3). None examined from Salebabu.

M. c. sanghirensis: Sangihe Is; skins of 4 ♂♂, 5 ♀♀, and 3 unsexed birds (BMNH, RMNH, SMTD).

	n	mean	s.d.	range
wing	12	231.4	5.19	224–240
tail	10	71.7	3.31	66–76
tarsus	11	66.7	3.18	62.0–71.7
mid-toe	10	38.9	1.38	36.5–40.8
mid-claw	11	20.4	1.47	18.7–22.2
bill (skull)	12	28.3	1.27	26.1–29.8
bill (nostril)	12	12.8	0.70	11.7–14.0
bill depth	5	8.3	0.38	7.9–8.7

Wing of population on Sangihe 225–237 mm (232.7 ± 6.66, n = 3), Siau 228–234 mm (230.8 ± 2.50, n = 4), Tahulandang 228–240 mm (234.0, n = 2), and Ruang 224–235 mm (229.0 ± 5.57, n = 3).

CHICK: wing 102–111 mm (106.2 ± 3.42, n = 5, *gilbertii*), 92–96 mm (93.5 ± 1.80, n = 3, nominate *cumingii*), 99 mm (n = 1, *dillwyni*), 113 mm (n = 1, *pusillus*). Tarsus 27.8–29.0 mm (28.2 ± 0.47, n = 5, *gilbertii*).

WEIGHTS
No information.

GEOGRAPHICAL VARIATION
Marked, both in colour and in size; no marked variation in relative size of wing, tail, leg, or bill between various populations, *M. cumingii* forming natural group within *Megapodius*. Variation in size appears to follow a stepped

cline, increasing from Sulawesi through Sabah, Palawan, Luzon and off-lying islands, and central Philippines to reach end in gigantic *tabon* of E Mindanao, gradually or suddenly smaller again from E Mindanao through the Sulu islands to Sabah and from E Mindanao through Talaud and Sangihe Is to Sulawesi. A cline of increasingly rufous pigments follows that of size, with rather olive population of Sulawesi showing rather restricted sandy-olive tinge, then colour gradually more extensive and more rufous-olive towards Luzon, then more deeply rusty rufous towards E Mindanao, Sangihe, and Talaud Is. Variation in depth of grey of head and body not parallel to other clines, but more or less an east-west cline, with darkest birds within Pacific reaches (especially Sangihe, Talaud, and E Mindanao), paler birds on islands bordering S China Sea.

M. c. gilbertii: from Sulawesi small, cinnamon-olive-brown on entire upperparts, dark or medium plumbeous-grey with faint olive tinge on underparts; occasionally, underparts strongly cinnamon (for example, in one bird from southeast peninsula), but variation individual, not geographical. Nominate *cumingii* from Sabah to Palawan has upperparts colder and purer greyish-olive-brown; rump to upper tail-coverts dark olive-grey, less brown, but some variation in colour, feather tips in some birds broadly olive-grey with dark grey of feather bases hidden, in others largely dark grey with olive-grey reduced; underparts medium plumbeous-grey, olive tinge faint or absent. On average, birds from Palawan slightly paler and larger than those from Sabah, but difference small. *M. c. gilbertii* often considered inseparable from nominate *cumingii* of Sabah, and difference in colour slight indeed, especially in worn plumage, but difference in size marked: wing of *gilbertii* 213 mm or less in 24 birds (220 mm or less in sample of 13 birds in Stresemann 1941), nominate *cumingii* 218 mm and over in ad (only two of 29 birds below 220 mm); also marked difference in other measurements, especially tail. *M. c. dillwyni* from Luzon, Mindoro, and Marinduque somewhat larger than nominate *cumingii* from Palawan, especially wing; colour rather different, upperparts extensively saturated drab-cinnamon, cinnamon-olive-brown, or rufous-yellow-brown, less dark than in *gilbertii*, less greyish than in nominate *cumingii*; underparts slightly paler medium grey, more distinctly washed olive- or cinnamon-brown, chin and throat light grey on feather bases. On Fuga and probably other northern islands, darker brown on upperparts than in typical *dillwyni* from Luzon and darker grey below, more like a large *gilbertii*, but provisionally included in *dillwyni*. In *dillwyni*, *pusillus*, and *tabon*, head and throat tend to be slightly less densely feathered, naked skin being more distinctly exposed.

M. c. tabon: from E Mindanao markedly large; upperparts strongly orange- or golden-rufous-brown or rufous-chestnut; cap brownish-black, side of head and neck and all underparts blackish-grey or dark plumbeous-grey, usually with some olive tinge visible on feather centres of breast and belly; shorter lesser upper wing-coverts often extensively grey; as in other races of *M. cumingii*, back to upper tail-coverts variable, either rufous like upperparts, or blackish-grey to blackish-brown, or mixture of rufous and black; *tabon* somewhat comparable with large black-and-rufous race *castanonotus* of Orange-footed Megapode *M. reinwardt*, but *tabon* more compactly build, less chestnut-brown above, mantle like remainder of upperparts (not dark grey), rump blackish (not maroon), and bare part colours less bright red and orange. *M. c. pusillus* as defined here a rather variable mixture between *dillwyni*, *tabon*, and probably nominate *cumingii*; perhaps to be split up further, but too few examined. Birds examined from Masbate, Cebu, Zamboanga peninsula (W Mindanao), and Basilan olivaceous rufous-brown on cap and upperparts, intermediate between *dillwyni* and *tabon*, more rufous (less olive) than nominate *cumingii*; rump variable, dark grey-brown, plumbeous-grey, blackish-brown, or a mixture of this; side of head and underparts dark plumbeous-grey with some rufous-brown or olive-brown tinge, darker than *dillwyni* or nominate *cumingii*, but less blackish than *tabon*. Size of birds from islands of Central Philippines apparently as on Luzon (but sample small), those of Zamboanga peninsula and Basilan larger, intermediate between *dillwyni* and *tabon*. Situation on Sulu islands and south to Maratua obscure; none examined and data of only three birds found in literature: type of *balukensis* Oberholser, 1924, from Balukbaluk (west of Basilan), type of *tolutilis* Bangs and Peters, 1927, from Maratua, and single bird from 'Sulu islands' in Blasius (1897). Judging from descriptions, Balukbaluk bird similar to Basilan population (rufous above, dark grey below) but small (though perhaps imm); 'Sulu' bird like *gilbertii*, but slightly paler (thus apparently near nominate *cumingii*), size rather small (but perhaps imm); Maratua bird darker and browner than nominate *cumingii* (thus apparently near *gilbertii*, but size rather larger, or near *pusillus*, but smaller); a single bird examined from Omadel (BMNH), south of Bum-Bum I. in E Sabah, similar in colour to nominate *cumingii*, but size rather small (though imm); perhaps a cline of decreasing size and decreasing depth of rufous and grey runs to southwest through Sulu islands from *pusillus* to nominate *cumingii*, while Maratua bird combines size of nominate *cumingii* with colour of *pusillus* or *gilbertii*; provisionally, Sulu and Maratua birds included in *pusillus*.

M. c. talautensis: distinct; plumage very dark, upperparts black-brown with maroon or chestnut tinge; crown black-brown, bordered dark grey at side and rear; rump to upper tail-coverts either brownish-black or black-brown mixed with some dark grey; underparts plumbeous-black, about as dark as *tabon* or slightly darker, feather centres more sooty black or faintly olive-black, olive tinge usually restricted to lower flank and thigh; lesser upper wing-coverts sometimes show much grey of bases; size large, as in *pusillus*. *M. c. sanghirensis* dark also, but paler than *talautensis* and distinctly smaller; upperparts slightly browner, chestnut-brown, less blackish or maroon; underparts paler, deep grey with olive tinge on belly, not as plumbeous-black as *tabon* or *talautensis*, more like underparts of *gilbertii* or *pusillus*, but upperparts distinctly darker than these latter two. In worn plumage, both *talautensis* and *sanghirensis* are duller and browner, more tinged with olive; palest birds of *sanghirensis* then hardly darker above than *gilbertii* in fresh plumage (for example, one from Ruang), but size distinctly larger.

Note: *M. cumingii* differs from all non-black *Megapodius* except Nicobar Megapode *M. nicobariensis*, Tanimbar Megapode *M. tenimberensis*, and Sula Megapode *M. bernsteinii* in upper mantle being of same colour as lower mantle and scapulars, rather than having a contrasting grey collar across the upper mantle. In relative proportions of extremities, it seems nearest to the New Guinea Megapode *M. decollatus*, Melanesian Megapode *M. eremita*, *M. bernsteinii*, *M. tenimberensis*, and Dusky Megapode *M. freycinet*.

Range and status

M. c. gilbertii: Sulawesi (N, central, and SE) and neighbouring islands of Togian group (Gulf of Tomini), Talisei, Tendile, and Lembeh (just off tip of north peninsula). *M. c. cumingii*: small islands off Sabah, from Omadel (near Bum-Bum, in east), via Malawali, Banggi, Balambangan, and Mantanani (in north), to Selingaan, Poffan, Gaya, Sapi, Manukan, Tiga, Kuruman, and Labuan (in west); on mainland of Sabah, mounds or birds recorded from Kudat peninsula (Indarasan, Kaniang R.), Maruda Bay coast (Sibayan, Matunggong R.) (Appell 1965), and Sandakan area (Ryves 1955); also (in SW Philippines) on Balabac, Cagayan Sulu, Ursula, Palawan, and (perhaps this race) Calauit, Tamalpulan, and Tanobon (in Calamian group). *M. c. dillwyni*: Luzon (but not certain whether this race in southeast), Marinduque, Mindoro, and northern islands of Palaui, Camiguin Norte, Fuga, Calayan, and Batan (at 20°26′N, rather near Taiwan); either this race or *pusillus* on Catanduanes, Ticao, Cresta de Gallo, Sibuyan, Romblon, and Tablas. *M. c. pusillus*: Masbate, Cebu, W Mindanao (Zamboanga peninsula, but probably west from *c*. 124°E), Basilan, and (probably this race) Samar, Leyte, Gigantes, Bantayan, Negros, Bohol, Dinagat, and Siargao; apparent intermediates between *pusillus* and nominate *cumingii* occur on Sulu Is (Balukbaluk, E Bolod, Jolo, Tara, Tawitawi, Manuk Manta, Sibutu, Tumindao, and Tres Islas) and (perhaps this race) on Maratua (and perhaps other islands off

E Borneo). *M. c. tabon*: E Mindanao, east of *pusillus*. *M. c. talautensis*: Talaud Is (Karakelong, Salebabu, and Kaburuang). *M. c. sanghirensis*: Sangihe Is (Sangihe, Siau, Tahulandang, and Ruang).

In Philippines, apparently not yet recorded from larger island of Panay or from medium-sized islands of Dumaran, Culion, Busuanga, Lubang, Burias, Biliran, Guimaras, Siquijor, and Camiguin; in Indonesia not from Bunguran or Natuna Is.

Status: *M. c. gilbertii*. Probably widespread, but locally scattered and secretive, in almost all parts of Sulawesi that are not actually deforested (MacKinnon 1978). Quite common in north-central Sulawesi (Watling 1983*a*).

M. c. cumingii: rare or restricted in range or locally extirpated on many islands. Still very common on Pulau Tiga, a small island of 7.5 km^2 off Sabah, with 212 birds per km^2 estimated from transect counts under the assumption that the species is evenly distributed over the island (Stuebing and Zazuli 1986). No recent observations of birds or mounds from Labuan, where already uncommon at end of nineteenth century. Locally common on Palawan; estimated population on Tabon (24 ha), off the east coast of Palawan, *c.* 100 birds in 1981 (W. Suter, *in litt.*).

M. cumingii ssp.: rare on the Philippine islands of Romblon, Tablas, and Sibuyan, though common on Cresta de Gallo, a small island off the coast of Sibuyan (McGregor 1905*a*). Reported as 'not really rare' on numerous islands in the Philippines, such as Siargao and Dinagat (DuPont and Rabor 1973).

Field characters

Length 31–44 cm. The only *Megapodius* sp. in its range. Upperparts brown; underparts, neck and head grey. Bare skin of head red; legs and feet dark. For comparison of *gilbertii* with sympatric Maleo *Macrocephalon maleo*, see that species.

Voice

Probably no duet, in contrast to, for example *M. bernsteinii*, *M. freycinet*, and *M. eremita*.

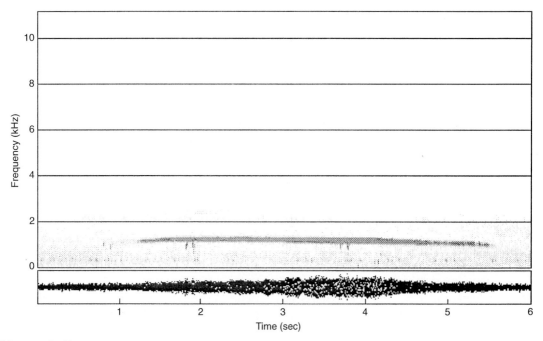

M. c. cumingii
J. Scharringa, Palawan, Philippines, Apr 1985.

Long, mournful whistle (see sonogram), probably produced by the ♂, like an air-raid alarm, lasting 4–5 sec, initially rising and subsequently descending in pitch 'ky-ouououououououououououou' (description from tape-recordings from Palawan made by J. Scharringa). Difficult to locate due to its indistinct and drawn-out beginning and ending. Regularly calls at night as do other *Megapodius* spp. Also described as 'mow', like a cat in distress (on Palawan and Sabah).

Habitat and general habits

In N and SE Sulawesi in hill and mountain forest at 250–2000 m (Stresemann 1941), in woods near the coast (for example, in Tangkoko-Batuangus Nature Reserve) and on small offshore islets (for example Pulau Molossing, off the north coast near Lolak: R. Dekker, personal observation). Most reports of the nesting of this species in Sabah refer to it as breeding on islands only, where the birds are common close to the sea and never seem to go far inland. However, several mounds have been reported from the mainland of Sabah, for instance in a small area of coastal jungle surrounded by mangrove swamps (Ryves 1955; Appell 1965; see 'Range and status'). In the Philippines in coastal scrub on small islands as well as in montane dipterocarp forests. At 2100 m on Mt. Apo (Hachisuka 1931–1932). On Palawan close to the sea.

Secretive; away from their mounds more often heard than seen. Very vocal at night.

Food

Includes worms, grubs, termites (Isoptera), seeds, and fruits. The crop of one specimen contained beetles.

Displays and breeding behaviour

Breeding behaviour not described. On Sulawesi (*gilbertii*), eggs often embedded against or in decaying roots of trees at a depth of approximately 30 cm. At each of two such nesting-sites, only one bird (a ♀) was seen (Stresemann 1941), suggesting that the ♀ is not accompanied by the ♂ while egg laying and that only one pair makes use of a particular tree. The ♀ or the ♂ also visits the tree on days when no eggs are laid (R. Dekker, personal observation). On one occasion, the whole process of egg laying took 2 h (MacKinnon 1978). In contrast to *gilbertii*, nominate *cumingii* builds communal mounds on Palawan and islands off Sabah. On Tabon off the east coast of Palawan a maximum of 14 birds of undetermined sex visited a large mound on a single day (W. Suter, *in litt.*). On Pulau Tiga, 24 of 88 birds (27%) encountered during surveys were paired, suggesting monogamy. Mounds are usually built near the seashore and used year after year; the temperature of the egg chamber of four mounds was constant over a certain period and measured 33–34 °C (Stuebing and Zazuli 1986).

Breeding

NEST: buries its eggs in mounds and between decaying roots of trees. Mounds are frequently formed round the foot of a large tree or dead stump. On islands, they are often situated near the seashore and largely consist of sand mixed with shells and sticks. On Pulau Tiga, 88% of 41 mounds were located within 50 m of the beach in the forest, while 98% were within 110 m; only one was found well into the interior of the island. The height of these 41 mounds ranged from 0.16 to 1.21 m (average 0.69 m), with a circumference ranging from 4.2 to 28.6 m (average 22.2 m); the distance between them was usually greater than 100 m (Stuebing and Zazuli 1986). Other reports mention mounds up to 1.8 m in height and 6–7 m in diameter (McGregor 1905a). In Sabah, mounds measure approximately 0.9 m in height and 18 m in outside circumference (Ryves 1955). Large communal mounds up to 1.65 m high have been reported from the Sulu islands. In Sulawesi, eggs are said to be laid in mounds, although recent reports and personal observations indicate that they are more commonly laid between decaying roots of trees.

EGGS: similar in shape and coloration to eggs of *Macrocephalon maleo*, *Eulipoa wallacei*, and other *Megapodius* species (see under *M. pritchardii*). SIZE: *M. c. gilbertii*. From Sulawesi: 76–82 × 48.6–50 mm (80 × 49.5, n = 4, Schönwetter 1961); 79–83 × 49.4–52.8 mm (81.1 ± 1.53 × 50.6 ± 1.30, n = 7, Meyer and Wiglesworth 1898; Riley 1924; RMNH; R. Dekker, unpublished data). *M. c. cumingii*: 71.9–80.1 × 45.9– 48.4 mm (76.5 × 47.1, n = 8, Sabah and Palawan, Steadman 1991); 74–84 × 46–51 mm (78.7 × 48.7, n = 12, Schönwetter 1961); 72.1–75.7 × 45.5–47.9 mm (74.4 × 46.5, n = 3, Pulau Tiga and Megalum, Steadman 1991); 78 × 47 mm (n = 1, Labuan, Smythies 1981); 79.3 × 48.9 mm (n = ?, Sabah, Ryves 1955). *M. c. dillwyni*: 76–87 × 45–52 mm (81.0 ± 4.32 × 48.6 ± 2.76, n = 7, Luzon, Oustalet 1881); 81.6 × 48.5 mm (n = 1, Luzon, Steadman 1991). *M. c. talautensis*: 78.7 ± 2.08 × 48.2 ± 0.40 mm *(n* = 3, Karakelong, Meyer and Wiglesworth 1898). *M. c. sanghirensis*: 77.5–78.7 × 45–48.2 mm (78.0 × 46.9, n = 4, Schönwetter 1961). *M. cumingii* ssp: 73.6–89.5 × 43.9–54.1 mm (80.2 × 48.9, n = 44, Mindanao, Steadman 1991); 78.0 × 51.0, 85.5 × 52.5 mm (n = 2, Mindanao, Schönwetter 1961); 75.9–85.0 × 46.6–50.8 mm (80.3 × 48.8, n = 7, Philippines, Steadman 1991); 78–82.5 × 47–52.5 mm (80.3 ± 1.92 × 50.7 ± 2.17, n = 5, Philippines, Hachisuka 1931–1932.) WEIGHTS: 110 and 114 g (n = 2, N Sulawesi, R. Dekker, unpublished data).

BREEDING SEASON: in Sulawesi eggs recorded in Nov, Dec, Mar, and May. In Sabah eggs are said to be found practically all year round. Other data indicate seasonality: on Pulau Tiga, 90% of 41 mounds were active from Aug through Oct, while the proportion dropped to 73% during Dec and Jan, at the onset of the rainy season (Stuebing and Zazuli 1986). On Palawan, eggs in June, July, and Aug, on Tanobon and Cagayan in Dec, on Fuga in May (Dickinson *et al*. 1991). Two chicks in the collection of the RMNH from N Sulawesi from Apr and Sept, another from C Sulawesi from July. Incubation period depending upon temperature. One fresh egg found between decaying roots hatched after 63 days in a Maleo hatchery (see Chapter 9) in volcanic soil.

Local names
Tabon, tavon, tabun, or taboun (Philippines, general), ou-cong (Calayan, Camiguin Norte, in far north), ayam tambun (Sabah), menambum (small islands off Sabah), moleo-kitjil, maleo utan, kokokoren, or kokotjoden (N Sulawesi), eoa (Kaburuang, Talaud), keoe (Sangihe), moyo (Sulawesi).

References
Low (1851), Sharpe (1875, 1877), Tweeddale (1877), Meyer (1879), Schlegel (1880*b*), Oustalet (1881), Guillemard (1886), Blasius (1888, 1896, 1897), Whitehead (1888), Ogilvie-Grant (1893, 1897), Meyer and Wiglesworth (1898), Vorderman (1898*a*), McGregor (1905*a*), Oberholser (1924), Riley (1924), Bangs and Peters (1927), Coomans de Ruiter (1930), Hachisuka (1931–1932), Ripley (1941), Stresemann (1941), Hoogerwerf (1949), Ryves (1955), Schönwetter (1961), Appell (1965), DuPont and Rabor (1973), MacKinnon (1978), Smythies (1981), Watling (1983*a*), Stuebing and Zazuli (1986), White and Bruce (1986), Holmes (1989), Dickinson *et al*. (1991), Steadman (1991), Roselaar (1994).

Sula Megapode *Megapodius bernsteinii* Schlegel, 1866

Megapodius Bernsteinii Schlegel, 1866. *Nederlandsch Tijdschrift voor de Dierkunde*, **3**, p. 251. (Sula Is; type from Mangole.)

PLATE 6

Other names: Sula Scrubfowl.

Monotypic. Synonym: *perrufus* Neumann, 1939, Peleng (Banggai Is).

Description
PLUMAGES
ADULT: sexes similar. Forehead, crown, and very short slightly pointed crest on nape medium grey with olive or brown tinge, often somewhat warmer rufous-brown on central crown, duller greyish olive-brown on forehead, along side of rear crown, and at side and tip of crest. Lore covered with scanty short bristle-like feathers, rather restricted patch round eye bare; lower cheek, chin, throat, and area round ear covered with short dull brown-grey feathers, bare skin just visible. Neck dark greyish-brown, merging into dark greyish olive-brown on upper mantle, side of breast, and chest. Lower mantle and scapulars gradually warmer rufous olive-brown towards rear, merging into rufous-brown on upper wing-coverts and deep rufous-brown or chestnut-brown on tertials. Back, rump, and

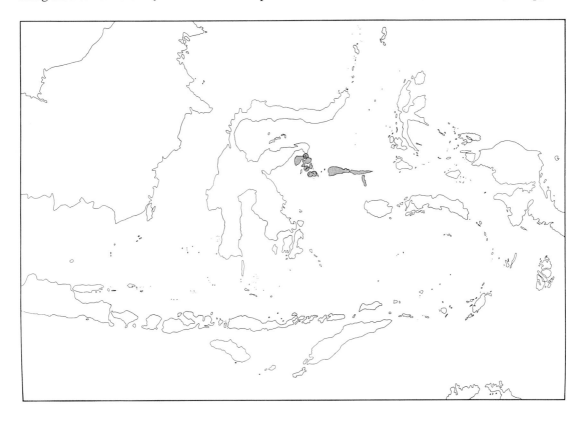

Sula Megapode *Megapodius bernsteinii*

upper tail-coverts variable, either blackish or dark rufous-brown, often contrastingly darker and duller than tertials; when rufous, less bright rufous-chestnut than, for example, rump of Orange-footed Megapode *M. reinwardt*. Underparts rufous-brown, tinged olive on breast, more rufous on flank and side of belly; thigh greyish-brown, under tail-coverts dark or sooty rufous-brown. Tail blackish chocolate-brown. Outer webs of flight feathers and greater upper primary coverts rufous cinnamon-brown, inner webs dark greyish-brown; bases of inner secondaries gradually darker rufous-brown to chestnut-brown towards tertials. Under wing-coverts and axillaries olive-brown, shorter coverts rufous-brown. Some influence of bleaching and wear: in fresh plumage, front part of body warm olive-brown, gradually shading to bright rufous-brown on rear body and chestnut-brown on tertials; if worn, general tinge of body more greyish-olive, less rufous.

CHICK: no information.

IMMATURE: only one in first imm plumage examined: colours markedly less saturated than in ad, less rufous or olive, more buffish; outer scapulars and some wing-coverts with trace of pale buff tip or subterminal bar; underparts pallid buffish-grey; appearance of body colour rather like the Tanimbar Megapode *M. tenimberensis* or Nicobar Megapode *M. nicobariensis*; outer primaries still juv, relatively short, p10 with pointed tip; length of p10 to insertion in skin 79–81 mm ($n = 2$, in ad 97–118 mm, average 103.8 ± 6.13, $n = 14$); tail often still juv, short, 54 mm ($n = 1$). In second imm plumage, like ad, but outer primaries from second generation, inner from third, innermost sometimes fourth; p10 relatively shorter than in ad, tip slightly more pointed, length to insertion in skin 87–104 mm (97.2 ± 6.24, $n = 6$), tail 52–56 mm (54.3 ± 1.44, $n = 5$).

BARE PARTS
ADULT, IMMATURE: iris brown or chestnut. Bare skin round eye brown or greyish-black. Bill horn-colour, brown, fuscous-brown, or reddish-black, base in some birds black. Leg and foot red-brown, red, or bright-red in ad, but blackish in imm, scutes at upper surface of toes in some birds dark horn, grading to horn-black near tips of toes.

CHICK: no information.

MOULTS
Primary moult serially descendant. Of birds examined, only two were younger imm: a ♂ from Nov with p1–p5 of second generation, p6–p8 growing, and p9–p10 of first generation (juv); one undated ♀ had outer three primaries (p8–p10) and probably tail still juv, body a mixture of first and second imm plumage (second and third generation of feathering), p3–p7 of second generation (p7 still growing), p1–p2 of third (both growing).

Remaining 15 birds examined all older imm or ad with two moult series in each set of primaries, except for one undated bird which had p1, p4, and p9–p10 growing and thus had three active moulting series. Of these 15 birds, only two without collecting data, but others all from restricted period of Aug–Dec and, hence, little information about timing of moult in population; none of the birds had moult recently started, all having an inner moult series reaching to (p3–) p4–p6 and outer series reaching to p9 or p10. All three birds examined from Aug were in active primary moult; of three from Sept and three from Oct, one in each month had primary moult suspended; in Nov and early Dec moult suspended in three of four birds.

MEASUREMENTS
ADULT: Sula and Banggai Is; skins of 14 ♂♂, 5 ♀♀ and 2 unsexed birds (BMNH, MZB, RMNH, SMTD, ZMA).

	n	mean	s.d.	range
wing	21	197.0	6.95	186–210
tail	17	60.5	3.25	54–68
tarsus	20	61.1	2.87	57.0–66.2
mid-toe	14	35.9	1.48	33.3–37.8
mid-claw	17	18.3	0.85	17.2–19.4
bill (skull)	16	26.3	0.93	24.9–28.1
bill (nostril)	18	11.9	0.65	10.8–13.1
bill depth	10	8.1	0.43	7.3–8.6

Wing. Banggai group: Peleng, 203–210 mm (205.5 ± 3.19, *n* = 4). Sula group: Taliabu, 187–197 mm (193.1 ± 3.83, *n* = 5); Mangole, 190–199 mm (194.1 ± 2.53, *n* = 8); Sanana (Sula-Besi), 206–210 mm (208.0, *n* = 2). On Peleng, 195–206 mm (*n* = 8); Taliabu, 182–188 mm (*n* = 14); Mangole, 185–193 mm (*n* = 8) (Neumann 1939).

CHICK: no information.

WEIGHTS
No information.

GEOGRAPHICAL VARIATION
Rather slight. Neumann (1939) separated birds from Peleng (Banggai Is) from typical *bernsteinii* from Sula Is as '*perrufus*', being larger, and deeper and more uniform rufous. Part of Neumann's original series examined for present work (SMTD), and indeed one of three Peleng birds markedly deep and extensive rufous on upperparts, as well as extensively dark rufous on belly and flank when compared with four birds from Taliabu (Sula Is), which latter were mainly dull greyish-olive; however, two other birds from Peleng were more or less intermediate between these, as were four from Mangole (Sula Is), though these latter were rather olive on underparts. In Leiden (RMNH), Amsterdam (ZMA), and Tring (BMNH), small series of birds from Mangole and Sanana showed more or less same extent of variation in colour, independent of locality, and apparently colour differences mainly caused by abrasion, colour being deeper and more extensively rufous when plumage freshly moulted, more olive-grey when worn. In size, birds from Taliabu and Mangole in western and central Sula Is smaller indeed than birds from Peleng in Banggai Is, further west, but birds from Sanana at the eastern end of the Banggai–Sula chain as large and almost as rufous as birds from Peleng on the western end of the chain. As it seems impractical to join the populations at the opposite ends of the geographical range of *M. bernsteinii* into a single larger-sized race, with a smaller nominate race in between, '*perrufus*' not recognized here.
  Note: structurally closest to the Philippine Megapode *M. cumingii*; in size, near *M. c. gilbertii*, but in view of aberrant colour (upperpart colours extending to underparts; intensifying of rufous tinge towards rear of body) and red legs considered a separate species.

Range and status
Probably on all islands in the Sula and Banggai Archipelago (between Sulawesi and the Moluccas). Older records from Mangole (collected prior to 1863, in 1864, 1897), Sanana (collected in 1864 and prior to 1876), Peleng (eight specimens collected in 1938), and Taliabu (14 specimens collected in 1938). More recently reported from Taliabu, Seho, and Mangole (Davidson *et al.* 1994), Peleng, Bangkalan, Banggai, Labobo, and several smaller islands south of Banggai near Salue (Indrawan *et al.* 1992).
  Status: vulnerable. However, probably still widespread and locally common on many islands (Indrawan *et al.* 1992; Davidson *et al.* 1994).

Field characters
Endemic to the Sula and Banggai Is, where it is the only megapode. Medium-sized, ground-living bird which appears tail-less. Plumage entirely rufous-brown, legs red. Betrays presence by its noisy behaviour while walking or scratching for food among dry leaves.

Voice
Duet (see sonogram) initiated (probably by the ♀) with two notes 'keyou' 'keyouw', both rising in pitch, and with the second note descending at the end and lasting slightly longer than the first. The partner joins in on the first or second note with a single 'weyou-prrrrrouroutou-tou-tou', initially rising in pitch but with the second part maintained at one level, slightly lower than the first. (Description from tape-recordings of University of East Anglia Taliabu Expedition 1991.) Similar in structure to duet of Dusky Megapode *M. freycinet* and Forsten's Megapode *M. forstenii*, in which species it seems that the ♂ begins the duetting. It cannot therefore be excluded that the ♂ may sometimes initiate the duet in *M. bernsteinii* also.

Sula Megapode *Megapodius bernsteinii*

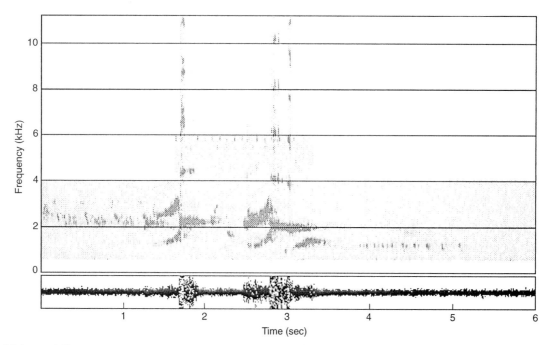

M. bernsteinii
University of East Anglia Taliabu Expedition, Taliabu, Indonesia, Sept–Oct 1991.

Regularly calls at night, as do other *Megapodius* spp.

Habitat and general habits
In all forest types including primary lowland forest, bamboo-dominated forest, dry coastal scrub, as well as secondary vegetation at lower altitudes. Recorded up to 450 m altitude at Mt. Bungkuko, Labobo I. Observed as single birds and in groups of up to three or five individuals (Indrawan *et al.* 1992; Davidson *et al.* 1994). Regular observations of two birds working together at mound suggest pair-bond and monogamy (M. Indrawan, personal communication).

Food
Probably omnivorous, similar to other megapodes.

Displays and breeding behaviour
Mounds visited almost daily (M. Indrawan, personal communication), which indicates that mounds are maintained or used by more than one pair.

Breeding
NEST: mound, sometimes mainly consisting of sand, found in various habitats, including cultivated areas.

EGGS: similar in shape and coloration to eggs of *Macrocephalon maleo*, *Eulipoa wallacei*, and other *Megapodius* spp. (see under *M. pritchardii*). SIZE: no information. WEIGHT: no information.

BREEDING SEASON: eggs and a chick recorded in Dec (Indrawan *et al.* 1992), which indicates that eggs are laid at least between Oct and Dec.

Local names
Kaelong (Peleng, Banggai and Labobo), lambeta (Labobo), tambun (Bowokan Is, south of Banggai).

References
Schlegel (1866*a*, 1880*a*), Oustalet (1881), Ogilvie-Grant (1893, 1897), Hartert (1898*c*), Neumann (1939), Eck (1976), White and Bruce (1986), Holmes (1989), Yong (1990), Indrawan *et al.* (1992), Davidson *et al.* (1994).

Tanimbar Megapode *Megapodius tenimberensis* Sclater, 1883

Megapodius tenimberensis P. L. Sclater, 1883. *Proceedings of the Zoological Society of London*, 1883, p. 57. (Lutu and Kirimun, Tanimbar Is.)

PLATE 8

Monotypic.

Description
PLUMAGES

ADULT MALE: a large, pale, diluted-coloured megapode. Forehead, crown, upperparts from lower mantle to upper tail-coverts, and upper wing pale brownish-olive, near colour of upperparts of the Philippine Megapode *M. cumingii* of N Borneo, or the Nicobar Megapode *M. nicobariensis*, but slightly more olive and colour less saturated. Nape with very short broad or slightly pointed crest. Feathered forehead, crown, and nape contrast with mostly naked rest of head and neck, which shows scanty reduced grey feathers on chin and neck only, as well as some bristles on lore. Hindneck medium grey; upper mantle pale brownish-olive, like lower mantle and scapulars, but some medium grey of feather bases visible; no sharp boundary between greyish-olive upper mantle and brownish-olive lower mantle. Rump to upper tail-coverts medium grey, feather tips suffused olive to variable extent. Chin, throat, and front and side of neck medium ash-grey, slightly darker than pale ash-grey of *M. nicobariensis*, remainder of underparts and underwing pale ash-grey with extensive pale olive tinge, like *M. nicobariensis*, but vent somewhat paler still, pallid greyish-olive. Flight feathers blackish-grey, outer webs of flight feathers suffused olive-brown; tail brownish-olive.

ADULT FEMALE: none examined, but probably similar to ♂.

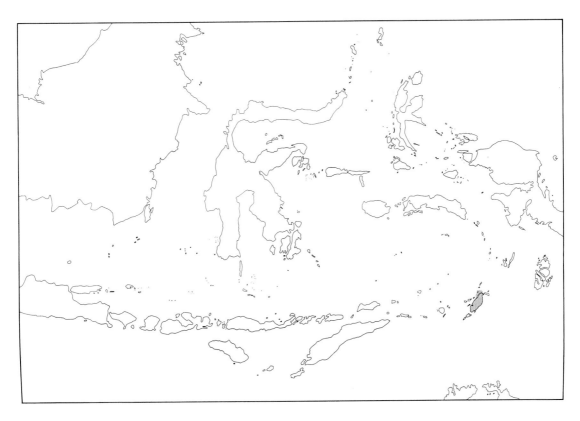

Tanimbar Megapode *Megapodius tenimberensis*

CHICK: none examined.

IMMATURE: none examined; p10 undoubtedly shorter and more pointed at tip than in ad (length of ad p10 to insertion in skin 136–144 mm, $n = 2$).

BARE PARTS
ADULT: iris dark brown. Bare skin of head reddish. Bill tip pale yellow, basal half of upper mandible dusky, of lower mandible greyish-horn. Front of tarsus black, but red on tibio-tarsal joint; back of tarsus red; foot black (Sclater 1883), in other specimen legs and feet reported as being red (Ogilvie-Grant 1893).

MOULTS
Only three ad specimens examined for moult of primaries: in one from Sept two moult series in each wing, of which outer active (p7 growing) and inner suspended (p1 new), p2–p3 and p8–p9 old (BMNH); both other birds, without collecting date, each with three series in each wing: in one, p1–p2 and p8 growing, p10 just completed, p3–p4 and p9 old; in other p1, p4–p5, and p10 growing, p2–p3, p6–p7, and p9 old (SMTD).

MEASUREMENTS
ADULT: skins of 2 ♂♂ and 2 unsexed birds (BMNH, SMTD).

	n	mean	s.d.	range
wing	4	255.0	7.26	246–263
tail	4	88.5	5.45	81–94
tarsus	4	73.0	3.26	70.2–76.8
mid-toe	4	40.5	1.76	38.2–42.3
mid-claw	4	22.8	0.91	21.5–23.6
bill (skull)	4	30.4	1.19	29.4–31.8
bill (nostril)	4	13.6	0.93	12.4–14.5
bill depth	4	8.8	0.68	7.8–9.2

WEIGHTS
No information.

GEOGRAPHICAL VARIATION
None reported.

NOTE
At first sight surprisingly similar to *M. nicobariensis*, being olive above (though less saturated and buffish than *M. nicobariensis*) and pale grey below. It differs from *M. nicobariensis* in not showing a pale grey line from above eye to nape and by proportionally longer tail and somewhat shorter tarsus and toes; in proportions it is close to *M. cumingii* and the Dusky Megapode *M. freycinet*. The geographical range of *M. tenimberensis* is entirely surrounded by the Orange-footed Megapode *M. reinwardt*, and therefore it is sometimes considered to be a race of that species (for example, in White and Bruce 1986), but, while various populations of *M. reinwardt* differ mainly in size and hardly in colour, *M. tenimberensis* is very different: it cannot be considered to form a diluted-coloured race of *M. reinwardt* as it does not show, for instance, the long pointed crest, long tail, and relatively short tarsus, toes, and claws of that species. *M. tenimberensis* probably has a relict distribution; when *M. reinwardt* expanded its range, in part as a colonist reaching uninhabited small islands, in part wiping out or swamping existing populations of small-sized megapodes in areas it reached (as may have happened on Vogelkop peninsula), it did not manage to get a foothold on the Tanimbar Is, which were already inhabited by the powerful *M. tenimberensis*, which is with the Vanuatu Megapode *M. layardi* the largest of all *Megapodius* spp. except for Australian races of *M. reinwardt*.

Range and status
Restricted to the Tanimbar Is. In Dec 1991, regularly recorded and five or six mounds seen along a trail over a distance of 10–12 km in the southeast part of Tanimbar (K. Monk, *in litt.*).

Field characters
Unmistakable. Length *c*. 43 cm. The only megapode on the Tanimbar Is. Generally brownish-olive and grey with reddish legs; short crest.

Voice
No information.

Habitat and general habits
The above-mentioned observations were all in primary forest, 4 km or more inland. No information is available on general habits, food, and breeding behaviour, but likely to be similar to other *Megapodius* spp.

Breeding
NEST: mound, 1.5–2.0 m high (K. Monk, *in litt.*).

EGGS: similar in shape and coloration to eggs of *Macrocephalon maleo*, *Eulipoa wallacei*, and other *Megapodius* species (see under *M. pritchardii*). SIZE: 78.9 × 49.1; 78.8 × 49.9 mm ($n = 2$, RMNH). Note: both eggs are small for a *Megapodius* of this size. Since they were obtained from the local market on Tanimbar, it cannot be excluded that they were imported from elsewhere and belong to another species; for example, the Moluccan Megapode *Eulipoa wallacei*. WEIGHTS: no information.

BREEDING SEASON: no information.

Local names
None reported.

References
Oustalet (1881), Sclater (1883), Salvadori (1891), Ogilvie-Grant (1893, 1897), White and Bruce (1986).

Dusky Megapode   *Megapodius freycinet* Gaimard, 1823

Megapodius Freycinet Gaimard, 1823. *Bulletin Général et Universel des Annonces et de Nouvelles Scientifiques*, **2**, p. 451. (Waigeu I.)

PLATE 7

Other names: Dusky Scrubfowl, Sooty Scrubfowl.

Polytypic. Three subspecies. *Megapodius freycinet freycinet* Gaimard, 1823, Waigeu and small islands around it, Gag, Gebe, Kofiau, and Misol. (Synonym: *Alecthelia urvillii* Lesson, 1826, Gebe; *alecthelia* Reichenbach, 1862.) *Megapodius freycinet oustaleti* Roselaar, 1994, Batanta, Salawati, and small islands off W Vogelkop peninsula (New Guinea). *Megapodius freycinet quoyii* G. R. Gray, 1861, Morotai, Rau, Ternate, Tidore, Mare, Kajoa, Halmahera, Bacan, and islands of Obi group; unknown what race inhabits Tifore and Batang Kecil.

Description
PLUMAGES
ADULT: sexes similar. Entirely deep black; feathers of crown, mantle, scapulars, lesser and median upper wing-coverts, and entire underparts fringed dark plumbeous, giving crown and body slight bluish-grey tinge; somewhat duller and more sooty black on chin, throat, round neck, and on flight feathers; dull sooty black feather centres partly visible on underparts; vent and feathering of thigh dark sooty-grey to dull black. Forehead and lore bare, except for short and scanty black bristle-like feathers; skin round eye fully bare; area round ear and lower cheek covered with rather scanty short and narrow black feathers; chin, throat, and all foreneck either entirely covered with short black feathers, leaving skin virtually invisible, or (perhaps in some populations only) feathers more scanty, sooty-grey. Feathers of nape slightly elongated, forming short broad crest, which ends in slightly pointed tip (latter

Dusky Megapode *Megapodius freycinet*

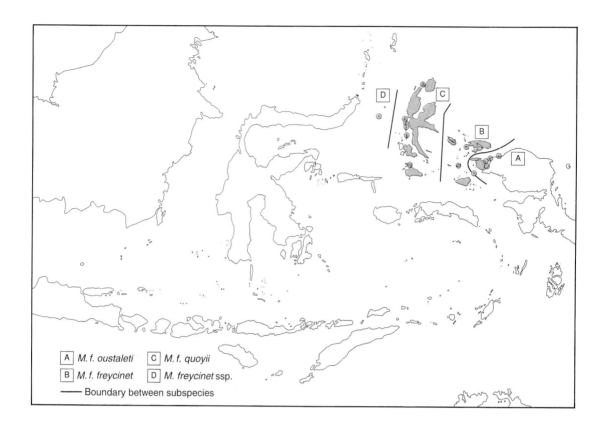

A — *M. f. oustaleti*
B — *M. f. freycinet*
C — *M. f. quoyii*
D — *M. freycinet* ssp.
—— Boundary between subspecies

more distinct in some populations, virtually absent in others). In worn plumage, black duller and slightly brownish, plumbeous feather fringes worn off; lower scapulars and tertials sometimes slightly tinged olive-green.

CHICK: central forehead, crown, hindneck, and upper mantle sooty fuscous-brown, merging into ochre-brown on side of forehead and round eye and this in turn merging into buffish- or isabelline-ochre on lower cheek, chin, and throat. Lower mantle, scapulars, lesser and median upper wing-coverts and tertials black, a variable number of feathers rather narrowly but usually contrastingly barred rufous-cinnamon (if fresh) or buff (if worn). Rump to upper tail-coverts variable, fuscous-black to rufous-chestnut or mixture of this. Chest, flank, and belly buff-brown with slight olive-grey tinge, merging into isabelline-buff on vent. Primaries and primary coverts uniform greyish-black; secondaries and greater upper wing-coverts black with cinnamon to pale fringe along tips and with some traces of rufous bars subterminally on outer webs. Second generation of primaries starts to grow from innermost primary (p1) onwards when juv (first generation) wing-tip (p5–p8) not fully grown, juv p9–p10 hardly started to grow, and chick still small; at the same time, barred feathers of mantle, scapulars, and longer tertials replaced by uniform black feathers. Extent of barring on upperparts apparently not dependent on locality; small chicks with extensive barring recorded from Morotai, Halmahera, Ternate, Bacan, Obi, and Misol, those with bars restricted to some scapulars and wing-coverts from Halmahera, Ternate, Bacan, and Misol. Chicks with extensive barring and chestnut-brown rump rather similar to chick of Moluccan Megapode *Eulipoa wallacei*, but general colour of *E. wallacei* browner, less sooty, rump brighter, more cinnamon, under wing-coverts white, not brown-grey, and flight feathers more strongly patterned; see also *E. wallacei* species account.

IMMATURE: in first imm plumage, like ad, but black of body slightly duller, plumbeous feather-fringes present but less distinct (absent when plumage worn), and forehead, lore, cheek, and area round ear more fully feathered. Outermost primaries still juv, relatively much shorter than second generation primaries (if not yet replaced by third generation, second generation p1 also relatively short); tip of p10 tapering to narrow and sharply pointed but often somewhat frayed tip; length of p10 from tip to insertion in skin 90–110 mm (97.0 ± 7.87, $n = 5$), in ad 120–138 mm (129.1 ± 5.25, $n = 15$); some secondaries still juv, short, dark brown with rufous fringes or marks; tail sometimes still juv, 52–65 mm (60.2 ± 6.71, $n = 3$). In second imm plumage, outermost primaries are from second generation, relatively shorter than neighbouring third generation and (if present, on innermost primaries) fourth generation; tip of p10 more pointed than in ad, length 105–122 mm (111.9 ± 6.11, $n = 8$); tail 73–82 mm (76.1 ± 3.28, $n = 8$; in ad, mainly 79–87 mm, occasionally 74–93 mm). No small-sized birds of Salawati population included in data above; p10 in ad 118 mm ($n = 1$), in second imm 106 mm ($n = 1$), tail in second imm 64 mm ($n = 1$).

BARE PARTS

ADULT, IMMATURE: iris brown, dark brown, chestnut, orange-brown, or (once, Waigeu) red. Bare skin of head and neck dull red, dark blood-red, or dark vinous-red at side of head and dark flesh-colour on throat; skin at base of bill black. Bill dark horn-brown, black-brown, olive-brown, blackish-olive, or black, gradually paler distally, tip and cutting edges yellow, yellowish-horn, yellow-brown, olive, ochre, or pale brown. Lower tibia and tarsus dark olive-green, dark greenish-ochre, olive-brown, olive-black, greyish-black, or black, occasionally with some red at joint with toes (once, Halmahera), front of tarsus sometimes darker than side and rear; toes dark slate, olive-black, or black; occasionally, legs yellow (twice, Waigeu, but over 10 others from Waigeu black). Unknown whether bare part colours depend on sexual activity, sex, age, or locality; tarsus recorded as (mainly) olive in birds from Morotai, Ternate, Halmahera, Waigeu, Misol, Kamuai (Schildpad Is), and Salawati, among which are ad and imm of both sexes, (mainly) black in birds from Ternate, Halmahera, Bacan, Waigeu, and Sorong I. which were mostly recorded as ad; variation perhaps seasonal.

CHICK: iris grey-brown or dark brown. Bill yellowish-green with horn tip, greyish, olive-black, or black. Leg and foot grey-green, greyish, or black.

MOULTS

Primary moult serially descendant. In chick, moult to first imm plumage starts with shedding of innermost primary (p1) when juv wing tip (p5–p8) still growing and juv outermost primaries (p9 and p10) have just started to grow. When this first moult series of second generation primaries reaches p6–p8 (on average p7.1, $n = 8$) and (now fully grown) outer primaries are still juv (first generation), a second moult series starts with shedding of p1; thus, two moult series active then. When first series has completed with regrowth of p10 and second series active half-way on wing, a third series starts with p1, fourth generation of primaries appearing; this situation continued in ad, where a moult series is followed by a next one when p5–p8 (on average p6.3 ± 0.93, $n = 24$) is reached. Unlike some other megapodes, not more than two series active in each set of primaries in this species, apparently because moulting season is long or moult is rather rapid; thus, always either two or three generations of primaries in each wing (two when one series just completed with regrowth of p10 and a second series active at about p4; three when series active at, for example, p2 and p8), and in each series innermost primary relatively oldest, outermost new or growing. During adverse feeding conditions or when sexually active, moult of both series may temporarily be suspended, moult halted when primary fully grown but next old one not shed, to be resumed again in both series when situation favourable again. Suspension of primary moult uncommon in imm: of 30 examined, only 10% showed suspended moult; of 60 ad, 22% had moult suspended, a low value for megapodes, pointing to long moult season (and short breeding season) for each individual. Timing of moult and duration of primary replacement difficult to establish; many birds examined, but from many different islands, which may each have different moulting season; also, likely to be much variation among individuals of each population. With all areas lumped, most small chicks (not necessarily just hatched) examined from May to July (11 birds) and mid-Nov to Jan (nine birds), only two in between; chicks just starting primary moult with p1 recorded Feb (2), Apr (2), July (1), Aug (2), and Dec (2), perhaps at age of about one month; ad and older imm with moult suspended occur Feb (1), Mar (3), May (1), Jun (1), Aug (3), Oct (1), and Dec (1), ad and older imm just starting new moult series with p1 mainly recorded Jan–Apr (34) and Aug–Oct (21), but also 12 in May–June and nine in Nov–Dec; as peak laying periods assumed to be Mar–Apr and Oct–Nov (concluded from chick data above), active moult and suspension of ad moult do not appear to agree with assumption that active moult and laying do not overlap. No evidence for simultaneous flight-feather moult found in ♀ from Misol by Stresemann (1913).

MEASUREMENTS

ADULT: *M. f. quoyii*. Ternate, skins of 4 ♂♂, 1 ♀, and 8 unsexed birds (RMNH, ZMA).

	n	mean	s.d.	range
wing	13	241.1	5.65	234–252
tail	9	86.6	3.53	82–93
tarsus	9	72.6	3.31	68.0–77.0
mid-toe	8	42.4	1.20	40.8–43.5
mid-claw	9	23.3	1.60	21.3–26.2
bill (skull)	9	32.1	1.63	29.8–34.1
bill (nostril)	9	13.9	1.23	12.5–15.0
bill depth	6	9.2	0.71	8.5–9.9

Morotai, Rau, Halmahera, Tidore, Mare, Kajoa, and Bacan; skins of 18 ♂♂, 20 ♀♀, and 13 unsexed birds (BMNH, RMNH, SMTD, ZMA, ZMB).

	n	mean	s.d.	range
wing	46	231.5	5.77	219–241
tail	43	82.7	2.55	78–87
tarsus	39	72.0	3.25	66.0–78.5
mid-toe	37	39.9	1.70	36.3–44.2
mid-claw	40	23.1	1.48	19.6–26.0
bill (skull)	39	30.9	1.20	28.2–33.8
bill (nostril)	33	13.8	0.69	12.5–14.9
bill depth	26	8.9	0.42	8.2–9.6

One bird from Halmahera (Paris Museum) recorded with wing 258 mm, tail 112 mm (Oustalet 1881). Sexual difference slight, e.g. wing: ♂ 219–241 mm (230.1 ± 6.33, n = 17), ♀ 222–240 mm (230.2 ± 5.00, n = 17).

Obi group of islands; skins of 4 ♂♂ and 3 ♀♀ (BMNH, RMNH).

	n	mean	s.d.	range
wing	7	224.4	6.83	215–237
tail	6	78.9	2.58	75–83
tarsus	6	69.4	2.66	66.2–72.2
mid-toe	6	40.0	2.20	36.4–42.5
mid-claw	6	21.4	2.13	19.1–25.0
bill (skull)	6	29.4	1.96	27.3–31.8
bill (nostril)	6	13.8	0.65	13.0–14.5
bill depth	3	9.3	0.30	9.0–9.6

M. f. freycinet: Gebe, Gag, and Waigeu; skins of 8 ♂♂, 2 ♀♀, and 2 unsexed birds (BMNH, RMNH, ZMB); wing includes data of 9 ♂♂ and 6 ♀♀ from Rothschild *et al.* (1932), De Schauensee (1940b), and Gyldenstolpe (1955).

	n	mean	s.d.	range
wing	27	229.9	4.45	222–238
tail	10	77.5	1.87	73–81
tarsus	10	70.4	2.35	68.0–74.0
mid-toe	9	39.7	1.92	37.0–43.0
mid-claw	10	20.5	0.80	19.4–21.8
bill (skull)	10	31.4	1.65	27.4–32.4
bill (nostril)	10	14.1	0.53	11.8–15.0
bill depth	9	9.1	0.39	8.5–9.7

Sexual difference slight, e.g. wing: ♂ 222–238 mm (229.8 ± 4.41, n = 16), ♀ 222–235 mm (229.2 ± 4.77, n = 8).

Misol; skins of 5 ♂♂, 5 ♀♀, and 3 unsexed birds (BMNH, RMNH).

	n	mean	s.d.	range
wing	12	224.8	6.10	214–232
tail	8	75.9	2.76	72–80
tarsus	12	67.5	3.59	63.8–71.5
mid-toe	11	38.4	1.28	36.9–41.2
mid-claw	12	20.0	1.52	17.8–22.8
bill (skull)	12	29.8	1.74	27.4–32.0
bill (nostril)	12	13.3	0.83	11.8–14.5
bill depth	4	9.0	0.19	8.7–9.1

Tarsus once 76.2 mm, excluded from range. Data from literature partly low: wing, 203, 205, 215, 225 mm (Stresemann 1913; De Schauensee 1940b), perhaps due to inclusion of some imm or of some birds from offshore islands. Wing of birds from Kamuai (half-way between Misol and Salawati): 212, 229 mm (De Schauensee 1940b).

M. f. oustaleti: Salawati, Sorong, Tsiof, and Jef Maän, just west of Vogelkop peninsula; skins of 4 ♂♂, 2 ♀♀, and 1 unsexed bird (RMNH); also data included of 3 ♂♂, 2 ♀♀, and 2 unsexed birds from Oustalet (1881; recalculated due to different measuring technique) and De Schauensee (1940b).

	n	mean	s.d.	range
wing	14	210.8	5.82	203–223
tail	10	72.8	2.95	69–78
tarsus	10	64.7	2.54	61.0–69.3
mid-toe	7	37.1	1.91	34.5–40.3
mid-claw	7	18.4	0.86	17.4–19.8
bill (skull)	7	29.7	1.59	27.2–32.6
bill (nostril)	7	13.1	0.86	12.0–14.1
bill depth	6	8.5	0.57	8.2–9.3

On Batang Kecil, mean wing of 3 birds 210 mm (Salvadori 1882).

In summary, sample size, mean, and standard deviation for wing and tarsus.

	Wing			Tarsus		
	n	mean	s.d.	n	mean	s.d.
Ternate	13	241.1	5.65	9	72.6	3.31
Morotai, Rau	6	229.3	8.94	6	69.7	3.56
Halmahera	23	233.1	4.97	19	73.0	3.07
Tidore, Mare, Kajoa	4	226.5	2.65	4	68.7	3.01
Bacan	14	230.6	5.50	10	72.7	2.08
Obi	7	224.4	6.83	6	69.4	2.66
Gebe, Gag	5	227.8	3.90	5	69.9	1.95
Waigeu	22	230.6	4.42	5	71.0	2.81
Misol	12	224.8	6.10	12	67.5	3.59
Schildpad I.	2	220.5	12.0	–	–	–
Salawati	6	211.7	6.25	6	66.2	1.95
Sorong, Tsiof, Jef Maän	4	210.5	7.05	4	62.4	1.11

CHICK: wing of small chick 103–116 mm (112.4 ± 3.21, n = 14), tarsus 28–35 mm (32.5 ± 1.88, n = 14).

WEIGHTS

ADULT: ♂ 640 g, ♀ 660 g (Bacan, RMNH); ♂ 600, 610, 700 g and ♀ 610, 670 g (Waigeu, Rothschild et al. 1932).

CHICK: newly hatched, 82 g (RMNH).

GEOGRAPHICAL VARIATION

Slight in colour, marked in size. *M. f. quoyii* of northern Moluccan islands from Morotai south to Halmahera and Bacan large, especially birds of Ternate markedly so, tail and claws relatively long. Nominate *freycinet* from Waigeu, Gag, Gebe, Kofiau, and Misol rather large also, but tail and claws relatively short; in *quoyii*, tail 78–93 mm (in only five of 48 birds below 81 mm), in nominate *freycinet*, tail 72–81 mm (in only three of 17 birds over 78 mm); in *quoyii*, middle claw 19.6–26.2 mm (in only six of 49 ad birds below 21.5 mm), in nominate *freycinet*, middle claw 17.8–22.8 mm (in only four of 16 birds over 21.5 mm); scattergram of tail against middle claw fully separates the two races (see Roselaar 1994). Birds inhabiting Obi group included in *quoyii*, as relative proportions similar to that race, though absolute measurements overlap strongly with nominate *freycinet*; slightly smaller size than typical *quoyii* not due to influence of nominate *freycinet*, but to some introgression of characters of small Forsten's Megapode *M. forstenii forstenii* from Ceram; this influence also visible in colour, at least four of seven Obi birds showing some olive tinge on scapulars, tertials, back, and rump; this tinge is usually very faint, but quite distinct in one bird, upperpart colour of latter being almost similar to darkest Ceram birds. Birds from Misol also somewhat smaller than Waigeu, Gebe, and Gag birds, but relative proportions of both similar and no direct evidence in structure or colour that introgression of characters from *M. forstenii* or *M. freycinet oustaleti* occurs.

M. f. oustaleti from Batanta, Salawati, and small islands off W Vogelkop peninsula much smaller than both other races, wing rarely over 220 mm (in two of 14 birds); in nominate *freycinet*, over 220 mm, except for three of 37 birds (these three all from Misol; also some low data from the literature for Misol, but these perhaps include imm). Proportionally intermediate between *quoyii* and nominate *freycinet*, but relatively short middle claw similar to nominate *freycinet* and bill relatively heavier than in both these races. Small size perhaps due to character displacement with large Orange-footed Megapode *M. reinwardt*, which occurs on Vogelkop peninsula and which is occasionally reported within range of *M. f. oustaleti*, for example on Batanta, near Sorong, and in Dore Hum Bay (north coast of Vogelkop peninsula at c. 131°30′E). In sample of black-backed *M. f. oustaleti*, no influence of size or structure of *M. reinwardt*, but secondary intergradation occurs, as some intermediates are known, which were excluded from data in Measurements. Of nine birds from 'Sorong' examined in RMNH (unknown whether refers to Sorong I. or Sorong proper on opposite mainland of New Guinea), four pure *M. reinwardt*, inseparable from other Vogelkop birds, two pure *M. f. oustaleti* (four such birds from 'Sorong' also in Salvadori's collection, Genova, and some similar birds collected by De Schauensee on Sorong I.); one bird similar in size and structure to *M. reinwardt*, legs (in skin) pale, but crest short and body medium plumbeous-black except for extensive olive tinge on mantle, scapulars, and upperwing; two birds largely dark plumbeous-black (like *M. freycinet*), but with trace of olive on upperparts, legs apparently dark, size and proportions intermediate between *M. freycinet* and *M. reinwardt*; these latter three birds closely similar to New Guinea Megapode *M. decollatus* from Japen and Mamberamo basin in size and appearance, but proportions differ: average wing 221 mm (*M. decollatus*, n = 8, also 221 mm), tail 78.3 mm (*M. decollatus* 73.5 mm), tarsus 63.8 mm (*M. decollatus* 67.3 mm), middle toe 18.9 mm (*M. decollatus* 16.8 mm), bill to skull 30.3 mm (*M. decollatus* 27.8 mm), thus tail relatively long and tarsus relatively short, as can be expected from hybrids with *M. reinwardt*.

Three birds collected on Batang Kecil (near Tifore, between Halmahera and Sulawesi) markedly small (see 'Measurements'), like *M. f. oustaleti*; none from here examined for present work, and not known whether small size is due to influence of Philippine Megapode *M. cumingii gilbertii*, to immaturity, or if a truly stable small-sized race inhabits this tiny isolated island. A similar small bird reported from 'Pulu Penang, north-west coast of New Guinea' (Cabanis and Reichenow 1876) is a normal nominate *freycinet* in first imm plumage and with wing (164 mm) still juv (ZMB).

Range and status
M. f. quoyii: Morotai, Rau, Halmahera, Ternate, Tidore, Mare, Kajoa, Bacan, Obi Bisa, Obi Latu, Obi Major. *M. f. freycinet*: Waigeu and offshore islets (Boni, Saonek), Gag, Gebe, Kofiau, Misol, Kamuai. *M. f. oustaleti*: Batanta, Salawati, islands in Sele Strait (Sakamun, Arar, Jef Maän, Tsiof), and islands off the coast of NW Vogelkop peninsula opposite Salawati (Sorong, Pulu Hum) east to Dore Hum Bay and Ramoi. Unknown what race inhabits Tifore and Batang Kecil.

'Very common' on Ternate around 1875 (Von Rosenberg 1878). Common on Halmahera in 1930s and 1950s (Heinrich 1956; G. A. L. de Haan, on museum labels); on Misol, common and widespread in 1950s (Ripley 1960).

Field characters
Length 34–41 cm. Medium-sized megapode with length dependent on race; smallest on Salawati, largest on Ternate. Looks entirely black, with large feet and short but conspicuous crest. Sympatric with *M. reinwardt*, which is mainly brownish-grey with orange legs, on Batanta and Sorong (see 'Geographical variation'). Also sympatric with Moluccan Megapode *Eulipoa wallacei* on Misol, Bruijn's Brush-turkey *Aepypodius bruijnii* on Waigeu, and Red-billed Talegalla *Talegalla cuvieri* on NW Vogelkop. For identification, see these species.

Voice
Duet (see sonograms). Initiated (probably by the ♂) with a laughing 'kejowowowowowowow'

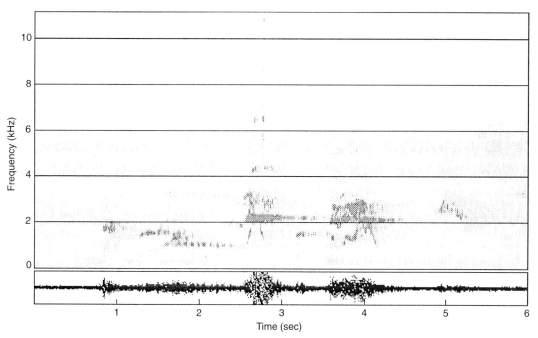

M. f. freycinet
K. D. Bishop, Misol, Indonesia.

Dusky Megapode *Megapodius freycinet*

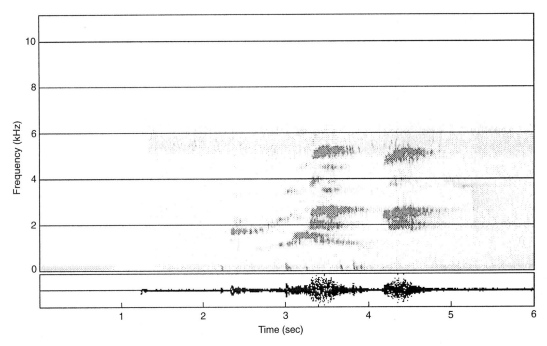

M. f. quoyii
K. D. Bishop, Halmahera, Indonesia.

(Misol) or 'keyouououarr' (Halmahera), initially rising and subsequently descending in pitch. The partner joins in shortly after with two notes: 'keya keyauw' (Misol) or 'keyou keyou' (Halmahera), both rising in pitch and the second note slightly shorter than the first. (Description from tape-recordings from Misol and Halmahera made by K. D. Bishop.) Similar in structure to duet of Sula Megapode *M. bernsteinii*, Melanesian Megapode *M. eremita*, and *M. forstenii*. Often heard at night as in other *Megapodius* spp.

Habitat and general habits

On Halmahera in areas such as mangroves, edges of swampy woodland, and sago swamps, much wetter habitat than reported for most other *Megapodius* spp. (Stresemann 1941). Shy.

Food

Stomach of one specimen contained snails and beetles (Cabanis and Reichenow 1876).

Displays and breeding behaviour

No information.

Breeding

NEST: mound, consisting of earth mixed with leaves and sticks, sometimes built against a rotten log, once reported as 1.5 m high and 2 m in diameter. On Bacan recorded at 600 m altitude. On Misol mounds situated on the edge of beaches as well as in the centre of the island (Ripley 1960). EGGS: similar in shape and coloration to eggs of *Macrocephalon maleo*, *Eulipoa wallacei*, and other *Megapodius* species (see under *M. pritchardii*). SIZE: 85.6–92.8 × 51.6–57.3 mm (89.6 ± 2.35 × 54.7 ± 1.90, $n = 11$, Halmahera, RMNH); 90.8–93.6 × 55.2–55.8 mm ($n = 2$, Bacan, RMNH); 76–92 × 48.5–54.2 mm (84.6 × 52.6, $n = 7$, Schönwetter 1961). WEIGHTS: no information.

BREEDING SEASON: on Waigeu, adult in breeding condition collected in Oct; specimens just out of breeding condition shot in Dec (Ripley

1964). On Misol, eggs laid Oct–Nov. Data of small chicks (collections and literature) recorded as: Morotai, Sept (1); Halmahera, July (1), Dec (2); Ternate, Jan (1), Dec (1); Misol, early Jan (several), May (2), June (1), July (1), Aug (1); Waigeu, Nov (1), Dec (1); Bacan, May (1), Jun (1); Obi, July (2). Thus, possibly two hatching peaks, from May to July (–Sept) and from Nov to Jan. See also under 'Moults' for possible relation between moult and reproductive period.

Local names
Manesaque, momkirio, or mankirio (Waigeu and Boni), blévine (Gebe), maleo (Ternate).

References
Gaimard (1823), Gray (1861), Reichenbach (1862), Schlegel (1866b, 1880a), Cabanis and Reichenow (1876), Von Rosenberg (1878), Oustalet (1881), Salvadori (1882), Ogilvie-Grant (1893, 1897), Vorderman (1898b), Rothschild and Hartert (1901), Stresemann (1913, 1941), Rothschild *et al.* (1932), Mayr (1938, 1941), Mayr and De Schauensee (1939) De Schauensee (1940b), Gyldenstolpe (1955), Heinrich (1956), Ripley (1960, 1964), Schönwetter (1961), Rand and Gilliard (1967), White and Bruce (1986), Holmes (1989), Roselaar (1994).

Biak Megapode *Megapodius geelvinkianus* A. B. Meyer, 1874

Megapodius geelvinkianus A. B. Meyer, 1874. *Sitzungsberichte der Mathematisch-Naturwissenschaftlichen Classe der Kaiserlichen Akademie der Wissenschaften, Wien*, **69** (1–5), p. 74. (Mafor [= Numfor] and Misory [= Biak]; a syntype from Korido on Supiori examined, other type-specimens lost.)

PLATE 7

Monotypic.

Description
PLUMAGES

ADULT: sexes similar. Crown and broad short crest on nape black with slight dark plumbeous tinge. Forehead and lore bare, except for sparse and short bristle-like dull black feathers. Region round eye fully bare. Feathers of hindneck dull black, normal-shaped, remainder of head and neck from chin, lower cheek, and region round ear to lower neck covered with scanty and short sooty grey to dull black feathers. Head rather like that of the Dusky Megapode *M. freycinet*, but short crest less inclined to end in short point and throat and neck distinctly less feathered, though some variation occurs in both species. Mantle, shorter and outer upper wing-coverts, chest, and breast black, feathers with dark plumbeous fringes, showing as slight blue tinge when plumage fresh, similar to *M. freycinet*. Rear body different from *M. freycinet*, however: scapulars, back, rump, tertials, and longer inner upper wing-coverts dusky brownish-olive, appearing almost olive-black in some lights, upper tail-coverts with slightly more chocolate-brown tips; underparts down from flank and belly dark plumbeous-grey with slight olive tinge, gradually paler greyish-olive towards vent, darker and more sooty again on lower flank and under tail-coverts; whole body (but not head and neck) closely similar to the Micronesian Megapode *M. laperouse*; colour of head and body about intermediate in tinge between *M. freycinet* and the Melanesian Megapode *M. eremita*. Tail black with brown tinge; flight feathers greyish-black (if fresh) or brownish-black (if worn), outer webs of secondaries tinged dark brownish-olive. Under wing-coverts and axillaries sooty-black, greyer on longer coverts.

CHICK: no information.

IMMATURE: none in first imm plumage examined. Undoubtedly like ad, but tail and outer primaries still juv, short; juv p10 sharply

Biak Megapode *Megapodius geelvinkianus*

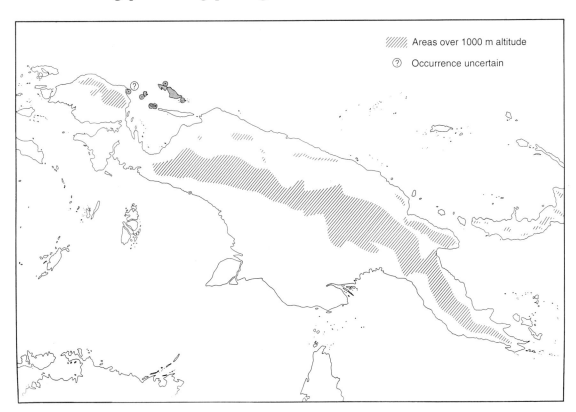

pointed. In second imm plumage, like ad, but outer primaries and tail of second generation, still somewhat shorter than ad; second generation p10 somewhat pointed, less rounded than in ad; length from tip to insertion in skin 94–102 mm (97.5 ± 3.42, n = 4), in ad 102–114 mm (106.5 ± 5.72, n = 5).

BARE PARTS
ADULT, IMMATURE: iris brown, dark brown, or chestnut. Bare skin of head and neck cherry-red or bright flesh-red, slightly duller flesh-red in imm. Bill yellow or dull orange, base of upper mandible (except cutting edges) dark horn-brown to black. Leg and foot cherry-red or red, scutes of lower front or all front of tarsus in many birds dark greyish-horn, black-brown, or black, upper surface of toes usually black; soles sometimes grey, in at least some imm (and perhaps ad) front of tarsus horn-black and rear red-brown or brown-red.

CHICK: no information.

MOULTS
Primary moult serially descendant. Ad and imm primary moult closely similar to *M. freycinet*, with two moult series and two or three generations of primaries in each wing; none of six older imm examined had moult suspended, and only two of eight ad; next series started with shedding of p1 when previous series had reached p4–p8 (on average p6.3 ± 1.03, n = 6) in imm, p4–p7 (p5.6 ± 0.92, n = 8) in ad. Majority of specimens examined undated; singles had just started an innermost series with p1 while an outer series active on p6 and p8 in Nov and Mar, both series more advanced in birds from Feb (2), Apr (1), May (1), and July (1), moult suspended in single bird from Feb.

MEASUREMENTS
ADULT: Biak, Supiori, and Aifundi Is (just north of Supiori); skins of 1 ♂, 2 ♀♀, and 8 unsexed birds (RMNH, SMTD, ZMA).

	n	mean	s.d.	range
wing	11	189.4	5.37	178–197
tail	11	61.2	1.85	58–64
tarsus	11	61.5	2.14	59.3–64.9
mid-toe	11	30.8	1.51	28.9–33.5
mid-claw	11	16.9	0.83	15.5–18.0
bill (skull)	11	26.2	1.13	24.2–27.7
bill (nostril)	11	12.2	0.59	11.0–12.9
bill depth	8	8.2	0.40	7.6–8.5

A single bird from Manokwari (E Vogelkop peninsula) had wing 197 mm; tail 67 mm; tarsus 58.5 mm; toe 32.3 mm; claw 18.5 mm; bill to skull 27.6 mm; bill to nostril 13.1 mm; bill depth 7.6 mm (RMNH).

Numfor; skins of 3 ♀♀ and 1 unsexed bird (RMNH, ZMB), as well as data on wing from Rothschild *et al.* (1932).

	n	mean	s.d.	range
wing	6	197.2	4.21	190–210
tail	4	66.9	2.46	64–70
tarsus	4	61.3	1.99	60.0–64.2
mid-toe	3	33.8	0.89	33.1–34.8
mid-claw	3	17.1	0.60	16.5–17.7
bill (skull)	3	25.9	1.66	24.7–27.8
bill (nostril)	3	12.5	0.36	12.1–12.9
bill depth	3	8.2	0.49	7.9–8.8

A single ♂ and ♀ from Mios Num had wing respectively 210, 204 mm; tail 71.5, 67 mm; tarsus 59, 61.5 mm; toe 37.0, 35.6 mm; claw 17.5, 16.8 mm; bill to skull 24.5, 26.9 mm; bill to nostril 11.7, 11.8 mm; bill depth 8.2, 8.7 mm (RMNH).

CHICK: Numfor, wing 123.5 mm (Rothschild *et al.* 1932), apparently not recently hatched.

WEIGHTS
ADULT: ♀ 525 g (Numfor) (Rothschild *et al.* 1932).

GEOGRAPHICAL VARIATION
Slight, apparently mainly due to some introgression of characters of the New Guinea Megapode *M. decollatus* in some populations of *M. geelvinkianus*. Birds from Biak and Supiori are typical *M. geelvinkianus*. Three out of five birds from Numfor are similar to Biak–Supiori birds, but two others are slightly browner-olive above, tinge as dark as typical *M. geelvinkianus* but somewhat warmer brown, less dark olive; also, Numfor birds larger in all measurements, except tarsus and bill. A single bird from E Vogelkop peninsula, labelled 'Dorei' (= Manokwari) collected 12 Jan 1869 (RMNH) similar in colour to typical *M. geelvinkianus*, size similar to Numfor birds; may have been shipped to Manokwari from Numfor, but perhaps *M. geelvinkianus* occurs on coast and islands along E Vogelkop, for example on Mansiman in Dorei Bay, similar to situation in W Vogelkop, where *M. freycinet* occurs on islets and coast; otherwise, birds from E and W Vogelkop, including 'Dorei', are Orange-footed Megapode *M. reinwardt*. Two birds examined from Mios Num, west of Japen in Geelvink Bay, are much less typical *M. geelvinkianus*: both are considerably larger than Biak–Supiori birds; upperparts of one bird are rather extensive olive-brown and underparts medium grey, lighter than in *M. geelvinkianus*, other bird much darker, however, similar to typical *geelvinkianus*. Both are completely intermediate in size between typical *M. geelvinkianus* and *M. decollatus* from Japen, and the paler bird is also intermediate in colour. Thus, characters of Mios Num population point to secondary hybridization of *M. geelvinkianus* and *M. decollatus*, and Numfor population (with its slightly larger size than typical *M. geelvinkianus* and occasional browner coloration) probably also shows some introgression of *M. decollatus* characters. Crest of Mios Num birds is broad and short, and head and neck are mostly naked, similar to Numfor and Biak–Supiori birds, unlike *M. decollatus*. No evidence for occurrence of *M. geelvinkianus* on Japen; single bird collected there by Meyer a chick and thus unidentifiable; of two birds collected by Beccari and described by Salvadori (1882), one typical of Japen population of *M. decollatus*, the other slightly darker and browner above, like some Numfor birds, and thus perhaps hybrid and not a pure *M. geelvinkianus*; many other

birds collected on Japen are all typical *M. decollatus* (for example, ten birds in Rothschild and Hartert 1901 and two skins in RMNH).

Recently, birds observed on Biak were typical *M. geelvinkianus*, but those on Owi, southeast of Biak, showed bluish instead of red facial skin and black or greyish rather than (partly) red legs, while local people confirmed occurrence on Biak of two species, one red-legged, one black-legged (D. Gibbs, *in litt.*); not known whether characters of birds showing dark bare parts are due to immaturity of typical *M. geelvinkianus*, whether influence of (black-legged) *M. decollatus* is now more extensive than in the past (hybridization being now more widespread on Geelvink Bay islands resulting in variable mixture of characters), or whether in fact another (undescribed) species is involved.

Note: *M. geelvinkianus* usually considered a race of *M. freycinet*, both showing rather similar forehead-feathering, crest-shape, bill-colour, and colour of feathering of head and front part of body. *M. geelvinkianus* markedly smaller than most populations of *M. freycinet*, but *M. f. oustaleti* of Batanta, Salawati, and western shores of Vogelkop peninsula bridge the gap. The Biak Megapode differs from *M. freycinet* in brighter colour of skin of head and neck, mainly red legs, and olive tinge of upperparts; also relative proportions rather different from *M. freycinet*, showing, for instance, relatively long tarsus and strong bill. In some aspects of body colour and proportions, nearer to *M. laperouse*, the Polynesian Megapode *M. pritchardii*, and Forsten's Megapode *M. forstenii* than to *M. freycinet*, and therefore *M. geelvinkianus* separated from *M. freycinet*, though all these species probably more closely related to each other than to any other *Megapodius*.

Range and status
Islands in the Geelvink Bay: Biak and offshore islands (for example, Owi, Supiori), Mios Korwar (Aifundi Is), Numfor, Pulu Manim, Mios Num; records from Japen doubtful (see 'Geographical variation'). Rosenberg collected a specimen at 'Dorei' (= Manokwari) on the mainland of New Guinea; whether this is a straggler from a nearby island or a wrongly labelled specimen remains unclear (see 'Geographical variation' and Chapter 3).

Small numbers reported from Owi and Supiori in 1991 (D. Gibbs, *in litt.*).

Field characters
Length 30 cm. The only megapode on Numfor, Mios Num, Biak and surrounding islets. Much smaller than *M. freycinet*, from which it differs in having the legs partly or mostly reddish and the bare skin of the head bright red instead of dull pale reddish. Might be sympatric with *M. decollatus* on Japen, which species is larger, generally brownish-grey and with dark instead of reddish legs. Possibly sympatric with much larger Orange-footed Megapode *M. reinwardt* in the area near Manokwari, which species has orange legs and rufous-brown upper parts.

Voice
Very vocal. 'Keyeew-kyu-kyu' with emphasis on the first note; less varied, shorter and harsher than *M. freycinet* on Halmahera (D. Gibbs, *in litt.*).

Habitat and general habits
On Supiori in secondary scrub near river; on Owi in dry scrub in the centre of the island and in the remains of evergreen forest. Heard in logged forest near Warafri, E Biak. Shy and wary (D. Gibbs, *in litt.*). No further information on habitat-preference, general habits, food, or breeding behaviour, but probably similar to other *Megapodius* spp.

Breeding
NEST: no information. Most probably builds mounds or buries its eggs between decaying roots of trees.

EGGS: probably similar in shape and coloration to eggs of *Macrocephalon maleo*, *Eulipoa wallacei*, and *Megapodius* species (see under *M. pritchardii*). SIZE: no information. WEIGHT: no information.

BREEDING SEASON: no information.

Local names
None reported.

References
Meyer (1874a), Oustalet (1881), Salvadori (1882), Ogilvie-Grant (1893, 1897), Rothschild and Hartert (1901), Rothschild *et al.* (1932), Mayr (1938, 1941), Gyldenstolpe (1955), Ripley (1964), Rand and Gilliard (1967), Diamond (1985).

Forsten's Megapode *Megapodius forstenii* G. R. Gray, 1847

Megapodius Forstenii G. R. Gray, 1847. *The Genera of Birds*, **3**, p. 491, pl. 124, Longman, London. (No type locality = Ceram, substantiated by Stresemann, 1914a, *Novitates Zoologicae*, **21**, p. 41.)

PLATE 8

Polytypic. Two subspecies. *Megapodius forstenii forstenii* G. R. Gray, 1847, Ceram and surrounding islands; *Megapodius forstenii buruensis* Stresemann, 1914, Buru.

Description
PLUMAGES
ADULT: *M. f. forstenii*. Sexes similar. Central forehead, all crown, and short broad crest on nape dusky olive-brown. Lore and patch round eye bare; forehead bare (apart from strip of feathers on upper centre and some short dark brown bristle-like feathers), cheeks and area round eye thinly covered with short dusky olive-brown feathers; amount and extent of feathers and bristles on forehead and side of head similar to the Dusky Megapode *M. freycinet*. Chin, throat, and entire neck virtually completely covered with short black-brown feathers. Upper mantle medium plumbeous-grey, sometimes slightly tinged brown; lower mantle, scapulars, tertials, and all upper wing-coverts

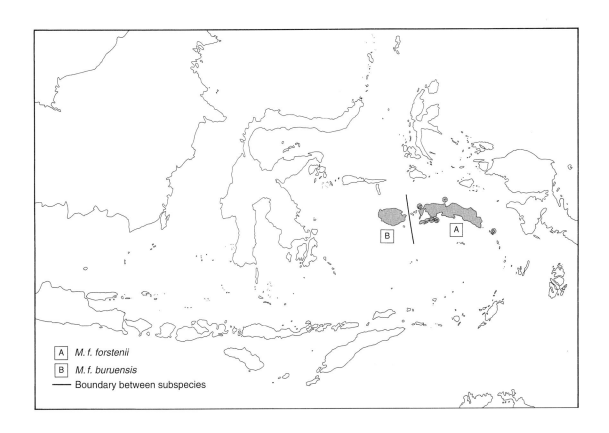

A *M. f. forstenii*
B *M. f. buruensis*
— Boundary between subspecies

dusky olive-brown; upperparts distinctly more extensive olive-brown than in the Biak Megapode *M. geelvinkianus*, brown of latter more olive and darker; tinge of *M. forstenii* near tinge of upperparts of the Philippine Megapode *M. cumingii gilbertii*, but latter slightly browner and tinge extends over upper mantle; extent of olive-brown on upperparts of the Orange-footed Megapode *M. reinwardt* equal to *M. forstenii*, but tinge in former distinctly paler, and even in more olive-tinged variants still more rufous than in *M. forstenii*. Back, rump, and upper tail-coverts dark umber-brown, darker and less olive than scapulars. Underparts medium plumbeous-grey, sometimes slightly washed olive-brown on chest, flank, and thigh; breast and belly more distinctly tinged olive-brown, mid-belly and vent dusky olive-grey. Flight feathers and tail greyish-black, outer webs tinged olive-brown (more markedly so on inner secondaries).

CHICK: upperparts including upper wing dark brown, tinged olive on cap and deep rufous on rump; mantle, scapulars, tertials, and inner wing-coverts contrastingly barred buff; buff bars reduced towards outer coverts. Side of head, chin, and throat cinnamon-buff, remainder of underparts dull cinnamon-brown, tinged olive on chest and flank, more buffish on mid-belly. Flight feathers fuscous-black.

IMMATURE: in first imm plumage, like ad, but some secondaries, tail, and outer primaries still juv, shorter than in ad; juv tail 58–63 mm ($n = 2$); juv outermost primary (p10) strongly pointed, less rounded than in ad, length from tip to insertion in skin 88–92 mm ($n = 2$; in ad, 108–122 mm, mean 113.3 ± 4.40, $n = 15$). In second imm plumage, like ad, but outer primaries from second generation still shorter and p10 more pointed than in ad (but less so than juv), length 101–115 mm (110.1 ± 5.24, $n = 5$).

BARE PARTS
ADULT, IMMATURE: iris brown, dark brown, or reddish-brown. Bare skin of head and neck plumbeous- or greyish-black, in at least some birds 'partly red' on side of head (Siebers 1930). Bill apparently variable, in some uniform, in others with darker base and paler tip, perhaps depending on season; recorded on labels of skins as dark yellow-olive with paler dirty yellow tip, dark yellow with olive culmen, dirty flesh with darker base, or uniform dirty yellow, dark grey, brown-black, or black. Leg and foot variable also: olive-green, grey-olive, dark olive, dark olive-brown, grey with yellow rear, grey, dark grey, blackish-olive, or black (Siebers 1930; BMNH, RMNH).

CHICK: iris grey or grey-brown. Bill greenish-grey or dark brownish-olive. Leg and foot dark brownish-olive or greenish-grey with darker toes (Siebers 1930).

MOULTS
Primary moult serially descendant. In both ad and older imm, next series starts with shedding of p1 (innermost primary) when previous series has reached about p4–p7 (on average, 5.6 ± 0.76, $n = 13$). Usually two, occasionally three series active in each wing; however, six of 17 ad show seemingly irregular moult, asymmetrical between wings, with sequence in some short series seemingly ascendant, others descendant; these either older birds, or birds with regrowth of some feathers after accidental loss (but unlikely that this happens more often in *M. forstenii* than in other species). Apparently a clear moulting season in population: during Feb–Aug, moult active in nine of 12 ad (suspended in three, Apr–May, when three others active); in Sept–Jan, moult active in two of ten ad (suspended in eight); all imm (Feb, Apr, Aug, and twice Oct) had active primary moult. Including birds from unknown collecting date, 37% of 27 ad had primary moult suspended.

MEASUREMENTS
ADULT: *M. f. forstenii*. Ceram, Tudju Is just north of Ceram, Ambon, and Haruku; skins of 7 ♂♂, 12 ♀♀, and 7 unsexed birds (BMNH, RMNH, ZMA).

Forsten's Megapode *Megapodius forstenii*

	n	mean	s.d.	range
wing	26	210.0	5.38	201–225
tail	20	70.3	2.69	65–74
tarsus	24	66.3	2.91	61.0–70.7
mid-toe	21	38.2	1.58	35.4–40.7
mid-claw	24	19.7	0.92	17.9–21.4
bill (skull)	24	29.8	1.15	27.6–32.5
bill (nostril)	23	13.0	0.60	11.8–14.3
bill depth	18	8.4	0.36	7.6–8.8

Ceram and Ambon, wing: 190–207 mm (201.3 ± 6.17, n = 6) (Stresemann 1914a, excluding data from BMNH and RMNH; probably includes some imm). No differences between islands, e.g. wing: Ceram and Tudju Is 201–215 mm (208.4 ± 4.19, n = 10), Ambon 202–225 mm (210.7 ± 5.95, n = 15), Haruku 213 mm (n = 1). Virtually all birds had wing below 215 mm, tail below 74 mm, tarsus below 70 mm; those with wing over 215 mm (n = 3) all from from Ambon and perhaps wrongly-labelled birds from Buru (see below). Sexes probably similar, e.g. wing: ♂ 203–215 mm (–225) (212.0 ± 7.14, n = 7), ♀ 201–213 mm (–217) (208.7 ± 4.48, n = 12). Gorong (Ceramlaut Is), unsexed (n = 2): wing 213, 213 mm; tail 70, 72 mm; tarsus 66, 69 mm; bill to skull 28.1, 29.6 mm, to nostril 12.9, 13.8 mm (BMNH).

M. f. buruensis: Buru, skins of 5 ♂ ♂, 3 ♀ ♀, and 1 unsexed bird (BMNH, RMNH); wing includes data from Stresemann (1914a, excluding BMNH and RMNH birds).

	n	mean	s.d.	range
wing	20	225.8	5.68	215–235
tail	8	77.4	4.53	72–83
tarsus	9	71.0	3.92	67.4–75.5
mid-toe	8	41.1	2.22	38.2–44.0
mid-claw	9	23.6	1.78	21.3–26.9
bill (skull)	9	30.2	1.17	28.8–32.2
bill (nostril)	9	13.0	0.51	11.8–13.6
bill depth	8	8.3	0.34	7.9–8.9

Wing: 214–233 mm (219.4 ± 6.14, n = 8) (Siebers 1930), 230–235 mm (n = 4) (Salvadori 1882). Thus, wing of *buruensis* 215 mm and over, birds with shorter wing (one from RMNH in Stresemann 1941a, two in Siebers 1930) being imm with wing not fully grown; in nominate *forstenii*, wing below 215 mm, and larger ones of doubtful origin; two large birds from Ambon, cited in Oustalet (1880), doubtful also, the one in Paris Museum ('wing 241 mm, tail 120 mm') being even too large for *buruensis*, the other in BMNH (coll. A. R. Wallace, 'wing 233 mm') at present has wing of 206 mm.

CHICK: shortly after hatching, wing of nominate *forstenii* 104 mm, tarsus 30.0 mm (n = 1); wing of *buruensis* 106 and 106 mm (n = 2).

WEIGHTS
ADULT: *M. f. forstenii*. Ceram: ♀ 650 g (Stresemann 1914a). *M. f. buruensis*. Buru: 620, 735, 752 g (Stresemann 1914b).

GEOGRAPHICAL VARIATION
Rather slight in colour, marked in size. *M. f. buruensis* from Buru distinctly larger than nominate *forstenii* from Ceram and surrounding islands, except for bill size; especially claws markedly longer, perhaps pointing to difference in digging or in inhabited substrate. Colour of upperparts of *buruensis* greener than in nominate *forstenii*, mantle and scapulars more greenish-olive, less olive-brown, back to upper tail-coverts more olive-brown, less dark umber-brown. Within nominate *forstenii*, birds from Ceram, Ambon, and Haruku inseparable in size and colour; birds of Tudju Is (just north of Ceram) and of Gorong (southeast of Ceram) both darker on head and body than Ceram, head, upper mantle, and underparts slightly darker and purer grey, less tinged olive, lower mantle and scapulars colder and more restricted dark olive, much like *M. geelvinkianus*; darker tinge perhaps due to some introgression of characters of *M. freycinet*. Average wing of Tudju I. birds somewhat shorter than on Ceram (203.5, n = 2), on Gorong somewhat longer (213.0, n = 2); samples for each small and not known whether differences in size and colour constant. One bird from Ambon (RMNH) has upperparts completely plumbeous-black, like *M. freycinet*, but underparts with much olive, similar to normal *M. forstenii*.

Note: sometimes considered a race of *M. reinwardt* (for example, in White and Bruce 1986), but markedly different: crest short and broad (*M. reinwardt* long and pointed), legs

Forsten's Megapode *Megapodius forstenii*

and bare skin of head mainly dark (*M. reinwardt* yellow or orange), colour of body much darker (though distribution of olive-brown colour similar to *M. reinwardt*), tail relatively much shorter, tarsus, toes, and claws much longer. In structure, close to *M. freycinet* and (especially) *M. geelvinkianus*, differing from these mainly in reduction of black pigments.

Range and status
M. f. forstenii: central Moluccan islands of Ambon, Pulau Pombo (off Ambon), Haruku, possibly Saparua (R. Dekker, personal observation), Gorong (Ceramlaut Is), Ceram, Nusa Tulun and Pulu Aleï (Tudju Is, off N Ceram), and Boano (C. Moeliker, personal communication). *M. f. buruensis*: restricted to Buru.

Status: moderately common in lowland primary and secondary forest and especially scrubby growth in Manusela National Park, Ceram (Bowler and Taylor 1989). Still common on Buru (M. M. J. van Balgooij, personal communication).

Field characters
Length 30–39 cm. Medium-sized *Megapodius* restricted to the central Moluccan islands where it is sympatric with the Moluccan Megapode *Eulipoa wallacei*. Generally brownish-grey with short crest and olive, greyish, or blackish legs. *E. wallacei* is smaller, much more brightly coloured, and lays its eggs not in mounds, but in burrows at communal nesting grounds.

Voice
Duet (see sonogram), rather similar to the Sula Megapode *M. bernsteinii*, Melanesian Megapode *M. eremita*, and *M. freycinet*. Initiated, probably by the ♂, 'keyou-ou-ou–ou-ou', rising in pitch on the first note and subsequently de-

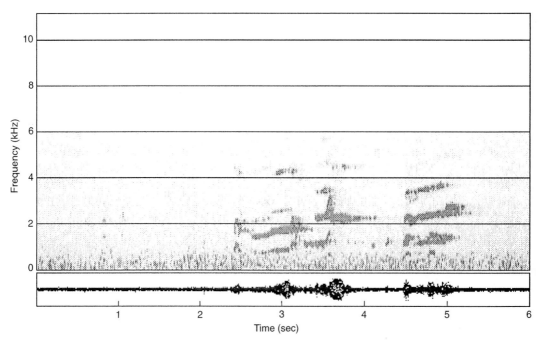

M. f. forstenii
M. Argeloo, Haruku, Indonesia, Jan 1992.

scending in pitch on all others, with the second part slowing down. The partner joins in shortly after, on the first note, with two (or sometimes three) notes 'kyou' 'keyou' ('keyou'), the first one slightly shorter and higher in pitch than the second, but neither note rising or descending in pitch (unlike some other *Megapodius* spp.) (Description of duet of nominate *forstenii*, Haruku, R. Dekker, personal observation). Often heard at night as in other *Megapodius* spp. Vocalization of birds from Ceram said to be different from those of Buru (Stresemann 1914a). Also a soft 'kruk-kru(k)' when disturbed near mound.

Habitat and general habits

Primary and secondary forest. Most frequently in scrubby secondary forest, especially in coastal scrub. Common in wooded country with light undergrowth on Buru, from the coast up to 1900 m at Mt. Fogha; in forests near the coast and in the hills on Ceram; also on small offshore coral islets (Stresemann 1914a). Mound up to 1550 m in altitude on Buru and 1165 m on Ceram (Stresemann 1914a, b). Seen singly or in small groups of three to five birds (Bowler and Taylor 1989). Some birds spend the night on small islands two to three kilometres off shore, as in *M. eremita*.

Food

Undescribed. Most probably similar to other megapodes; will include invertebrates, fruits, and seeds.

Displays and breeding behaviour

No information.

Breeding

NEST: mound. Composed of forest litter and sand depending upon location. Mounds close to the beach (as on Haruku) consist mainly of sand, those further inland (as on Ceram at 200 m altitude) mainly of leaves and sticks (R. Dekker, personal observation). On Buru, the mounds found by Stresemann (1914b) consisted of earth and stones with few leaves, 3–4 m in diameter and 1–1.50 m high; others, however, are flat, more spread out and situated between buttresses which separate them into several parts (Martin 1894; R. Dekker, personal observation).

EGGS: similar in shape and coloration to eggs of *Macrocephalon maleo*, *Eulipoa wallacei*, and other *Megapodius* species (see under *M. pritchardii*). SIZE: *M. f. forstenii*: 80–85 x 55 mm ($n = 5$, Ambon, Blasius and Nehrkorn 1883); 80–85 × 50–55 mm (Haruku, Martin 1894); 73–84 × 47–51 mm (78.5 × 49.0, $n = 15$, Schönwetter 1961). *M. f. buruensis*: 81–93 × 50–56 mm (85.9 ± 3.86 × 52.9 ± 1.96, $n = 9$, Buru, Siebers 1930); 75.5–83.4 × 44.7–50.8 mm (79.4 × 48.9, $n = 6$, Buru, Steadman 1991). WEIGHTS: no information.

BREEDING SEASON: *M. f. forstenii*. Eggs in May on Ceram; small chicks collected in Sept on Ceram. *M. f. buruensis*: eggs in Feb, Mar, and Apr; chicks collected in Mar, May, and June.

Local names

Kèhô (Buru), maléhu (Ambon), maleo (Haruku, also used for *Eulipoa wallacei*), moma, muma, momalo, uma, maleune, ilaun, memai, and maja (Ceram).

References

Wallace (1863), Oustalet (1880, 1881), Schlegel (1880b), Salvadori (1882), Blasius and Nehrkorn (1883), Ogilvie-Grant (1893, 1915), Martin (1894), Rothschild and Hartert (1901), Stresemann (1914a, b), Shufeldt (1919), Toxopeus (1922), Siebers (1930), Mayr (1938), Schönwetter (1961), White and Bruce (1986), Bowler and Taylor (1989), Holmes (1989), Jones and Banjaransari (1989), Steadman (1991).

Melanesian Megapode *Megapodius eremita* Hartlaub, 1867

Megapodius eremita Hartlaub, 1867. *Proceedings of the Zoological Society of London*, 1867, p. 830. ('Echiquier' = Ninigo Is.)

PLATE 7

Other names: Melanesian Scrubfowl, Bare-faced Scrubfowl, Hartlaub's Scrubfowl.

Monotypic. Synonyms: *M. brenchleyi* G. R. Gray, 1870, Gulf [= Ugi] I.; *M. hueskeri* Cabanis and Reichenow, 1876, New Hannover; *M. rubrifrons* Sclater, 1877, Admiralty I.

Description
PLUMAGES
ADULT: sexes similar. Lore and forehead up to above eye bare, apart from scanty hair-like bristles, rather sharply demarcated from densely feathered crown. Crown and very short broad crest on nape dusky olive-brown, bordered by dark plumbeous-grey at side of crown and at rear of crest; some or much plumbeous tinge of feather bases sometimes visible on crown. Side of head, chin, throat, and side and front of neck bare, apart from rather scanty and reduced dark grey feathers; feathering on hindneck slightly denser. Upper mantle dark plumbeous-grey, often rather sharply demarcated from uniform dusky olive lower mantle, scapulars, and inner and longer upper wing-coverts; shorter and outer upper wing-coverts dark plumbeous-grey, colour gradually shading to blackish (if feathering fresh) or dusky brown (if worn) of upper primary-coverts and flight feathers. Rump and upper tail-coverts dusky black-brown or olive-black, some feathers sometimes indistinctly tipped dull rusty-brown. Chest, side of breast, and upper flank dark plumbeous-grey, slightly tinged blue when plumage fresh, feather centres slightly blacker; remainder of underparts dark grey with slight olive tinge, latter more pronounced towards

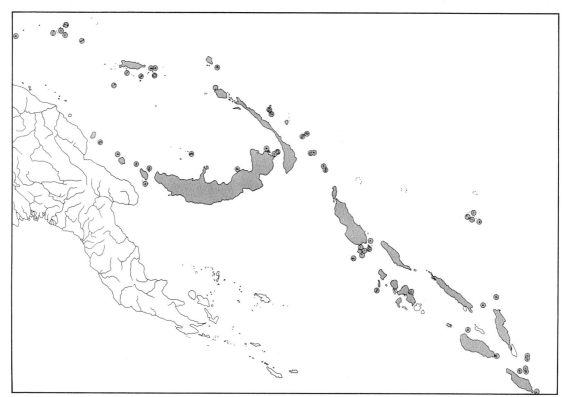

rear of flank, thigh, vent, and under tail-coverts. Tail feathers greyish-black (if fresh) or blackish-brown (if worn); axillaries and under wing-coverts dusky grey.

CHICK: rather blackish, like chick of Dusky Megapode *M. freycinet*, but crown and mantle slightly tinged red-brown, rump and upper tail-coverts more maroon, and underparts slightly paler greyish rufous-brown with more isabelline of feather-bases visible; black-brown of lower mantle, scapulars, back, tertials, and inner upper wing-coverts barred light rufous-brown.

IMMATURE: in first imm plumage, like ad, but upperparts backwards from lower mantle as well as flank and belly sometimes browner, less dusky olive; in all birds, tail, some secondaries, and outer primaries still of first (juv) generation, shorter and narrower than in ad and than neighbouring feathers of second generation. Juv outer primaries (p9–p10) short, tapering, tip pointed; juv p8, if present, shorter than p9 (longer in older birds); length of juv p10 from tip to insertion in skin 85–98 mm ($n = 2$; in ad, 117–133 mm, average 125.6 ± 6.34, $n = 8$). In second imm plumage, like ad, but tail and outer primaries of second generation, still shorter than in ad; length of p10 111–125 mm (117.6 ± 5.56, $n = 5$), of tail 62–74 mm (67.5 ± 4.49, $n = 5$).

BARE PARTS
ADULT, IMMATURE: iris light brown, golden-brown, brown, dark brown, red-brown, dark brown-red, or red. Bare skin of head and neck deep pink-red, light red, bright red, or brick-red; crown sometimes more strawberry-red, side of head dull purple-red or strawberry-red, throat dull pink. Bill-tip dull yellow or golden-yellow; base dull green, fuscous-green with black wash, brown-olive, or greyish yellow-horn. Leg and foot greenish-yellow, greenish-brown, yellowish grey-horn, olive, olive-brown, dusky olive-green, blackish-olive, dark ash-grey, dark grey, or fuscous-black; toes often darker than tarsus, but soles paler greenish, yellowish, or pinkish; tarsus of ♀ on average slightly paler than in ♂.

CHICK: iris light yellow, grey-brown, or chocolate-brown. Bare skin round eye, ear, and on chin pinkish-brown. Bill dark fuscous-grey to black, base pink-brown or dull brown. Leg and foot yellowish-black, grey-black, or dark grey-brown; toes blacker, soles dull orange or yellow.

MOULTS
Primary moult serially descendant. Apart from chicks which have just started moult with series of second generation primaries from innermost (p1) outwards, generally two series of moult in each wing, with moult active or suspended on two centres; thus, two generations of primaries in each wing (when inner series reached, for example p4, and outer had just completed with regrowth of p10) or three generations in each wing (when inner series reached, for example p1–p2, an earlier series to, for example p7–p8, and outermost feathers of a still older generation); exceptionally (in one of 21 birds), three moult centres active on primaries. In both ad and older imm, series in wing 4–8 (6.1 ± 1.13, $n = 13$) feathers apart. Of seven imm, primary moult active in 86%; of 18 ad, primary moult active in 72%, remainder suspending moult. Too few dated birds examined to be certain whether a distinct moulting season in population exists; of birds with known collecting date, moult active in Jan (two birds), Mar (1), May (4), June (1), July (2), Aug (2), Nov (1), and Dec (1); moult suspended in birds from Feb (1), June (2, of which one an egg-laying ♀), and July (1). Chicks examined from Jan (1), June (2), Aug (2), and Nov (2), but none of these had started moult.

MEASUREMENTS
ADULT: New Britain ($n = 18$), New Ireland ($n = 3$), New Hannover ($n = 2$), and one each from islands of Duke of York, Watom, Mioko, and Credner; skins of 6 ♂♂, 12 ♀♀, and 9 unsexed birds (BMNH, RMNH, SMTD, ZFMK, ZMB).

Melanesian Megapode *Megapodius eremita*

	n	mean	s.d.	range
wing	27	221.5	7.96	205–235
tail	27	72.3	3.30	65–78
tarsus	23	66.4	3.04	62.8–71.8
mid-toe	18	35.7	1.98	33.0–39.0
mid-claw	21	19.9	1.23	18.0–22.1
bill (skull)	23	26.9	1.42	24.3–29.3
bill (nostril)	24	11.8	0.85	10.5–13.3
bill depth	22	8.2	0.49	7.3–8.9

Wing mainly 216–230 mm. Sexual differences slight, e.g. wing, ♂ (205–) 219–231 mm (221.6 ± 9.29, $n = 6$), ♀ (210–) 216–230 mm (220.6 ± 5.41, $n = 12$); tarsus, ♂ 63.5–71.8 mm (67.0 ± 3.89, $n = 6$), ♀ 63.2–69.0 mm (65.6 ± 2.09, $n = 10$). Presumed hybrids from Karkar and Bagabag: wing, ♂ 229 mm, ♀ 215, 225, 232 mm; tail, ♂ 73 mm, ♀ 68, 72 mm; tarsus, ♂ 65 mm, ♀ 61, 62, 63 mm; bill, ♂ 28 mm, ♀ 27, 27, 28 mm (Diamond and LeCroy 1979).

Solomon Is: Bougainville to Guadalcanal; skin of 3 ♂♂ and 3 ♀♀ (BMNH, SMTD).

	n	mean	s.d.	range
wing	6	223.6	5.50	217–230
tail	6	73.0	3.54	69–77
tarsus	6	70.4	2.34	68.0–73.8
mid-toe	4	36.4	0.97	35.7–37.8
mid-claw	4	20.7	1.36	19.3–22.3
bill (skull)	5	27.6	0.70	26.8–28.7
bill (nostril)	4	11.8	0.45	11.2–12.2
bill depth	4	8.4	0.17	8.2–8.6

San Cristobal, skins of 1 ♂, 1 ♀, and 1 unsexed bird (BMNH): wing 225–241 mm (average 231.0), tail 76–83 mm (78.8), tarsus 76.2–78.5 mm (77.6), toe 38.0 mm ($n = 1$), claw 22–24 mm (22.8), bill to skull 28.1, 30.2 mm ($n = 2$), to nostril 11.8, 13.8 mm ($n = 2$), bill depth 8.3 mm ($n = 1$). Purdy and Pigeon Is (Admiralty group), skins of 2 ♂♂ (BMNH, ZMB): wing 201, 208 mm; tail 72, 73 mm; tarsus 61.8, 67.0 mm; middle toe 33.0, 36.7 mm; middle claw 18.2, 20.1 mm; bill to skull 26.0, 26.5 mm; bill to nostril 10.3, 11.5 mm; bill depth 7.0, 7.1 mm.

Sexes combined, wing and tarsus of birds from (1) Ninigo Is, (2) Manus, Rambutyo, and other islands of Admiralty group, (3) Long I., (4) Umboi (Rooke) I., (5) New Britain, Vitu, and Duke of York Is, (6) New Ireland, New Hannover, and Mussau (Squally) I., (7) Tabar, Lihir, Tanga, Feni, and Nissan groups of islands, (8) Bougainville and its offshore islands (Mono, Shortland, Alu, Poporang, Fauro, Oema), (9) Choiseul and Santa Isabel, (10) central Solomons from Vella Lavella to New Georgia, (11) Guadalcanal and its offshore islands (Pavuvu, Savo, Komachu, Beagle), (12) small islands in E Solomons (Ramos, Ndai, Santa Ona), (13) San Cristobal (Mayr 1938).

	Wing				Tarsus			
	n	mean	s.d.	range	n	mean	s.d.	range
(1)	18	210.3	6.13	198–219	–	–	–	54–63
(2)	18	214.5	5.70	202–220	c. 14	59.4	–	57–65
(3)	10	221.1	4.51	215–230	10	62.8	–	55–69
(4)	3	213.0	4.36	210–218	3	61.7	2.08	60–64
(5)	22	218.6	6.02	212–235	–	c.62.5	–	58–68
(6)	12	225.6	6.37	217–235	c.11	65.4	–	58–70
(7)	29	221.4	5.28	213–231	29	63.6	–	56–67
(8)	12	221.2	7.39	211–235	10	63.7	3.65	60–72
(9)	16	221.9	9.07	206–233	–	–	–	–
(10)	16	219.2	9.74	200–237	–	–	–	–
(11)	13	222.2	5.93	214–233	7	65.1	2.27	64–70
(12)	11	216.8	4.58	208–224	11	66.2	3.31	59–70
(13)	4	237.0	7.75	226–244	4	70.8	4.27	68–77

CHICK: wing 86–100 mm (94.5 ± 4.80, *n* = 6); tarsus 26 mm (*n* = 1) (BMNH). Live birds (New Britain, K. D. Bishop, unpublished data): wing 87–98 mm (92.4 ± 3.45, *n* = 17), bill 6.9–8.9 mm (7.5 ± 0.56, *n* = 17), tarsus 21.0–28.5 mm (26.1 ± 1.80, *n* = 16).

WEIGHTS
ADULT: New Britain, New Ireland, and off-lying smaller islands: 570–660 g (Heinroth 1902); ♂ 500+ g (*n* = 1), ♀ 395 g (*n* = 1) (AMNH); ♀ 550–580 g (561.7 ± 12.11, *n* = 6) (Wolff 1965). San Cristobal: ♂ 700, 750, 750 g, ♀ 925 g (Mayr 1938). Bougainville: ♂ 615, 640 g (*n* = 2), ♀ 735 g (*n* = 1). Presumed hybrids between this species and the New Guinea Megapode *M. decollatus* from Karkar and Bagabag: ♂ 750 g (*n* = 1), ♀ 580, 700, 780 g (*n* = 3) (Diamond and LeCroy 1979).

CHICK: 46–73 g at hatching (56.9 ± 6.59, *n* = 16, New Britain, K. D. Bishop, unpublished data); 59 g (*n* = 1, Meyer 1930*a*).

GEOGRAPHICAL VARIATION
Rather marked, both in colour and size, but without clear trend and hence no race within *M. eremita* recognized. Size, as expressed in wing length, on most islands *c.* 220 mm, but somewhat smaller on islands from Admiralty group and further westwards, wing *c.* 210–215 mm, and on small islands N and E off Malaita and San Cristobal; somewhat larger on Choiseul and Santa Isabel, wing *c.* 225–227 mm, and (especially) on San Cristobal; according to Mayr (1938), birds largest in mountain areas of larger islands, smallest on smaller islands, but no difference, for example between Bougainville (wing *c.* 220 mm) and its offshore islands (wing *c.* 222.5 mm), between Guadalcanal (wing *c.* 221 mm) and smaller islands nearby (wing *c.* 223 mm), or between larger and smaller islands in Bismarck group of islands, and difference in size between large birds on San Cristobal and smaller ones on nearby Santa Ona probably not an effect of island size but due to isolated position of San Cristobal birds. Birds from Ninigo (type locality of *eremita*, at western end of geographical range) dark, mantle and scapulars dull olive; birds from nearby Admiralty group similar, but those from Ndai ('Gower') and Ontong Java off E Solomons (at opposite end of geographical range) also dark, and birds from Nissan (NW of Bougainville) and Ramos (near Ndai) scarcely paler. Population from San Cristobal markedly paler, grey of head and underparts lighter, mantle and scapulars lighter and more rufous-brown, less dull olive, crown paler and brighter, throat more densely feathered; birds from nearby Santa Ona and probably those of Ugi ('Gulf') I. (type locality of '*brenchleyi*') still more rufous; on the other Bismarck and Solomon Is, colour intermediate between these dark dull olive and more rufous-brown extremes, with birds of Bismarck group tending to be nearer typical *eremita* and those of larger Solomon Is near '*brenchleyi*' (Mayr 1938). In sample examined, some variation in colour of upperparts, but differences individual, not geographical, and no constant difference in colour between Bismarck and Solomon birds. On Karkar and Bagabag Is, between New Britain and mainland New Guinea, populations occur which are intermediate between the Melanesian and New Guinea Megapode, but nearer to the former (Mayr 1938; Diamond and LeCroy 1979).

Range and status
Range: Wuvulu, Ninigo ('Echiquier') Is, Hermit I., and Admiralty group (Manus, Purdy group, Pigeon, Lou, Pak, Tong, Rambutyo); Mussau ('St Matthias', 'Squally'), Emirau, Kung, New Hannover, Dyaul, New Ireland, New Britain, and smaller islands off N and E New Britain (Garove in Vitu group, Lolobau, Watom, Matupi, Mioko, Credner Is, Duke of York); Siassi I., Sakar, Umboi ('Rooke'), Tolokiwa, Long, and Crown Is; Tabar, Lihir, Tanga, Feni, Pinipel, and Nissan groups of islands; Buka, Bougainville, and islands off S Bougainville (Mono = 'Treasury', Shortland, Alu, Poporang, Whitney, Fauro, and Oema); Choiseul, Santa Isabel group, Ontong Java, Ramos, and Ndai ('Gower') Is;

New Georgia group (Vella Lavella, Ganongga = 'Simbo', Navoro, Kolombangara, Rendova, Keru, New Georgia), Guadalcanal and off-lying islands (Pavuvu, Savo, Beagle, Komachu), Malaita, and San Cristobal with off-lying islands (Ulawa, Olu Malau group, Ugi = 'Gulf', Santa Ona = 'Santa Anna').

Status: locally common to abundant. In the late nineteenth century, many thousands of megapodes were reported at a single nesting ground on Guadalcanal (Wolff 1965). The number of eggs laid at communal nesting grounds on New Britain is extremely high. At Garu, an estimated 30 000 eggs are laid per breeding season (Kisokau 1976). At the same nesting ground, an estimated 15 000 eggs were harvested between Apr and Dec 1971 (Downes 1972), and 22 489 eggs between 31 May and 9 Sept 1978, with c. 800 eggs on 31 May alone (Bishop 1980). At Pokili, 500 eggs were collected on a single day in 1978. The number of birds during the peak of the egg-laying season in June 1978 was estimated to be 2000 at Garu and 53 000 at Pokili (Broome et al. 1984). During the non-egg-laying season the birds leave the vicinity of the nesting grounds and probably move from the north coast, of New Britain to the drier south coast, although there are no data to confirm this (Broome et al. 1984). Flocks of over 60 individuals have been observed flying high in a westerly direction (Bishop 1980). The number of birds at a communal nesting ground on Savo on 15 Aug 1981 between 06.20 and 07.20 h was 171–198 adults, probably all ♀♀ (Roper 1983). Still common on Nissan and Pinipel Is, N Solomons, in 1982 (Harding 1982), and on New Georgia in 1986–1988 (Blaber 1990).

Field characters

Length 35–39 cm. A medium-sized very dark megapode with naked skin of head and neck dull pinkish-red, restricted to islands northeast of New Guinea. On Karkar and Bagabag Is forms mixed population with *M. decollatus*, which has bright olive-brown instead of blackish upperparts and more conspicuous crest.

Voice

Duet (see sonogram), starting with a short, lightly laughing 'keyououourrrr', initially rising and subsequently descending in pitch. The partner joins in shortly after with two notes 'keyou keyourr' of which the first note rises only slightly, if at all, in pitch, while the second note initially rises and subsequently descends a little (description by R. Dekker from tape-recordings from New Britain made by K. D. Bishop). Similar in structure to duet of the Sula Megapode *Megapodius bernsteinii*, *M. freycinet*, and Forsten's Megapode *M. forstenii* but at a lower pitch and shorter. According to K. D. Bishop (unpublished data), it is the ♂ who makes the 'kiaow kiaow' sound (described as 'keyou' above) and starts the duet, immediately answered by the ♀ who gives the 'low, rather tremulous whinnying chuckle'. However, based on similarity with other *Megapodius* spp., it would be expected that the ♂, not the ♀, produces the 'keyououourrrr' sound, as this is typical for the ♂ in other *Megapodius* spp., such as the Nicobar Megapode *M. nicobariensis* (R. Dekker, personal observation).

The most distinctive call, thought to be the territorial call, is a series of loud, slightly nasal, resonant braying cries, lasting 1–5 sec, uttered by the ♂ only (K. D. Bishop, unpublished data). Vocalizations also described as a low 'ko–ko–ko–ko'; a loud 'kio! kio! kio!'; a long dragging 'kio—o—o—o'; a low 'kro—kro—ko—ko' and 'kio..kio..kio' (New Britain, Kisokau 1976). An individual of the hybrid population on Karkar produced a high down-slurred 'kee-ya' (Diamond and LeCroy 1979).

Habitat and general habits

On New Britain mainly in lowland rainforest, sometimes in swamp forest, garden regrowth areas, and hill forest; virtually absent from palm plantations, and clear-felled seedling areas (Broome et al. 1984). Restricted to tall secondary and primary forest fringing the

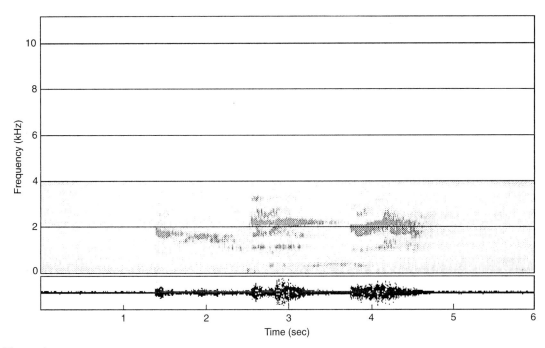

M. eremita
K. D. Bishop, New Britain, July 1986.

coast up to 5 km inland (below 50 m altitude) on S Bougainville. Elsewhere recorded up to 10 km inland and 600 m altitude. On Guadalcanal, San Cristobal, Ugi, and Ulawa mainly in coastal and lowland forest, but also in hill forest. Normally seen singly or in pairs when away from the nesting grounds. Roosts in trees, often communally, at a height of 4–15 m. Small flocks make roosting flights of several kilometres from S Bougainville to small offshore islands at a height of approximately 3–5 m above sea-level. Chicks become active at dusk as in other megapodes.

Food

Based on stomach contents: plant material, seeds, fruits, snails, earthworms, millipedes, centipedes, caterpillars, beetles, beetle larvae, grasshoppers, crickets, as well as a crawfish. Has been observed jumping up and plucking small cauliflorous fruits from the lower trunk of a small Fig tree, *Ficus* sp. (Bishop and Broome 1979).

Displays and breeding behaviour

At the onset of the breeding season the birds are very active, clearing the burrows of forest floor debris. On Savo, most birds arrive at the communal nesting ground between 05.30 and 06.00 h and stay there for 1.5–3.5 h. Probably the ♀♀ are alone responsible for the digging and are not accompanied by the ♂♂. When digging, one to 12 strokes (usually three to five) are given by each foot alternately (Roper 1983). At the site the birds continually utter short cries. Aggressive behaviour occurs when birds get too close to neighbouring burrows; this shows considerable variation and includes loud cries, bobbing of the lowered head, and raising and flapping of wings. Occasionally a bird flies over its opponent or both individuals flap into the air pecking at one another. No aggressive display seen resulted in the defending bird losing its burrow. The duration of 16 interactions, from the moment digging ceased until it was recommenced, was in the range of 1–20 sec, with a mean of 5.4 ± 1.23 sec (Roper 1983). No clear instances of courtship, for example, copulation, ever observed at the nesting ground (Bishop and Broome 1979). Hybridizes with *M. decollatus* on Karkar and Bagabag Is (Diamond and LeCroy 1979; see also Chapter 3).

Breeding

NEST: the Melanesian Megapode lays its eggs in burrows at volcanic soils (for example New Britain and Savo) and sun-exposed beaches (on a small islet east of Ganongga and on Watom), between decaying roots of trees (Wuvulu I., and the Solomon Is of Bougainville, Choiseul, San Cristobal, and Malaita), in mounds (Ninigo Is, Wide Bay, the E Solomons, and probably S New Britain), and even in man-made rubbish pits (Liga, New Ireland). Nesting grounds of this species are the largest and most spectacular communal laying grounds reported so far. The Pokili nesting ground on New Britain covers an area of 2–4 km^2 and carries tens of thousands of burrows often 1 m apart and occasionally even running into each other. The Garu nesting ground, also on New Britain, has an estimated 11 776 burrows in an area of 67 ha (Broome *et al.* 1984). Other nesting grounds on New Britain are Garili and Peesi in the northwest and Matupit I. and Vudal in the northeast. Communal nesting grounds are not known from S New Britain, where the species is said to build mounds. This, however, needs to be confirmed. On Ganongga (= Simbo), a volcanically heated nesting ground contained at least 200 separate burrows, which varied in shape from vertical holes with loose dirt in the bottom to tunnel-shaped holes. The dimensions of individual burrows varied considerably, from 25 to 90 cm in diameter, and 30–90 cm deep (Sibley 1946). At a geothermal nesting ground on Savo, eggs were deposited at a depth of 88 ± 3.4 cm ($n = 21$) at a temperature of 33.0 °C (Roper 1983, which should be referred to for illustration). Communal nesting grounds, heated geothermally or by the sun, are known also from Duke of York I., New Ireland, the Admiralty Is (for example, Lou I.), and Bougainville.

EGGS: similar in shape and coloration to eggs of *Macrocephalon maleo*, *Eulipoa wallacei*, and other *Megapodius* species (see under *M. pritchardii*). SIZE: 71.3–79.8 × 44.4–50.0 mm (76.9 × 47.5, $n = 17$, Bismarck Archipelago, Steadman 1991); 79.0 ± 2.00 × 47.0 ± 2.00 mm ($n = 4$, New Ireland, Oustalet 1881); 75–82 × 45–48.6 mm ($n = 9$, New Britain); 75–84 × 46–50 mm (79.6 ± 3.78 × 48.1 ± 1.57, $n = 7$, Watom); 73.5–84.4 × 43.9–51.4 mm (78.0 × 47.8, $n = 47$, Solomon Is, Steadman 1991); 78.2 ± 0.79 × 47.3 ± 0.06 mm ($n = 6$, Savo, Roper 1983). WEIGHTS: 84.5–107.8 g (101.9 ± 7.92, $n = 7$, Watom); 85–93 g (89.7 ± 4.16, $n = 3$, New Britain); 89–116 g (98.7 ± 8.62, $n = 9$, Bismarck Archipelago); 67–86.4 g ($n = $?, Bismarck Archipelago).

BREEDING SEASON: during the dry season from Apr to Nov (Dec) on Garu and Pokili, N New Britain, with a peak in June–July. Birds absent from the nesting grounds during the wet season from Dec until late Mar (Bishop 1980; Broome *et al.* 1984). Egg laying possibly in two waves, from Apr to June and from late Sept to Nov (Bishop and Broome 1979). Other reports do not suggest seasonality on New Britain. Seasonality possibly on Watom where eggs have been collected in June, July, and Oct (for example, Meyer 1930*a*). In the Bismarck Archipelago, eggs and chicks in Aug (Reichenow 1899). On Hakuola, Ontong Java, Solomon Is, egg collected in Jan (Bayliss-Smith 1972). Incubation period dependent on incubation temperature; varies between 6 and 9 weeks, once even 10 weeks (for example, Meyer 1930*a*).

Local names

Kaiyo (Nissan and Pinipel), apagei (Ninigo), kiau, malige tsiuti (Bougainville), kakiau (Duke of York), kihau (Guadalcanal), aupwau (S. Cristobal), kokoko (Ugi), ngiok, disin, nee-ock, or mulong (New Britain), lápi (Ganongga = Simbo), ngero (Savo), angiok or giok (Bismarck Archipelago generally), kho'io (Santa Isabel).

References
Hartlaub (1867), Gray (1870), Cabanis and Reichenow (1876), Sclater (1877), Oustalet (1881), Woodford (1888), Meyer (1890b), Ogilvie-Grant (1893, 1897), Bolau (1898), Dahl (1899), Reichenow (1899), Rothschild and Hartert (1901, 1914b, c), Heinroth (1902), Mayr (1930, 1938, 1945), Meyer (1930a, b, 1933), Pockley (1937), Sibley (1946), Cain and Galbraith (1956), Schönwetter (1961), Wolff (1965), Gilliard and LeCroy (1967a), Bayliss-Smith (1972), Downes (1972), Kisokau (1976), Schodde (1977), Bishop (1978, 1980), Bishop and Broome (1979), Diamond and LeCroy (1979), Harding (1982), Roper (1983), Broome et al. (1984), Coates (1985), White and Bruce (1986), Blaber (1990), Steadman (1991), Webb (1992).

Vanuatu Megapode *Megapodius layardi* Tristram, 1879

Megapodius layardi Tristram, 1879. *Ibis*, 1879, p. 194. (Efaté, Vanuatu.)   PLATE 7

Other names: New Hebrides Scrubfowl, Tristram's Scrubfowl.

Monotypic.

Note on nomenclature
When Tristram described *M. layardi* in 1879 from Vaté (= Efate), Vanuatu (formerly New Hebrides), he deliberately ignored the *epitheton specificum 'brazieri'* proposed by Sclater in 1869 and based on an egg sent by John Brazier from Vavua Lavu (= Vanua Lava), Banks Is, to the Zoological Society of London, since there was no proof that the bird he described was the same as the Banks Is' megapode. Later, when it became clear that *brazieri* and *layardi* referred to the same species and thus were synonyms, the senior synonym was 'found inadmissible' as its description was based on an egg (see, for example, Ogilvie-Grant 1893, p. 459).

According to the International Code of Zoological Nomenclature (ICZN), 3rd edition, 1985, a name is valid also when based on a particular product of a species (in this case an egg). So, the senior synonym *brazieri* is valid and thus has priority over *layardi* (Article 23 of the ICZN). However, in this case we propose not to apply the 'Principle of Priority' but to have the unused senior synonym suppressed (Article 79c of the ICZN), in order not to disturb stability or cause confusion, and to maintain existing usage of the junior synonym *layardi* which has been used in literature ever since (for example, Ogilvie-Grant

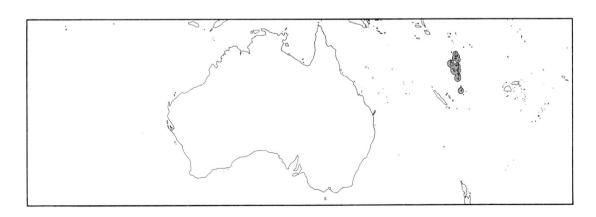

1897; Sharpe 1900; Shufeldt 1919; Amadon 1942; Dekker 1990c).

Description

PLUMAGES
ADULT: sexes similar. Head and neck more fully bare than in any other *Megapodius*, covered with a few short fine black bristles on forehead, crown, cheek, and side and rear of neck only, as well as some very scanty short feathers on chin and throat, and a thin strip of black feathers down central hindneck. However, a dense sharply delimited patch of black feathers springs down from rear crown and nape, forming a short broad crest; whole crest *c.* 2.5–3.5 cm long and *c.* 2 cm wide. Upperparts entirely plumbeous-black, slightly tinged bluish (but generally less bluish than most populations of the Dusky Megapode *M. freycinet*); lower scapulars, tertials, secondaries, and most longer wing-coverts slightly tinged olive, most pronounced on tertial-coverts which are dull greenish-black. Rump and upper tail-coverts sooty black with slight but distinct olive-brown tinge. Chest, flank, and under tail-coverts plumbeous-black; remainder of underparts black with plumbeous feather fringes, but black slightly tinged olive-grey (especially if plumage worn) and some grey feather-bases just visible on mid-belly and vent, underparts appearing slightly less blackish than in *M. freycinet*. Flight feathers, tail, axillaries, and under wing-coverts black, small coverts along bend of wing grey.

CHICK: buff-brown with dark brown barring on wings and back; abdomen pale buff (Bregulla 1992).

IMMATURE: similar to adult but overall duller.

BARE PARTS
ADULT: iris brown or dark brown. Bare skin of head and neck red. Bill pale yellow, horn, or brown. Leg and foot bright yellow, claw horn-coloured.

CHICK: no information.

IMMATURE: iris light brown. Bill dark olive-brown. Leg and foot yellowish-ochre with blackish-brown patches especially on the toes.

MOULTS
Primary moult serially descendant. Two moult series in each wing in five birds examined, but moult suspended in one from May, two from July, and one from Sept, active only in another bird from July. A single bird from Sept had one moult series only, reaching to p6 and then suspended; outer feathers contrastingly worn, perhaps second or third imm generation.

MEASUREMENTS
ADULT: data on skins of 6 ♂♂ and 2 ♀♀ from Efate and Espiritu Santo (BMNH, ZMB).

	n	mean	s.d.	range
wing	8	246.2	5.46	240–257
tail	8	89.6	3.42	85–95
tarsus	8	75.1	3.55	69.0–80.2
mid-toe	6	38.5	0.77	37.0–39.2
mid-claw	8	23.0	0.83	22.2–24.5
bill (skull)	8	27.3	0.76	26.0–28.6
bill (nostril)	8	11.8	0.56	10.7–12.3
bill depth	8	8.5	0.49	7.9–9.0

In ♂♂ of this sample, wing 240–257 mm (246.2 ± 5.64, $n = 6$), tarsus 74–80 mm (76.7 ± 2.17, $n = 6$); in ♀♀, wing as in ♂♂, 241 and 251 mm, but tarsus only 69.0 and 71.6 mm. Birds examined by Amadon (1942) from Efate, Epi, Mai (=Emae), Malekula, and Pentecost (central islands group) measured: wing, ♂ 237–256 mm (248.4 ± 6.54, $n = 9$), ♀ 236–251 mm (243.1 ± 6.20, $n = 8$); tarsus, ♂ 64–69.5 mm (mean 66.4), ♀ 59–67 mm (mean 63). On Ureparapara, Gaua, and Valua (Banks Is group), 1 ♂ and 3 ♀♀ had wing 252–260 mm (256.3 ± 4.04, $n = 3$), tarsus 64–71 mm (67.2 ± 2.99, $n = 4$) (Amadon 1942). Apparently, tarsus measured by Amadon in different way than in sample above, but large size of Banks Is birds is evident.

CHICK: no information.

WEIGHTS
CHICK: approximately 60 g at hatching (Bregulla 1992).

GEOGRAPHICAL VARIATION
Birds from Banks Is slightly larger than those from elsewhere in Vanuatu, but series of skins not large enough to rule out individual variation (Amadon 1942). If a larger data set of measurements of Banks Is birds becomes available and size difference cited under Measurements above proves to be valid, the population of Banks Is and the population of other Vanuatu islands may deserve recognition as two separate races.

Note: usually considered to be close to the Melanesian Megapode *M. eremita*, the bare forehead of *M. eremita* being present in *M. layardi* in an exaggerated form. However, it differs considerably in proportions from *M. eremita*, especially tail of *M. layardi* being relatively longer (approaching that of the Orange-footed Megapode *M. reinwardt* in relative length), toes, claws, and especially bill relatively shorter. Also, Santa Cruz Is, which may form a stepping stone between the distribution of *M. eremita* on the Solomon Is and *M. layardi* on the Banks Is, are apparently not inhabited by any megapode. Instead of this, relationship of *M. layardi* may have been with *Megapodius andersoni* Gray 1861, a possible megapode seen on New Caledonia by Anderson on Captain Cook's second voyage, described as a blackish-brown grouse-like bird with bare legs. (For discussion on *M. andersoni* and fossil remains of *Megapodius molistructor* from New Caledonia, see Chapter 3.)

Range and status
Endemic to Vanuatu (formerly New Hebrides). Reported (from north to south) from Ureparapara, Valua, Vanua Lava, Gaua (the Banks Is), and from Espiritu Santo, Malo, Oba, Maewo, Pentecost, Ambrim, Malekula, Paama, Lopevi, Epi, Tongoa, Tongariki, Emae, Mataso, Nguna, Mau, Efate (the central islands). Reported from Tana in the southern islands by Bennett (1862), confirmed by eggs in BMNH (Steadman 1991).

According to Layard and Layard (1878), common on Ambrim and Espiritu Santo, but becoming rare on Efate due to predation by introduced cats and pigs. However, still reported to be widespread and common as recently as 1977.

Field characters
Length *c.* 42–45 cm. The only megapode on Vanuatu. Unmistakable, generally blackish with bare parts of head and neck red; legs yellow.

Voice
Vocal, especially at dusk and dawn. Usually a subdued or loud hoarse clucking (Bregulla 1992), described by Layard and Layard (1878) as a hoarse croak. Also a loud, often repeated, two-syllabled slurred 'took-tooorrr' with the first syllable short and the second drawn-out and decreasing in volume. ♂ and ♀ are reported to call in duet. While feeding, both sexes produce soft short notes at irregular intervals (Bregulla 1992).

Habitat and general habits
A shy bird of the lowland forest, also found at moderate altitudes in deep densely wooded ravines. Usually occurs singly or in pairs, also roosts in trees in pairs. Probably monogamous. May visit offshore islands to roost as in *M. eremita* (Bregulla 1992).

Food
Invertebrates such as worms and snails as well as seeds and fruits. On labels of specimens from Efate, Layard mentioned 'small hard seeds' and 'small helices' (snails).

Displays and breeding behaviour
Breeding behaviour mostly undescribed but seemingly similar to other *Megapodius* spp. Mounds are used communally by several pairs for several years. The whole process of digging and egg laying in burrows around the base of large decaying trees takes between 2 and 4 h (Bregulla 1992).

Breeding

NEST: on Banks Is it builds very large mounds, but on many islands of Vanuatu it buries its eggs in holes around the base of large, decaying forest trees (Brazier 1869; Bregulla 1992). According to local information, eggs are also laid in burrows at communal sites on islands with active volcanoes (Bregulla 1992).

EGGS: similar in shape and coloration to eggs of *Macrocephalon maleo*, *Eulipoa wallacei*, and other *Megapodius* species (see under *M. pritchardii*). SIZE: 71.5–89.7 × 42.6–50.8 mm (81.8 × 47.4, n = 47, Tana, Espiritu Santo, Tongoa, Malekula, Efate, and Banks, Steadman 1991); 74.2–89.5 × 46.3–50.3 mm (79.5 × 47.4, n = 9, Schönwetter 1961); 77–85 × 43–47 mm (n = 20, Bregulla 1992); 86.4 × 48.2, 75.8 × 48.7 mm (Tongoa, n = 2); 90 × 50 mm (Efate, n = 1); 83 × 47 mm (Ambrim, n = 1). WEIGHTS: no information; 50–70 g given in Bregulla (1992) too low and certainly incorrect.

BREEDING SEASON: eggs on Banks Is in Aug and on Tongoa in Dec. According to local information, eggs laid every month of the year (Bregulla 1992).

Local names

Tarboush and malou (Efate), nemal (Espiritu Santo, label BMNH), also skraptak or sikraptak and namalau.

References

Gray (1861), Bennett (1862), Brazier (1869), Layard and Layard (1878), Tristram (1879), Oustalet (1881), Ogilvie-Grant (1893, 1897), Sharpe (1900), Shufeldt (1919), Amadon (1942), Mayr (1945), Schönwetter (1961), Parker (1967a), Diamond and Marshall (1976), White and Bruce (1986), Dekker (1990c), Steadman (1991), Bregulla (1992).

New Guinea Megapode *Megapodius decollatus* Oustalet, 1878

Megapodius decollatus Oustalet, 1878. *Bulletin hebdomadaire de l'Association Scientifique de France*, **21**, p. 248. (d'Urville = 'Tarawai' = Kairiru I.)

PLATE 8

Other names: New Guinea Scrubfowl, Brown Scrubfowl.

Monotypic. Synonyms: *jobiensis* Oustalet, 1881, Jobi (= Japen); *brunneiventris* A. B. Meyer, 1891, Astrolabe Bay; *huonensis* Stresemann, 1922, Heldsbach Coast, Huon Gulf.

Note on nomenclature

This species has until now been known as *M. affinis*, based on two specimens (syntypes) from Rubi (southwest shore of Geelvink Bay) described by Meyer (1874b) as 'similar in colour to *M. reinwardt*, but smaller and legs dark'. Both these specimens were in the collection of the Staatliches Museum für Tierkunde in Dresden (SMTD), but one of them was destroyed during World War II. The other one is still present, and on examination proved to be an Orange-footed Megapode *M. reinwardt* in first imm plumage, the legs being not dark but orange, as one would expect in a *M. reinwardt* skin, and the small size being due to immaturity. Two adults examined from Rubi were also true *M. reinwardt*, and these three birds were completely similar to *M. reinwardt* from the Vogelkop peninsula. Though one of the type specimens of *M. affinis* may have had darker legs than the other (Meyer described the legs of one of the two syntypes as 'black', the other as 'brown-black with red tinge', and probably the latter bird is the remaining one which now has legs orange in skin), it is quite likely that it also involved a true *M. reinwardt*. To avoid confusion it is better to avoid the use of the name

New Guinea Megapode *Megapodius decollatus*

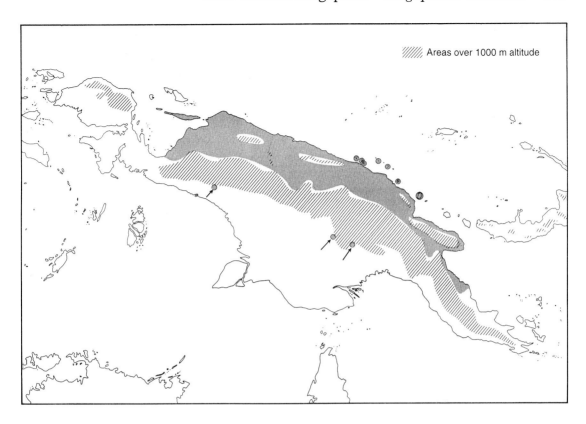

M. affinis for the birds inhabiting N New Guinea between Japen I. and Huon Gulf. We select this remaining syntype of *M. affinis* in the SMTD as lectotype of *M. affinis*; a synonym of *M. reinwardt*.

The next oldest name available is *M. decollatus*, described by Oustalet (1878) from d'Urville I. (= 'Tarawai' = Kairiru) off N New Guinea.

Description
PLUMAGES
ADULT: sexes similar. Rather like *M. reinwardt* in general colour of body and due to its pointed crest, but legs dark (not orange-red) and colours generally duller, more like those of Forsten's Megapode *M. forstenii* or Melanesian Megapode *M. eremita*. Forehead, side of head, chin, and front and side of neck covered with rather sparse and reduced grey feathers, bare skin showing more distinctly than in *M. reinwardt*, though forehead and side of head and neck less naked than in *M. eremita*. Crown and crest brown with olive tinge; crest pointed, as in *M. reinwardt*, but shorter. Hindneck, upper mantle, and underparts medium grey. Lower mantle, scapulars, upper wing-coverts, and tertials dark olive-brown, contrasting with grey of upper mantle; generally less rufous than *M. reinwardt*, more similar to *M. forstenii* or *M. eremita* or slightly brighter and paler; unlike *M. eremita*, no grey on lesser upper wing-coverts. Rump and upper tail-coverts dark brown or rufous-brown, not as chestnut-brown or maroon as in *M. reinwardt*. Unlike *M. reinwardt*, flank usually extensively olive-brown or rufous-brown; occasionally, brown tinge restricted to mid-flank (then more olive and less contrasting than in *M. reinwardt*; in latter species, brown mainly confined to rear of flank and thigh), but more frequently extends up to side or mid-belly and to chest. Wing and tail as in *M. forstenii*. Much variation in colour; grey of

head and underparts in some birds distinctly darker and more plumbeous than in others, brown of crown, upper side of body, and flank in some birds colder olive-brown, in others purer olive or more rufous; this variation in part individual or due to influence of bleaching and abrasion, in part geographical (see 'Geographical variation').

CHICK: upperparts fuscous-brown, distinctly barred black and buff on scapulars, tertials, and inner upper wing coverts; underparts dark greyish-brown, paler on cheek, chin, and mid-belly.

IMMATURE: in first imm plumage, like ad, but outer primaries, some secondaries, and sometimes tail still juv, much shorter than in ad, secondaries partly barred black and buff, juv outer primaries relatively shorter than second generation of inner primaries; p10 narrow, pointed at tip, length from tip to insertion in skin 91–99 mm ($n = 2$; in ad, 117–133 mm, average 125.2 ± 4.82, $n = 16$); juv tail 59 mm ($n = 1$). In second imm plumage, outer primaries of second generation, middle of third generation, inner of third or fourth; second generation primaries still shorter than in ad, tip rather pointed (less sharply than in juv, less broadly rounded than in ad); length of p10 111–118 mm (114.3 $\pm$ 3.18, $n = 3$), and tail also still somewhat shorter than in ad, 68–69 mm ($n = 2$).

BARE PARTS
ADULT, IMMATURE: iris light brown, reddish-brown, or dark brown. Bare skin of head and neck red, vermilion-red, or dull red. Bill either uniform greenish, green-brown, light-brown, dark brown, or plumbeous, or base green-brown to dull olive-brown with paler dull yellow, green-yellow, or horn-yellow tip. Leg and foot variable, either uniform yellow-brown, yellow-olive, grey-green, brown-green, dark olive-green, grey-yellow, olive-black, or black, or front of tarsus and upper surface of toes greyish-horn, dull olive-grey, or black with paler dull yellow, green-yellow, flesh-brown, or green-brown rear and soles. No relation between colour of leg and locality: greenish recorded in, for example, Humboldt Bay area, along Mamberamo and Sepik rivers, and along Huon Gulf, black(ish) from Humboldt Bay area, Kairiru, Karkar ('Dampier'), Mamberamo, and Huon Gulf.

CHICK: iris brown. Bill horn. Leg and foot dark cherry-red-brown.

MOULTS
Primary moult serially descendant. Usually two series of primary moult in each wing, primaries showing two generations (when earlier series has reached outermost feather—p10—and a later series active or suspended at about p3–p7), or three generations (when earlier series has reached, for example p8, later series p3, and outermost feathers of a still older generation), but six of 34 ad birds had three series active in each wing and one bird four. In birds with two series, active feathers five to seven feathers apart, but actively growing feathers closer in birds with three or four series, two to four feathers apart. Of the 34 ad, 17 had moult of all series active on all centres, ten had moult entirely suspended, remainder had one (or more) series active and one (or more) suspended. No clear moulting season apparent, especially because only half of ad dated; in dated birds, moult active in Feb (1), Apr (1), May (2), June (1), July (1), Aug (1), Sept (1), and Oct (1), partly suspended in Jun (1) and Nov (1), fully suspended in Jan (1), July (1), Aug (2), Oct (1), and Dec (1). Suspension may occur at any primary: up to p1 new (and p2 of previous series contrastingly old) in one bird, up to p2 in two, to p3 in nine, to p4 in three, to p5 in two, to p6 in three, to p7 in one, to p8 in five, and to p9 in six. In two older imm examined (May and undated), moult suspended in both, while second generation of primaries had reached p10 and third generation up to p3; in two birds in first imm plumage (Apr and undated) outer primaries still juv (first generation) with p9–p10 still growing, inner primaries of second generation, p5–p6 (Apr) or p3–p4 (undated) growing.

MEASUREMENTS

ADULT: Japen, skins of 1 ♂, 1 ♀, and three unsexed birds; some birds perhaps still imm (RMNH, SMTD).

	n	mean	s.d.	range
wing	4	212.8	11.32	200–227
tail	4	70.1	0.63	69–71
tarsus	5	62.6	2.95	60.0–67.4
mid-toe	4	35.9	2.02	33.6–38.5
mid-claw	5	19.3	1.63	17.7–21.5
bill (skull)	5	29.2	0.46	28.7–29.8
bill (nostril)	5	13.3	1.36	12.1–14.2
bill depth	5	8.7	0.90	7.7–9.8

Japen, wing of 5 ♂♂ and 5 ♀♀: 214–227 mm (221.8 ± 3.79, n = 10); Makimi (southern shore of Geelvink Bay), wing of 1 ♀ 222 mm (Mayr 1938).

Mamberamo and Sepik basins, southern slope of Snow Mts, shores of Astrolabe Bay and Huon Gulf, and Manam ('Vulcan') I., skins of 15 ♂♂, 11 ♀♀, and 9 unsexed birds (BMNH, RMNH, SMTD, ZMB).

	n	mean	s.d.	range
wing	33	223.8	5.56	213–236
tail	32	74.9	3.44	70–81
tarsus	34	67.7	2.09	64.4–72.1
mid-toe	31	37.6	1.50	34.6–41.0
mid-claw	32	19.8	1.46	17.2–22.3
bill (skull)	33	28.9	1.41	26.6–31.8
bill (nostril)	32	12.8	0.69	11.3–13.9
bill depth	27	8.4	0.37	7.7–9.4

Sexes closely similar, e.g. wing ♂ 219–235 mm (225.1 ± 4.59, n = 13), ♀ 213–236 mm (224.2 ± 7.59, n = 10); tarsus ♂ 65.5–72.1 mm (68.6 ± 2.05, n = 14), ♀ 65.3–70.0 mm (67.6 ± 1.67, n = 11). Wing from literature: lower Mamberamo basin 218–227 mm (222.2 ± 3.70, n = 5) (Von Bemmel 1947), upper Mamberamo basin and Humboldt Bay area 223–235 mm (228.5 ± 4.66, n = 8) (Mayr 1938); northern slope of Snow Mts 215–234 mm (220.3, n = 13) (Rand 1942b).

CHICK: Japen: wing 101 mm, tarsus 29.2 mm (SMTD).

WEIGHTS

ADULT: ♂ 715 g, ♀ 550, 692 g (NE New Guinea, Senckenberg Mus; ZFMK). Possible hybrids between this species and *M. eremita* from Karkar and Bagabag islands: ♂ 750 g, ♀ 580, 700, 780 g (Diamond and LeCroy 1979).

GEOGRAPHICAL VARIATION

Slight in size, more marked in colour. Variation in colour difficult to assess due to the strong change in colour in older skins, for example birds from Mamberamo R. collected 1920 markedly different in colour from skins from same locality collected 1939 and examined when still quite fresh (van Bemmel 1947). Slight increase in size of wing, tail, and middle claw eastwards on mainland New Guinea, but no apparent differences in length of tarsus, mid-toe, or bill; for example wing of birds from Mamberamo and Sepik basins east to Manam 213–225 mm (221.3 ± 3.97, n = 10), those from shores of Astrolabe Bay 214–236 mm (223.8 ± 6.60, n = 10), those from tip of Huon peninsula southward 218–235 mm (225.7 ± 5.55, n = 12); birds examined from Japen appear distinctly smaller, but probably some data of imm birds are included and sample is small; wing of Japen birds measured by Salvadori (1882) and Mayr (1938) not different from Mamberamo birds. In colour, *M. decollatus* rather resembles *M. reinwardt*, but in structure it is rather different: tail is markedly shorter (ratio of tail to wing 0.33–0.34 in various populations of *M. decollatus*, 0.38–0.41 in most populations of *M. reinwardt*, except *M. r. macgillivrayi*) and tarsus longer (ratio of tarsus to wing 0.295–0.305 in *M. decollatus*, 0.276–0.287 in most *M. reinwardt* populations, but 0.297–0.308 in populations of *M. reinwardt* from Vogelkop and just south of Snow Mts); also, toe and bill of *M. decollatus* relatively longer and heavier, crest shorter (up to c. 2 cm long), and head and neck often more extensively bare.

In NW New Guinea (Mamberamo basin to Humboldt Bay area; also southern slope of Snow Mts), grey of head and underparts averages slightly darker than in *M. reinwardt*; lower mantle, scapulars, upper wing, and tertials duller and darker olive-brown, less rufous; rump to upper tail-coverts duller brown, less maroon; flank rather extensive olive-brown or rufous-brown. Birds from Japen, Kairiru, and Manam on average darker than those of northwest mainland, mantle and scapulars slightly darker olive-brown, grey of head and underparts darker and more slaty, olive-brown of sides of body rather dull and restricted. In contrast to these, birds from shores of Astrolabe Bay and Huon Gulf on average brighter, with more extensive brown than those of northwest mainland, olive-brown of upperparts and sides of body somewhat tinged rufous, brown of flank more extensive, especially in birds from Astrolabe Bay, in which rather rufous-brown tinge often extends to side of breast, chest, and mid-belly. Birds from Karkar and Bagabag combine characters of *M. decollatus* and *M. eremita* (Mayr 1938; Diamond and LeCroy 1979; Coates 1985); on Karkar, two ♀♀ are both rather near *M. decollatus*, of four ♂♂, one is near *M. decollatus*, one is near *M. eremita*, and two are intermediate; birds of Manam further west are pure *M. decollatus*, birds of Long I. further east are almost pure *M. eremita* (Mayr 1938). Hybrids between *M. reinwardt* and Dusky Megapode *M. freycinet* from Sorong (western Vogelkop) and probably elsewhere resemble *M. decollatus* closely in colour, but are quite different in measurements.

Range and status
N New Guinea, mainly north of Central Range, from Japen, south and east shore of Geelvink Bay (east from Makimi), and Mamberamo basin east to Huon Gulf and Mambare R. basin in southeast, including Valif, Tarawai, Kairiru, Schouten and Manam islands; locally on south slope of Central Range in upper Setakwa basin, Moroka district, and perhaps elsewhere (shown by arrows). Identity of birds from Mt. Sisa and Lake Kutubu requires confirmation, but most probably this species. Forms mixed population with *M. eremita* on Karkar and Bagabag.

Common in the Memenga forests and along the Sepik R. (northeast New Guinea); also in the outskirts of the villages (Gilliard and LeCroy 1967b).

Field characters
Length 33–35 cm. Slightly smaller than *M. reinwardt* with which it lives sympatrically in S New Guinea in the area of the Utakwa, Mimika, Kapare, Wataikwa, and Setekwa Rivers (see Chapter 3). Can be distinguished by its dark instead of orange legs, bright olive-brown instead of rufous-brown upperparts, and sparse feathering of the throat.

Voice
Duet, probably initiated by the ♂, 'kok, nyacal, nailleue', ascending to a high screech. The partner joins in shortly after with a continuing, stuttering 'nu-nu-nu-nu-nu'. Duets produced at all hours of the day and night, but most prevalent on moonlit nights (Gilliard and LeCroy 1966).

Habitat and general habits
From sea level up to at least 1260 m at the upper Utakwa R. Abundant in swamp forest bordering the middle Sepik R. (Gilliard and LeCroy 1966). Eggs collected at an altitude of 1800 m at Tapu (Upper Ramu R. plateau, SE Bismarck Range).

Food
No information available. Probably omnivorous as in other megapodes.

Displays and breeding behaviour
Mounds are reported to be used by two, sometimes three, rarely four pairs for consecutive years (compare *M. reinwardt*). Judging from sound orientation, Gilliard and LeCroy (1966) found that pairs seemed to keep closely together at night (only rarely separating for more than 3–6 m) and to remain within *c.* 50 m of the mound. These observations suggest a strict pair bond and monogamy. Reported to hybridize with *M. eremita* on Karkar and Bagabag.

Breeding

NEST: mound similar in form and shape to that of *M. reinwardt* and made of sand, earth, and leaves. A low mound situated on nearly flat ground in tall original forest measured 5.4–6.0 m in diameter. A mound in the middle Sepik region measured 0.9 m in height and 3.6 m in diameter (Gilliard and LeCroy 1966). Sometimes lays its eggs in mounds of the Wattled Brush-turkey *Aepypodius arfakianus* and *Talegalla* spp.

EGGS: similar in shape and coloration to eggs of *Macrocephalon maleo*, *Eulipoa wallacei*, and other *Megapodius* species (see under *M. pritchardii*). SIZE: 77.5–81.0 × 47.5–50.7 mm (79.3 × 49.1, $n = 4$, N coast of New Guinea); 82–91 × 50.2–57.2 mm (86.0 × 53.0, $n = 3$, Huon Gulf); 80.0 × 48.7 mm ($n = 1$, Japen, Schönwetter 1961); 78.8–82.5 × 46.5–52.1 mm (80.4 × 49.9, $n = 3$, Papua New Guinea, Steadman 1991); 77.8–81.2 × 48.5–52.9 mm (79.3 ± 1.1 × 50.9 ± 1.5, $n = 7$, Mt. Sisa, S Highland Province, Dwyer 1981, data probably refer to this species; see 'Range and status'). WEIGHTS: 108–129 g (117.1 ± 7.4, $n = 8$, Mt. Sisa, S Highland Province, Dwyer 1981, data probably refer to this species; see 'Range and status').

BREEDING SEASON: according to local information, eggs are laid throughout the year (Gilliard and LeCroy 1966). However, eggs for sale in markets near Mt. Sisa between Jan and May indicate breeding during the wet season. Eggs found during same period at NE Lake Kutubu (early Feb), in the Trans-Gogol area, Madang Province (Mar or Apr), and at the Upper Ramu R. plateau, SE Bismarck Range (Apr). Occupied mound in middle Sepik R. area in Feb. Birds in breeding condition collected in Feb, Mar, June, and Oct.

Local names

Kinjow (Adelbert Range), pajie (lower Mamberamo R.), sirouqua (middle Sepik), kaube, oedóc, neyo (E Highlands), aeo (Mt. Sisa, S Highland Province).

References

Meyer (1874b, 1892), Oustalet (1878, 1881), Salvadori (1882), Ogilvie-Grant (1893, 1897), Rothschild and Hartert (1901), Stresemann (1922), Mayr (1930, 1938, 1941), Rothschild *et al.* (1932), Rand (1942b), Van Bemmel (1947), Schönwetter (1961), Ripley (1964), Gilliard and LeCroy (1966, 1967b), Rand and Gilliard (1967), Diamond and Terborgh (1968), Harrison and Frith (1970), Diamond (1972), Schodde (1977), Diamond and LeCroy (1979), Dwyer (1981), Coates (1985), White and Bruce (1986), Holmes (1989), Steadman (1991).

Orange-footed Megapode   *Megapodius reinwardt* Dumont, 1823

Megapodius reinwardt Dumont, 1823. *Dictionnaire des Sciences Naturelles*, **29**, p. 416. (Lombok.)

PLATE 8

Other names: Orange-footed Scrubfowl.

Polytypic. Five subspecies. *Megapodius reinwardt reinwardt* Dumont, 1823, Lesser Sunda Is from Nusa Penida and Lombok east to Babar and Damar, north to Salajar, Tukangbesi, and Lucipara Is; Moluccan islands from Banda and Watubela groups to Manuk and Kai group; W New Guinea from Vogelkop peninsula east to southern shore of Geelvink Bay and Triton Bay, S New Guinea south of Central Range from Triton Bay to southeastern tip; also north of Central Range east from Kumusi R. basin, and on islands off S New Guinea from Adi and

Orange-footed Megapode *Megapodius reinwardt*

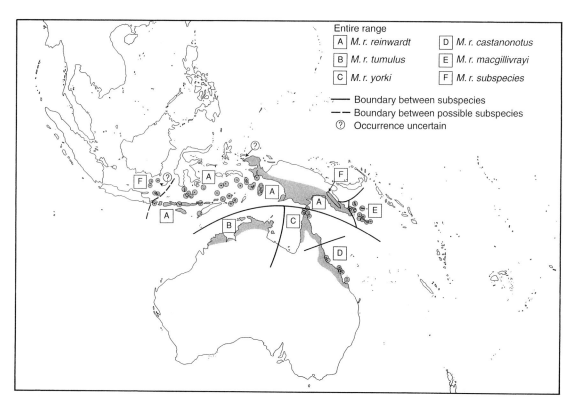
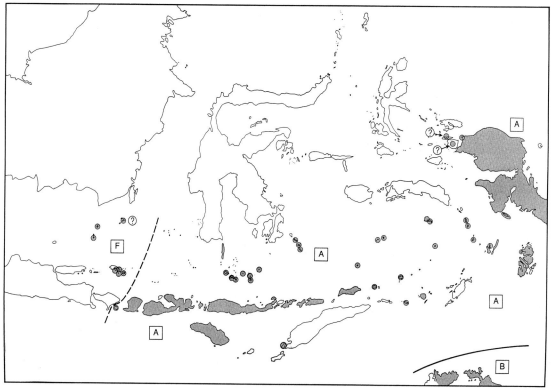

Orange-footed Megapode *Megapodius reinwardt*

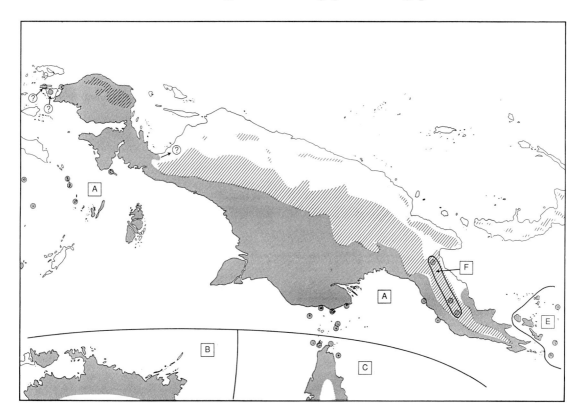

Aru group through Deliverance, Boigu, Saibai, Bet, Dungeness, and Daru to Yule and Vauga Is; occurs also (race uncertain) on islands of Kangean group, Solombo Besar, Karamian, and Matasiri (possibly), eastern Java Sea (synonyms: *rubripes* Temminck, 1826, Lombok; *duperreyii* Lesson and Garnot, 1826, Manokwari; *gouldii* Gray, 1861, Lombok; *amboinensis* Gray, 1861, Banda; *affinis* A. B. Meyer, 1874, Rubi, southwest shore of Geelvink Bay; *assimilis* Masters, 1876, Dungeness and Bet; *aruensis* Mayr, 1938, Aru Is); *Megapodius reinwardt tumulus* Gould, 1842, northern Western Australia and Northern Territory from York Sound east to Cape Arnhem, and on islands off Northern Territory from Melville and Croker east to Groote Eylandt (synonym: *melvillensis* Mathews, 1912, Melville I.); *Megapodius reinwardt yorki* Mathews, 1929, Cape York peninsula (N Queensland, Australia), south to Mitchell R. (in west) and to Cedar Bay and Bloomfield R. in east (south of Cooktown, near Weary Bay), including islands off Cape York north to Booby, Prince of Wales, Albany, Cairncross, and Haggerstone; *Megapodius reinwardt castanonotus* Mayr, 1938, E Queensland from Cape Tribulation to Yeppoon, including many offshore islands; *Megapodius reinwardt macgillivrayi* G. R. Gray, 1861, D'Entrecasteaux, Trobriand, and Woodlark Is, and Louisiade Archipelago, from Goodenough and Trobriand (Kiriwina) through Conflict and Bonvouloir groups to Tagula and Rossel Is; also (perhaps this race), mountains of SE New Guinea from about Wau to Astrolabe Mts.

Description
PLUMAGES

ADULT: (nominate *reinwardt*, Flores): sexes similar. Forehead, crown, and nape dark cinnamon-brown, sometimes tinged greyish-olive (especially on forehead and if worn); fringes of feathers often duller brown. Feathers on nape elongated, slightly curved upward at tip, forming pointed crest c. 1.5–2.5 cm long. Lores and ring round eye virtually bare; base of upper mandible covered with sparse bristle-like feathers; cheek and area round ear more densely covered with rather narrow and reduced dark grey or dark brown-grey feathers, leaving part of skin visible; feathering of chin, throat, and neck as on cheek, but denser still, less skin visible or none at all, especially central hindneck and foreneck usually fully covered. Upper mantle uniform medium grey, tinged plumbeous when fresh. Lower mantle, scapulars, upper wing-coverts, and tertials bright cinnamon-brown or rufous-brown, somewhat less dull and brown than crown; tips of feathers of outer mantle or of some coverts occasionally dark chestnut or maroon. Some variation in tinge of upperparts, mainly depending on bleaching and wear: in fresh plumage, fairly bright cinnamon or rufous; if worn, more olive-drab or clay colour, but still distinctly brighter, more cinnamon, and less olive than upperparts of, for example, Forsten's Megapode *M. forstenii* or the Philippine Megapode *M. cumingii gilbertii*; upperparts of the Sula Megapode *M. bernsteinii* rather similar, but tertials of latter species contrastingly more rufous, upper mantle not medium grey, and rump duller brown. Back, rump, and upper tail-coverts bright rufous-chestnut or light maroon. Underparts medium grey, tinged plumbeous when fresh, slightly more brownish when worn, especially on belly; lower flank and feathering of thigh rufous-brown or rufous-chestnut, but extent of rufous variable, in some birds confined to some feathers on outer thigh, in others extending along side of body to upper side of breast and to side of belly; vent light grey, some off-white of feather-bases shining through; under tail-coverts rufous-chestnut. Tail brown-black, slightly tinged chestnut; flight feathers greyish-black or brownish-black, outer webs tinged cinnamon-brown or rufous-brown, sometimes almost chestnut-brown. Under wing-coverts medium grey, axillaries rufous-brown.

CHICK: upperparts medium to dark rufous-brown, grey-brown, or olive-brown, rather indistinctly barred rufous-ochre and dark brown on scapulars, back, inner upper wing-coverts, inner secondaries, and tertials, or bars virtually absent; rump rufous-olive to maroon-brown; underparts pale greyish-olive, grey-brown, or dull rufous-cinnamon, paler on chin, cheek, and mid-belly; belly tinged rufous-isabelline. Variation in colour perhaps geographical, but samples small: chicks from Lombok and Flores are very pallid and olivaceous, faintly barred; those of Sumba, Kai, and Aru darker brown, faintly barred; those of Vogelkop peninsula darker brown, more distinctly barred; those of Astrolabe Mts (E New Guinea), Louisiade Is, and Trobriand darkest and well barred.

IMMATURE: in first imm plumage, like ad, but colour of body slightly less saturated, mantle and scapulars slightly more cinnamon, grey of underparts slightly paler, especially on belly, some traces of buff bars sometimes on scapulars, tertials, or flank; crest on nape shorter, 1–2 cm; head and neck more fully feathered, lores covered with scanty bristles, only ring round eye fully bare, no bare skin visible on neck. Tail, outer primaries, and some secondaries still from first (juv) generation, shorter, narrower, and more pointed at tip than in second generation; especially outermost primaries (p9–p10) sharply pointed, tip somewhat frayed, juv p8 (if present) shorter than p9 (in older birds, p8 longer than p9, except when in moult); absolute length of juv p10 (from insertion in skin to tip) in populations from Lesser Sunda Is and New Guinea 87–103 mm (96.1 ± 6.20, $n = 14$), in Australia 113–122 mm (118.9 ± 3.36, $n = 5$); in ad from Lesser Sundas and New Guinea, length of p10 119–145 mm (130.3 ± 7.26, $n = 25$), in Australia 137–155 mm (146.3 ± 6.32, $n = 7$). Juv tail in populations of Lesser Sundas and New Guinea 52–74 mm (66.2 ± 7.85, $n = 9$) (ad over 80 mm), in Australia 72–86 mm (79.3 ± 7.02; $n = 3$) (ad over 92 mm). In second imm plumage, similar to ad, but outer primaries still of second generation, slightly shorter and tip slightly more pointed than neighbouring third generation feathers or than ad feathers; length of p10 in populations from Lesser Sundas and New Guinea 108–121 mm (115.2 ± 4.34, $n = 8$), in Australia 125–140 mm (134.3 ± 8.14, $n = 3$); tail intermediate in length between juv and ad, length in Lesser Sundas and New Guinea 77–88 mm (81.6 ± 3.23, $n = 12$), in Australia 91–102 mm (97.3 ± 5.69, $n = 3$).

BARE PARTS

ADULT: iris light grey-brown, dull or bright coffee-brown, bright light brown, pale brown, umber-brown, sepia, chestnut, dark brown, black with red ring, light red-brown, or dark red. Bare skin of head and neck vinous-red, cherry-red, deep plum, dull crimson, deep red, reddish-umber, dull madder-brown, or reddish-black. Base of bill olive-grey, horn-olive, greyish-brown, dark grey, horn-brown, reddish-brown, or black, tip and cutting edges olive-yellow, brown-yellow, dirty yellow, yellow-brown, yellow-horn, ochre-yellow, orange-yellow, or orange. Leg and foot pale orange-yellow, reddish-yellow, yellow-red, orange, deep salmon, tomato-red, salmon-red, bright red, or vermilion, scutes of lower front of tarsus and upper surface of toes dark grey, slate-colour, red-brown, or black-brown.

CHICK: iris grey-brown with dark outer ring, pale brown, hazel, chestnut, or dark brown. Bare skin round eye orange. Bill dirty white, yellow-brown, or dark grey-brown with orange tip. Leg and foot flesh-orange, orange-yellow, reddish-yellow, or orange, front of lower tarsus and upper surface of toes dark brown or blackish-grey.

IMMATURE: iris brown or dark brown. Bare skin as ad. Bill base sooty-yellow, olive-horn, horn-olive, horn-brown, or black-brown, tip and cutting edges dull horn-yellow, lemon-yellow, dull orange, or lemon-red. Leg and foot dull salmon-pink, pale brick-yellow, horn-yellow with reddish soles, brown-yellow, reddish-orange, or vermilion, front of tarsus and upper surface of toes dusky olive, yellow-brown, brown, yellowish-black, or black.

MOULTS

Primary moult serially descendant. In ad, two moult series on each wing in virtually all birds (three series in two out of 91 birds examined), and two generations of primaries (when inner series active or suspended on central primaries and outer series just completed with regrowth of outermost primary, p10) or three generations present (when inner moult series has reached, for example p3, central series has reached e.g. p8, and p9 and p10 are from still older series). Both moult series generally separated by 4–8 (5.9 ± 0.92, $n = 48$) feathers, next series starting with p1 when previous series has reached p4–p8 ($p5.7 \pm 0.86$, $n = 25$). In all populations, moulting birds present throughout the year; birds suspending moult most frequent in Sept–Jan, when 52% of 23 dated birds examined suspended moult (in rest of year, 12.5% of 48 birds). Including undated birds, 31% of 99 ad had primary moult suspended, but 13% of 38 imm; in Lesser Sunda to Aru Is, 22% of 37 ad suspended moult, on mainland New Guinea 28% of 43, in Australia 58% of 19. Immature start moult of second generation of primaries with p1 when juv outermost primaries (p9–p10) and tail just started growth. Not known at what age moult started; unmoulted small chicks (not necessarily just hatched) examined mainly from Oct–Feb, imm just starting moult from Dec–June, and thus moult perhaps starting at age of about three months. A third generation of primaries starts when second generation reaches p5–p7 (–p9), on average at $p6.6 \pm 1.22$, $n = 18$, probably at age of about 8 months, or later when moult of second series suspended.

MEASUREMENTS

ADULT: *M. r. reinwardt*. Salajar, Tanahdjampea, Kalaotua, Nusa Penida (off Bali), Lombok,

Flores, Alor, and Wetar; skins of 6 ♂♂, 3 ♀♀, and 10 unsexed birds (BMNH, RMNH, ZMA, ZMB).

	n	mean	s.d.	range
wing	19	235.9	5.32	227–245
tail	17	89.9	4.63	84–99
tarsus	18	67.0	2.96	63.0–73.7
mid-toe	17	38.9	1.76	36.3–41.0
mid-claw	18	21.3	1.86	17.4–22.8
bill (skull)	18	29.7	1.33	27.2–31.7
bill (nostril)	18	13.7	0.96	12.1–15.5
bill depth	15	8.4	0.64	7.6–9.7

Wing. Djampea group, 233–242 mm (237.5 ± 3.70, $n = 4$); Lombok, Flores, and Sumbawa, 227–242 mm (234.4 ± 5.68, $n = 7$); Alor, Wetar, Romang, and Damar, 221–245 mm (236.1 ± 8.08, $n = 9$) (Mayr 1938). Tukangbesi I., 229–237 mm ($n = 3$) (Meise 1941). Semau (off SW Timor), ♂ 232–238 mm ($n = 3$), ♀ 228–232 mm ($n = 3$) (Hellmayr 1914). Kangean and Karamian islands, Java Sea (not certain whether these should be considered nominate *reinwardt*), 207–220 mm (213.2 ± 5.26, $n = 5$) (Meise 1941).

Sumba; skins of 2 ♂♂, 1 ♀, and 3 unsexed birds (BMNH, RMNH, SMTD, ZMB).

	n	mean	s.d.	range
wing	6	244.7	1.86	242–247
tail	6	96.9	2.87	94–102
tarsus	6	69.6	1.43	68.0–71.3
mid-toe	6	38.5	0.97	37.1–39.7
mid-claw	6	21.8	1.49	19.5–24.1
bill (skull)	6	30.2	1.14	28.7–31.1
bill (nostril)	6	13.7	0.37	13.2–14.1
bill depth	5	8.6	0.55	8.1–9.5

Wing. Sumba, ♂ 243–253 mm (250.2 ± 4.86, $n = 4$) (Mayr 1938). Birds from Babar large also, e.g. in 2 ♂♂ and 1 ♀: wing 241–250 mm (244.7), tail 91–96 mm (93.5), tarsus 67.9–71.9 mm (69.6), mid-toe 37.8–38.5 mm (38.2), mid-claw 18.8–22.3 mm (20.9), bill to skull 29.0–32.7 mm (30.5), bill to nostril 12.6–13.8 mm (13.3), bill depth 8.6–9.2 mm (8.9) (RMNH).

Banda, Kai and Aru islands; skins of 12 ♂♂, 11 ♀♀, and 2 unsexed birds (BMNH, RMNH, ZMA, ZMB).

	n	mean	s.d.	range
wing	25	231.8	5.54	219–245
tail	16	87.1	4.58	82–96
tarsus	21	66.5	2.80	63.0–72.5
mid-toe	18	37.3	1.57	34.7–39.5
mid-claw	20	20.5	1.44	17.4–24.0
bill (skull)	21	29.5	1.31	27.4–32.2
bill (nostril)	20	12.4	1.22	10.5–14.5
bill depth	10	8.5	0.54	7.9–9.3

Wing. Banda, 226–232 mm (229.7 ± 3.21, $n = 3$); South-East Islands (Kasiui, Tioor, Kilsuin, Kur), 219–232 mm (224.5 ± 3.38, $n = 10$) (Mayr 1938). Banda, 237 mm ($n = 1$); Kai, 219–235 mm (227.6 ± 5.34, $n = 8$); Aru, 223–245 mm (233.6 ± 5.06, $n = 16$) (BMNH, RMNH, SMTD, ZMA, ZMB).

Vogelkop peninsula south to southern shore of Geelvink Bay; skins of 10 ♂♂, 9 ♀♀, and 5 unsexed birds (BMNH, RMNH, SMTD).

	n	mean	s.d.	range
wing	24	238.6	6.26	226–250
tail	24	94.0	4.17	87–104
tarsus	24	70.9	2.79	66.5–75.2
mid-toe	23	40.6	1.79	37.8–43.5
mid-claw	24	20.5	1.44	16.9–23.5
bill (skull)	24	30.4	0.88	28.6–32.3
bill (nostril)	22	13.7	1.03	11.5–15.5
bill depth	20	9.3	0.59	8.4–10.5

SW New Guinea from Triton Bay to upper Lorentz (Noord) R.; skins of 12 ♂♂, 12 ♀♀, and 1 unsexed bird (BMNH, RMNH).

	n	mean	s.d.	range
wing	23	232.6	5.62	223–244
tail	13	87.5	3.72	82–92
tarsus	22	71.6	3.24	65.4–77.2
mid-toe	15	39.5	2.10	35.5–43.0
mid-claw	20	20.7	0.96	19.3–23.2
bill (skull)	20	30.5	1.55	26.8–33.2
bill (nostril)	20	13.8	0.62	12.8–15.1
bill depth	16	8.7	0.55	7.9–9.9

Orange-footed Megapode *Megapodius reinwardt*

S and SE New Guinea from Merauke area to Orangerie Bay; skins of 4 ♂♂ and 7 unsexed birds (BMNH, RMNH, ZFMK, ZMB).

	n	mean	s.d.	range
wing	11	239.4	7.04	230–252
tail	8	91.7	5.48	85–98
tarsus	10	68.1	2.97	65.6–72.5
mid-toe	8	38.8	2.01	36.0–42.0
mid-claw	9	20.1	1.48	18.5–23.0
bill (skull)	9	30.2	1.51	28.0–31.9
bill (nostril)	10	13.6	0.95	12.3–14.8
bill depth	8	8.5	0.38	8.0–9.1

Wing. Fly R., ♂ 226–243 mm (232.4, $n = 10$) (Rand 1942a), ♀ 228–244 mm (Mayr 1938). South coast of SE New Guinea, ♂ 232–247 mm (239.1 ± 5.38, $n = 10$), ♀ 235–246 mm ($n = 2$) (Mayr 1938), both sexes 228–245 mm (235.9 ± 6.01, $n = 11$) (Mayr and Rand 1937). North coast of SE New Guinea, 233–250 mm (240.0 ± 6.98, $n = 5$) (Mayr 1938). Dungeness I. (Torres Strait), ♀ 234 mm (Masters 1876).

M. r. tumulus: N Australia from Cobourg Peninsula to Groote Eylandt; skins of 2 ♂♂ and 6 unsexed birds (BMNH, RMNH, ZMB).

	n	mean	s.d.	range
wing	8	264.9	5.51	255–271
tail	7	109.6	3.47	105–115
tarsus	7	74.3	1.59	71.7–76.5
mid-toe	7	41.5	1.75	39.5–44.1
mid-claw	7	23.3	1.37	21.9–25.2
bill (skull)	7	31.2	0.98	29.4–32.4
bill (nostril)	7	14.2	1.14	12.9–15.7
bill depth	5	9.1	0.53	8.3–9.7

Northern Territory, wing: mainland 270–282 mm (277.2 ± 4.12, $n = 6$), Melville I. 252–272 mm (262.2 ± 7.55, $n = 6$) (Mayr 1938).

M. r. yorki: Cape York peninsula (Australia), south to Cooktown, and islands at north tip of peninsula (Booby, Prince of Wales, Cairncross); skins of 5 ♂♂, 4 ♀♀, and 10 unsexed birds (BMNH, RMNH, ZMB).

	n	mean	s.d.	range
wing	19	265.0	8.78	250–280
tail	15	103.1	4.56	94–113
tarsus	17	73.1	2.72	67.5–78.8
mid-toe	15	40.7	1.81	37.8–43.7
mid-claw	16	22.5	1.59	19.9–25.6
bill (skull)	16	30.8	1.38	28.7–32.6
bill (nostril)	15	14.3	0.70	13.2–15.6
bill depth	12	8.8	0.31	8.4–9.5

Wing. ♂ 240–274 mm (258.2 ± 12.21, $n = 6$), ♀ 243–266 mm (252.8 ± 8.26, $n = 6$) (Mayr 1938).

M. r. castanonotus: E Queensland (Australia) from Cairns to Mackay, including some offshore islands; skins of 2 ♂♂ and 3 unsexed birds (BMNH, RMNH).

	n	mean	s.d.	range
wing	5	261.2	5.26	255–268
tail	4	102.1	7.71	93–111
tarsus	5	73.7	3.14	69–77
mid-toe	4	41.9	0.57	41.3–42.5
mid-claw	5	24.1	2.04	21.6–26.4
bill (skull)	5	31.6	0.98	30.5–32.7
bill (nostril)	5	14.4	0.75	13.4–15.2
bill depth	5	8.9	0.39	8.6–9.4

Wing. Cairns area, 230–256 mm (247.1 ± 8.51, $n = 7$) (Mayr 1938).

M. r. macgillivrayi: D'Entrecasteaux and Louisiade Is; skins of 2 ♂♂, 1 ♀, and 1 unsexed bird (BMNH).

	n	mean	s.d.	range
wing	4	237.5	12.92	219–242
tail	3	83.7	4.01	79–87
tarsus	4	66.9	5.97	60.5–74.8
mid-toe	4	35.9	0.92	34.5–36.4
mid-claw	4	20.4	2.16	17.9–22.6
bill (skull)	4	28.7	0.87	27.5–29.6
bill (nostril)	4	13.0	0.95	12.0–14.1
bill depth	3	8.4	0.81	7.9–9.3

Wing, ♂ 227–245 mm (238.3 ± 5.54, $n = 10$), ♀ 220–246 mm (232.4 ± 10.34, $n = 7$); tarsus 64–72 mm (67.2 ± 2.75, $n = 14$) (Mayr 1938). Mountains of SE New Guinea; skins of 2 unsexed birds: wing 224–239 mm, tail 82–90.5 mm, tarsus 70.0–72.0 mm, mid-toe 38.5–41.3 mm, mid-claw 21.1–22.6 mm, bill to skull 30.4–32.7 mm, bill to nostril 12.5–14.6 mm, bill depth 9.7 mm (1) (BMNH).

CHICK: small chicks with p9–p10 not yet growing and not yet moulting, but not necessarily just-hatched: nominate *reinwardt*, wing 99–115 mm (107.9 ± 4.50, $n = 11$), tarsus 29.0–32.0 mm (30.7 ± 1.13, $n = 9$); one of Australian races, wing 118 mm; *macgillivrayi*, wing 100–112 mm (108.0 ± 6.93, $n = 3$), tarsus 30.5 ($n = 1$) (BMNH, RMNH, SMTD).

WEIGHTS
ADULT: *M. r. reinwardt*. ♀ 550 g (Flores, ZMA), unsexed 505–705 g (603 ± 100.0, $n = 3$, Sumba, R. Johnstone, *in litt.*), 810, 860 g (Vogelkop peninsula, New Guinea, Hartert 1930), 685, 750 g (Merauke area, Mees 1982; RMNH), ♀ 710 g (Sutter 1965). *M. r. tumulus*: ♂ 970–1000 g (Western Australia), ♀ 1000, 1000 g (R. Johnstone *in litt.*), 1091 g (Darwin, Northern Territory, Australia, Goodwin 1974). *M. r. castanonotus*: 900–1200 g (Cairns, Australia, Crome and Brown 1979).

CHICK: no information.

GEOGRAPHICAL VARIATION
Remarkably slight for a species inhabiting such a huge geographical range including many small and isolated islands; this may mean that present-day wide distribution is either evolutionarily recent, or that much more gene flow exists between various island populations than in other species of *Megapodius*, preventing populations from diverging. However, *macgillivrayi* differs markedly from others, and this race is considered separately below. Variation involves colour (mainly depth of olive or rufous of upperparts and upper wing) and size (as expressed in wing length). Most populations throughout Indonesia and Papua New Guinea have average wing length between 232 and 240 mm, only birds from Sumba and Babar larger, wing on average c. 244–245 mm, and those of Kangean, Solombo Besar, and Karamian Is in extreme northwest of geographical range smaller, average c. 213, but latter sample may include not full-grown imm birds; according to Meise (1941), population of Kangean and other northwest islands and of Sumba may deserve recognition as a separate race. Birds from Australia much larger, average wing c. 260–265 mm, crest long, c. 3 cm; based on colour, three races recognized here. Data on wing of Australian populations supplied by Mayr (1938) suggest that birds from mainland Northern Territory are largest, those of central E Queensland smallest, and those of Melville I. and Cape York area in between; however, no marked difference found between wing or other measurements of Australian races in the samples for present work examined. In measurements other than wing, all populations of *M. reinwardt* agree closely (apart from *macgillivrayi*), but tarsus, mid-toe, and length and depth of bill in populations from Vogelkop peninsula and in area south of Snow Mts relatively longer than in other populations. In colour, birds from Babar and (especially) Sumba average darker and more reddish-brown on upperparts than the lighter rufous-olive-brown of typical birds from other Lesser Sunda Is and islands in Java, Flores, and Banda seas; also, grey of head and underparts on average darker; as Babar and Sumba are widely separated, and smaller and paler forms occur on intervening islands, large size and dark coloration

on both islands may have been developed independently during a long isolation, or may be due to a wave of colonization from N Australia of which descendants survived on these islands only; both in colour and relative proportions, especially birds from Sumba are nearer to Northern Territory birds than to populations of Lombok or Flores. As not all Sumba and Babar birds are large and not all are dark, birds from both Sumba and Babar are provisionally included in nominate *reinwardt*.

Birds from Banda and from South-East Is (Watubela to Kai) are inseparable from nominate *reinwardt*, though geographically separated from typical nominate *reinwardt* from the Lesser Sunda Is by the Tanimbar Megapode *M. tenimberensis* on Tanimbar group, a different species which has nothing to do with *M. reinwardt*. About half of the birds from Aru Is are markedly more rufous on upperparts than are birds of the Lesser Sundas, especially tertials and upper wing-coverts being bright cinnamon-rufous or occasionally partly maroon; as remainder of birds are similar to nominate *reinwardt*, race 'aruensis' Mayr, 1938, not recognized here. Though on average populations of New Guinea are perhaps slightly darker, variation in depth of grey of head and underparts and in more rufous or more olive tinge of upperparts is similar to that of nominate *reinwardt* from Lesser Sunda Is; therefore, race 'duperreyi' as recognized for New Guinea birds by Mayr (1938) included in nominate *reinwardt* here.

In Australia, *tumulus* from northeast Western Australia and Northern Territory rather dark plumbeous-grey on side of head, upper mantle, and underparts and dark rufous-brown on lower mantle, scapulars, upper wing, and tertials, near Sumba birds; *yorki* from N Queensland is rather dark also, but upperparts are more cinnamon-brown. Some individual variation occurs, however, and birds from Booby and Prince of Wales islands in S Torres Strait are nearer to *tumulus*, while some (not all) from Cairncross I. are much darker; further north in Torres Strait, birds from Dungeness and Bet ('*assimilis*' Masters, 1876), are, as judged from original description, similar in colour to *yorki*, but much smaller, as is a bird examined from Saibai off S New Guinea; perhaps a cline of decreasing size exists northward through islands of Torres Strait, but for convenience the large birds from islands off Cape York peninsula are included in *yorki* and the small ones from Bet and Dungeness northward in nominate *reinwardt*. *M. r. castanonotus* from E Queensland (south of Cape Tribulation) is very dark plumbeous-black or sooty-grey on head, upper mantle, and underparts, distinctly chestnut-brown on lower mantle, scapulars, upper wing, rear of flank, and thigh, almost similar to the maroon of rump and upper tail-coverts; no difference in colour or size found between northern and southern populations (*contra* suggestion in Condon 1975), but only a few old specimens of each population examined.

M. r. macgillivrayi occurring on the islands off the southeast tip of New Guinea is markedly different from all other populations of *M. reinwardt*; general colour much darker, dark slate-grey with dull brownish-olive lower mantle, scapulars, inner upper wing-coverts, and tertials; rump and upper tail-coverts dark brown, less bright chestnut or maroon; crest shorter, broader, less pointed; forehead partly bare; lesser upper wing-coverts partly grey. In many characters, nearer to the Melanesian Megapode *M. eremita*, differing from latter especially in pale orange-red rather than blackish legs. According to Mayr (1938), who examined many specimens, *macgillivrayi* is a hybrid population between *M. reinwardt* and *M. eremita*, and this is supported by measurements of specimens examined: wing rather long, as in nominate *reinwardt*, but tail relatively shorter than in any population of *M. reinwardt*, intermediate between *M. reinwardt* and *M. eremita*, as are other measurements. In westernmost islands (D'Entrecasteaux group, nearest to mainland of SE New Guinea) upperparts still fairly pale rufous-brown, near nominate *reinwardt*, but in easternmost islands (Woodlark and Rossel) influence of *M. reinwardt* is very slight, birds being entirely plumbeous-black with rather limited dull olive on lower mantle, scapulars, inner upper wing-coverts, tertials, and flank, and black-brown on rump and upper tail-coverts; on intervening islands (Trobriand, Dugumenu, Bonvouloir group, Misima, Tagula) birds are either pale, dark, or intermediate (Mayr 1938). The type specimen from Duchateau I. (near Tagula) is dark, like birds from Rossel. Though the number of characters derived from *M. eremita* thus increases eastwards, size (as expressed in wing length) and leg colour appear to be constant throughout and similar to nominate *reinwardt*, and therefore *macgillivrayi* considered to be a race of *M. reinwardt* here. The chick of *macgillivrayi* is darker than that of other races of *M. reinwardt*, being brown with extensive black and rufous barring on upperparts.

The systematic position of the mountain population of SE New Guinea, from about inland of Huon Gulf to Astrolabe Mts, is obscure. In this area, birds occur which are superficially similar to *macgillivrayi*, appearing plumbeous-black with restricted cold olive-brown on upperparts, black-brown rump and upper tail-coverts, rather short broad crest, fairly extensive bare forehead and throat, and orange-red legs; they differ from *macgillivrayi* in relative proportions, however, showing relatively long tail and long and strong legs, claws, and bill; in proportions not similar to *M. eremita*, but near the New Guinea Megapode *M. decollatus*. These mountain populations are perhaps hybrids between *M. reinwardt* and *M. decollatus* according to size and proportions, but the general colour is much darker than that of the supposed parent species, which is difficult to explain.

Range and status

M. r. reinwardt: Lesser Sunda Is (Nusa Penida, Lombok, Sumbawa, Sumba, Komodo, Padar, Rindja, Flores, Solor, Lomblen, Pantar, Alor, Timor, Wetar, Semau, Romang, Damar, Sermatta, and Babar); islands in Flores and Banda Seas (Salajar, Tanahdjampea, Kalao, Bonerate, Sangi Sangiang, Kalaotua, Madu, Kakabia, Kaledupa, Tomea, Binongka, Gunung Api, Lucipara and Penju group, Manuk, Ai, Banda, Kasiui, Tioor, Kilsuin, Kur); Kai Is (Tual, Kai Kecil, Kai Besar), Aru Is (Wamar, Wokam, Kobroör, Maikoör, Trangan, Giabu-Langan); NW New Guinea (Vogelkop and Bombarai peninsulas east to Wandammen peninsula and Triton Bay); S New Guinea north to southern foothills of Central Range, from Triton Bay east to Orangerie Bay; mountains and north coast of SE New Guinea from Kumusi R. and foot of Astrolabe and Hydrographer Mts east to Milne Bay, and islands off S New Guinea (Adi, Deliverance, Boigu, Saibai, Dungeness, Bet, Daru, Yule, Vauga); also (perhaps this race), islands in Java Sea (Kangean group, Solombo Besar, Karamian, and Matasiri).

M. r. tumulus: northeast of Northern Territory (Australia) east from York Sound, and in Arnhemland (Northern Territory) north of *c.* 15°N, east to Cape Arnhem, including Cobourg peninsula and Melville, Croker, Woodah, Bickerton, and Groote I. *M. r. yorki*: Cape York area (N Queensland, Australia), in west occurs south to Mitchell R., in east occurs south to Cedar Bay and Bloomfield R. (near Weary Bay, south of Cooktown), and islands off N and E coast, including Booby, Prince of Wales, Nogo, Albany, Cairncross, and Haggerstone islands. *M. r. castanonotus*: E Queensland (Australia) from Cape Tribulation (north of Cairns) south to Byfield (near Yeppoon), including Fitzroy, Dunk, Barnard, Whitsunday, and Cumberland islands and probably others. *M. r. macgillivrayi*: D'Entrecasteaux Is (Goodenough, Fergusson, Normanby), Trobriand ('Kiriwina') Is, Marshall Bennett Is (Dugumenu), Woodlark Is, and Louisiade Archipelago: East I. in Bonvouloir group, Conflict group, and Misima ('St. Aignan'), Duchateau, Nimoa ('Pig'), Tagula ('Sudest'), and Rossel Is; also (perhaps this race) in mountains of SE New Guinea, from inland of Huon Gulf southeast to Astrolabe Mts.

Locally sympatric with the Dusky Megapode *M. freycinet* in W New Guinea and with *M. decollatus* along the southern slopes on the southwest side of the Snow Mts (see Chapter 3).

Orange-footed Megapode *Megapodius reinwardt*

Status: nominate *reinwardt*: no density figures available for Australia, but uncommon to common in suitable localities and sometimes abundant on small offshore islands. Twenty-three active and 19 inactive mounds in an area stretching approximately 2.5 × 1.5 km along the coast of Komodo (Lincoln 1974). Fairly common in some areas of Flores and Sumba, such as the Pengaduhahar area, though less common in the Tabundung area (Holmes 1989; Jones and Banjaransari 1990).

Field characters

Size dependent on race; nominate *reinwardt* c. 37 cm in length, Australian sspp. larger, up to 40 or 45 cm. Medium-sized darkish bird of the forest-floor with a small head, large, powerful feet and a short tail. Head with elongated, pointed crest. Neck and underparts grey, upper-parts usually dark olive-brown. Legs and feet orange. Very vocal, often duetting.

Voice

M. r. reinwardt: (Vogelkop peninsula, New Guinea): a rather far-carrying 'eukeu-keu-keu-keuw' or 'kokauw-krauw . . . . . . kokauw-krauw', a softer 'keurrr', 'oowooq-euweuw', and 'peutauw-teurrr', produced by individuals of unknown sex. When flushed, the ♂ produces a 'pie-wiet—wiet-iet' call. The ♀ has been heard to utter a soft 'kreu-kreu-kreu' (Hoogerwerf 1971). In E New Guinea, a descending high-pitched 'crou-o-o-o-ou', repeated two or three times, or an occasional 'croo'; when flushed, a soft high 'cu-cu-cu-cu-cu' (Coates 1985).

Duet more complex and melodious than that of other *Megapodius* spp. such as *M. eremita*, *M. forstenii*, and *M. freycinet*, but not known whether ♂ or ♀ initiates duet, nor which part of duet each member of pair produces. The duet (see sonogram) consists of a call, produced in its full length or with only the first and last note 'kli-au-(kiau)-kiau-kiau-kiau-kou

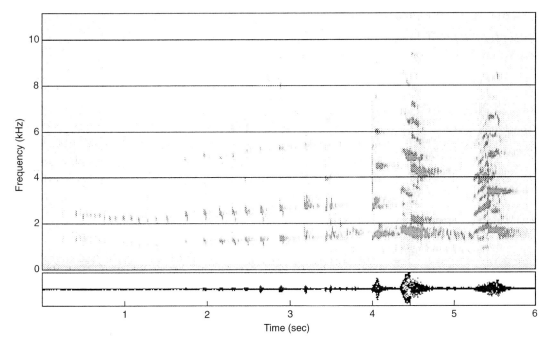

M. r. reinwardt
B. King, Flores, Indonesia, Aug 1987.

—kle-ou', of which the second part of the first note descends slightly in pitch and the second part of the last note clearly rises. The other component of the duet is a long, high-pitched, rolling but stammering 'krrrrr-uk-uk-uk-uk-krrrrr', descending in pitch, sounding somewhat like a Common Moorhen *Gallinula chloropus*. (Description by R. Dekker from tape-recording from Flores by B. King.)

M. r. castanonotus: duet given from a favourite roosting spot, heard at any hour of the day but mostly at night and during the breeding season (Aug–Jan). Early in the year the ♂ usually calls alone, but the ♀ gradually joins in later in the year and takes the initiative by Sept–Oct. During this period the ♀ starts with a series of loud clucks, the ♂ joins in with three chirrups followed by a far carrying, long drawn-out, descending, mournful, double note. Its function is thought to be mainly territorial (Crome and Brown 1979).

M. r. yorki: seldom heard during the day, but is noisy during the night (Macgillivray 1914); calls probably similar to the other sspp.

Habitat and general habits

Well adapted to secondary scrub and capable of recolonizing areas from which it has been exterminated. On Lombok, Sumbawa, and Flores it frequents bushy or wooded country near the coast (Rensch 1931*a*), but it has also been collected at 1000 m on Lombok (Hartert 1896). Forty-two mounds in an area of 250 ha along the coast of Komodo were all located in woodland or thick cover (Lincoln 1974). In SE New Guinea, favours various habitats, such as monsoon forest, swamp forest, secondary growth, scrub, and mangroves in the Vanapa River area, but less common or absent in primary hill forest. Common in all rain and monsoon forests in the Bensbach area, central S New Guinea. Often under 100 m (Mayr and Rand 1937). The occurrence in mangroves is surprising; otherwise this has been reported only for *M. freycinet*. Eggs collected as high as 1500 m on the northeast slope of Mt. Simpson, at Enaena, E Papua (Harrison and Frith 1970). Other reports mention altitudes of 600–1800 m (Diamond 1972). In the Vogelkop, NW New Guinea, the species prefers both heavy forest and secondary growth. *M. r. macgillivrayi* occurs in mountainous regions, for example on Goodenough and SE New Guinea. In E Australia, *M. r. castanonotus* is found in coastal ranges as well as further inland, for example in mixed mesophyll vine forest adjoining cleared areas (Crome and Brown 1979; Blakers *et al.* 1984). *M. r. yorki* and *M. r. tumulus* have also been recorded further inland.

Shy. Probably monogamous as it is usually seen in pairs. Coates (1985) suggested that single individuals are immatures. When disturbed either runs or noisily flies away. Capable of sustained flight and able to reach offshore islands. Territorial, with territories often maintained throughout the year (Blakers *et al.* 1984). A mound frequently forms the centre of a territory of 1–4 ha, with river beds demarcating the limits of the area (Lincoln 1974; Crome and Brown 1979). Digging behaviour similar to other *Megapodius* spp., but likely to be more powerful due to large size of some of its subspecies. In N Queensland, one bird was seen digging up a stone of almost 7 kg, moving it 70 cm (Coates 1985).

Food

One stomach contained insects, centipedes, snails, pebbles, and sand, another contained undetermined animal material and black cocoons with light green, brown-headed larvae. Stomach contents of two specimens from Kimberley, Western Australia, consisted of fruits, snails, and insects. Other reports mention: grubs, beetles, worms, the pupae of ants, young snakes, centipedes, millipedes, scorpions, and other invertebrates (Australia, Banfield 1913), seeds of Tamarind (Ashby 1922), fallen fruits, earthworms, snails, and centipedes (Lombok, Wallace 1869), and insects and snails (Macgillivray 1914).

Displays and breeding behaviour

Several reports of mound activity in nominate *reinwardt* suggest monogamy in this subspecies, as pairs work closely together (for example Gyldenstolpe 1955; Bergman 1963; Lincoln 1974; Coates 1985), as in *M. nicobariensis*. In New Guinea, Coates (1985) was of the impression that most digging activity at the mound took place from early to mid-morning, and was followed by a long break during which the birds behaved quietly. If confirmed, this would differ from the mound activities of the Nicobar Megapode *M. nicobariensis* which is recorded digging and egg laying at all hours of the day. On Komodo, Lincoln (1974) found that active mounds were used year after year and visited at least once every 4–7 days. Some birds worked at the mound for 2 or 3 days and than had a break of several days before returning. Most activity was early in the morning, although birds could be seen at the mound at all times of the day. Active mounds were never closer together than 100 m and were often separated by a natural barrier such as a river bed or rocky ridge. Inactive mounds were found as close as 15 m to active mounds (Lincoln 1974).

Communal use of mounds has been described for *castanonotus* (Crome and Brown 1979). In their study site at Clifton Beach (north of Cairns, Queensland), five out of 28 mounds were used by more than one pair. One particular mound was used partially successively, partially simultaneously by four different pairs over a period of at least 5 years, with a maximum of two pairs using the mound at any one time. The territories of these two pairs did not overlap. No confrontation or co-operation was ever seen at the mound. Observations of Crome and Brown (1979) show that though ♂ and ♀ often work together, gathering material from a radius of up to 25 m, and digging test and egg holes, the ♂ does most of the work at the mound, and may continue working when the ♀ is absent. Construction and maintenance continue all the year round, with increased activity just before the breeding season. *M. reinwardt* starts at the top of the mound and works its way down, head first, kicking leaves and sand backwards and upwards (Crome and Brown 1979). Before laying, test holes are dug by both ♂ and ♀. When ready to lay, the ♀ enters the hole, pointing her tail downwards and spreading her left wing while laying. This process takes 4–8 min, with the ♂ observing the ♀ from the edge of the hole. She then emerges slowly, covering the egg by scratching material from the sides of the hole, assisted by the ♂. The ♂ always returns at dawn the next day to add more material to the mound. Digging at the mound was also observed on days when no egg was laid, which is in contrast to the breeding behaviour of *M. nicobariensis* which seems to visit the mound on the day of egg laying only. Whether this difference is structural or depends on the timing of the breeding season needs to be further investigated. A single pair of *castanonotus* from N Queensland produced a total of 12–13 eggs at intervals of 9–20 days (average 13 ± 4 days) between 19 Aug and 8 Jan (Crome and Brown 1979). Copulation not described.

Breeding

NEST: mound, composed of sand, leaf litter and debris. Composition depends upon locality. Mounds in coastal areas are composed mostly of sand and may rely partially on solar heat. In shape they resemble miniature volcanic cones and may reach 4.5 m in height and 9 m or more in diameter (Macgillivray 1914). The largest mound on Saobi, Kangean, reported by Appelman (1938), measured *c.* 3.50 m in height with a circumference of *c.* 20 m. On Komodo they measured on average 0.9 m in height and 7.15 m in diameter, ranging from 0.3–3.0 m in height and 2.2–11.3 m in diameter (Lincoln 1974). Four mounds of *tumulus* from Port Keats, Northern Territory, Australia, measured 9.9–15.9 m in diameter at base, 31.5–50.1 m in circumference, and 1.70–2.55 m in height (Ashby 1922). Mounds are also constructed around a decaying tree-trunk (Lincoln 1974; Crome and Brown 1979) as described for other *Megapodius* species, *M. cumingii* and *M. nicobariensis*. However, Crome and Brown (1979) were doubtful about the function of this mound type as it was never found to contain an egg during their study.

EGGS: similar in shape and coloration to eggs of *Macrocephalon maleo*, *Eulipoa wallacei*, and other *Megapodius* species (see under *M. pritchardii*). SIZE: *M. r. reinwardt*. 84.0 ± 1.41 × 51.0 ± 1.41 mm (n = 2, Bonerate, Meyer and Wiglesworth 1898); 83.0–86.4 × 50.4–52.0 mm (84.3 × 51.2, n = 3, Komodo, Steadman 1991); 88 × 55 mm (n = 1, Komodo, Lincoln 1974); 86–88.5 × 54.6–55.4 mm (n = 2, Sumba); 78.6–88.4 × 48.6–55.1 mm (84.4 ± 2.36 × 52.3 ± 1.59, n = 53, Flores, RMNH); 80.1–90.8 × 49.2–54.9 (85.5 × 52.6, n = 10, Papua New Guinea, Steadman 1991); 83–87 × 48.5–55 (84.7 ± 2.08 × 51.7 ± 3.25, n = 3, central S New Guinea, Rand 1942a); 81.0–82.9 × 49.6–52.8 (82.3 ± 1.10 × 51.7 ± 1.79, n = 3, Mt. Simpson, E Papua New Guinea, Harrison and Frith 1970); 88.3 × 52.0 (n = 4, New Guinea, Hellebrekers and Hoogerwerf 1967); 75.9 × 52.1 mm (n = 1, Timor, RMNH, note small size; might be introduced egg of other *Megapodius* species); 81.6–88.9 × 51.3–54.5 mm (85.5 ± 2.98 × 52.7 ± 1.20, n = 5, Kangean, Lint 1967; possibly refers to nominate *reinwardt*). *M. r. macgillivrayi*: 79.5–90 × 52–56 mm (82.2 × 53.7, n = 8, Schönwetter 1961); 79.7–83.5 × 51.9–55.0 mm (81.2 × 53.0, n = 7, Fergusson I., Steadman 1991); 83.7–93.2 × 51.2–55.7 mm (88.2 × 54.2, n = 6, Trobriand Is, Steadman 1991). *M. r. yorki*: 81.3–92.0 × 48.8–55.9 mm (88.4 × 52.6, n = 29, Schönwetter 1961); 84.4–96.0 × 50.8–55.6 mm (88.7 × 53.6, n = 11, Queensland, Steadman 1991). *M. r. tumulus*: 86.1–93.4 × 52.8–57.4 mm (88.7 × 54.8, n = 9, Northern Territory, Australia, Steadman 1991). *M. reinwardt* ssp.: 85–93 × 50–55 mm (89.0 ± 3.37 × 52.8 ± 2.22, n = 4, Australia, Oustalet 1881, Meyer and Wiglesworth 1898); 84.8–89.0 × 52.0–55.3 mm (87.2 ± 2.18 × 53.4 ± 1.71, n = 3, Australia, Shufeldt 1919); 81.3–101.6 × 52.1–55.9 mm (n = ?, Shufeldt 1919). WEIGHTS: *M. r. reinwardt*. 126–140 g (132.5 ± 6.61, n = 4, Flores, J. Verheyen, *in litt*.); 115 g (n = 1, Komodo, Lincoln 1974); 111.5–130.5 g (121.3 ± 7.23, n = 5, Kangean, Lint 1967; possibly refers to nominate *reinwardt*).

BREEDING SEASON: *M. r. reinwardt*. Eggs from Komodo in Aug–Nov (Lincoln 1974), from Flores in Jan (n = 4), Feb (22), Mar (13), June (1), July (3), Aug (1), Oct (3), Nov (1), and Dec (3) (coll. RMNH), thus apparently no seasonality. Eggs from Kangean, probably referring to nominate *reinwardt*, in Apr and May. Three chicks from Wetar collected in Feb. New Guinea: on the northeast slopes of Mt. Simpson, eggs in Sept (Harrison and Frith 1970), on Pulau Adi, a chick collected in Dec, another one from S New Guinea in Feb. Mound activity in the Port Moresby area mostly during and shortly after the wet season from late Nov or early Dec to late Apr or early May (Coates 1985); on Vauga, a small coral island off Port Moresby, active mounds in Jan (Bell 1969). In the Fly River basin, breeding birds were taken in Aug, Oct, Nov, Dec, and Jan (Rand 1942a). *M. r. macgillivrayi*: chick collected in Mar on Trobriand Is. *M. r. yorki*: eggs from Cape York from Dec to Mar, chicks in Feb (Macgillivray 1914). *M. r. castanonotus*: breeding season from Aug to Jan, with that of a single pair lasting just over 20 weeks (Crome and Brown 1979). Eggs from Sept to Feb, a newly hatched chick from Sept (Bravery 1970; Campbell and Barnard 1917).

Local names
Gosong and wontong (Kangean and Komodo), djangul or danu (Aru), kwaar (Kai), mangoipé, manepreak (Manokwari, Andai), mangirio (west and southwest shore of Geelvink Bay), kata (Merauke area), oooregoorga (N Australia), koogerri (Australia).

References
Dumont (1823), Wallace (1869), Masters (1876), Oustalet (1881), Hartert (1896, 1902, 1904, 1930), Meyer and Wiglesworth

(1898), Finsch (1901), Rothschild and Hartert (1901, 1913, 1914*a*), Mathews (1910–1911), Banfield (1913), Hellmayr (1914), Macgillivray (1914), Ogilvie-Grant (1915), Campbell and Barnard (1917), Oberholser (1917), Shufeldt (1919), Ashby (1922), Meise (1930, 1941), Rensch (1931*a*), Rothschild *et al.* (1932), Junge (1937), Mayr and Rand (1937), Appelman (1938), Mayr (1938, 1941), Rand (1942*a*), Hoogerwerf (1949, 1971), Gilliard (1950), Gyldenstolpe (1955), Schönwetter (1961), Coomans de Ruiter (1962), Bergman (1963), Sutter (1965), Hellebrekers and Hoogerwerf (1967), Lint (1967, 1975), Rand and Gilliard (1967), Bell (1969), Bravery (1970), Harrison and Frith (1970), Diamond (1972), Goodwin (1974), Lincoln (1974), Condon (1975), Crome and Brown (1979), Clarke (1982), Mees (1982), Blakers *et al.* (1984), Coates (1985), White and Bruce (1986), Holmes (1989), Jones and Banjaransari (1990), Steadman (1991).

References

Aagaard, J. (1980). Home truths about Scrub Turkeys. *Tamborine Mountain Natural History Association*, December, 5.

Abbott, I. (1974). The avifauna of Kangaroo Island and causes of its impoverishment. *Emu*, **74**, 124–34.

Abdulali, H. (1964). The birds of the Andaman and Nicobar Islands. *Journal of the Bombay Natural History Society*, **61**, 483–571.

Abdulali, H. (1967). The birds of the Nicobar Islands, with notes on some Andaman birds. *Journal of the Bombay Natural History Society*, **64**, 139–90.

Abdulali, H. (1978). The birds of Great and Car Nicobar with some notes on wildlife conservation in the islands. *Journal of the Bombay Natural History Society*, **75**, 744–72.

Ackerman, R. A. (1981). Growth and gas exchange of embryonic sea turtles *(Chelonia, Caretta)*. *Copeia*, **1981**, 757–85.

Ackerman, R. A. and Seagrave, R. C. (1987). Modelling heat and mass exchange of buried avian eggs. *The Journal of Experimental Zoology*, **1**, (Suppl.) 87–97.

Alexander, W. B. (1923). A week on the Upper Barcoo, Central Queensland. *Emu*, **23**, 82–95.

Ali, S. and Ripley, S. D. (1969). *Handbook of the birds of India and Pakistan*, Vol. 2. Oxford University Press, Bombay.

Amadon, D. (1942). Birds collected during the Whitney South Sea Expedition, 49. Notes on some non-passerine genera, 1. *American Museum Novitates*, **1175**, 1–11.

Amadon, D. (1977). Megapodes of Australia. *Explorers Journal*, **55**, 178–80.

Anon. (1904). The Brush Turkey (from E. L. Bertling, *Avicultural Magazine*, 1904). *Emu*, **4**, 75.

Anon. (1905). The Brush Turkey (from E. L. Bertling, *Avicultural Magazine*, 1904). *Emu*, **4**, 143.

Anon. (1971). Mallee Fowl. *Nature Walkabout*, **7**, 41–5.

Anon. (1972). A new home for Brush Turkeys. *Zoonooz*, **45** (11), 13.

Anon. (1982). Annotated list of the birds recorded in the zones: Vanapa–Veimauri and Kanosia–Cape Suckling. *Papua New Guinea Bird Society Newsletter*, **195–6**, 29–37.

Anon. (1986). Een zware bevalling. *Dieren*, **3**, 106–8.

Anon. (1988*a*). Nicobar Scrubfowl faces extinction. *Oriental Bird Club Bulletin*, **7**, 9.

Anon. (1988*b*). Islanders exterminate goose with the golden eggs. *World Wildlife Fund News*, **48**, 3.

Anon. (1989*a*). Tradition serves the Maleo. *Voice of Nature*, **71**, 10–13.

Anon. (1989*b*). Malleefowl gets a helping hand. *Thylacinus*, **14**, 16–17.

Appell, G. N. (1965). Distribution of the megapode —as reported by the Rungus Dusun. *Sarawak Museum Journal*, **12**, 393–4.

Appelman, F. J. (1938). Iets over de 'Gosong'. *De Tropische Natuur*, **27**, 133–6.

Argeloo, M. (1992*a*). The Maleo—more than a symbol. *World Birdwatch*, **14** (1), 8–9.

Argeloo, M. (1992*b*). Problems of Maleo on Sulawesi. *Dutch Birding*, **14**, 54–5.

Argeloo, M. (1992*c*). *Maleo conservation project. Phase II*. Report, Institute of Taxonomic Zoology, University of Amsterdam.

Ashby, E. (1921). Notes on the supposed 'extinct' birds of the south-west corner of western Australia. *Emu*, **20**, 123–4.

Ashby, E. (1922). Notes on the mound-building birds of Australia, with particulars of features peculiar to the Mallee-fowl, *Leipoa ocellata* Gould, and a suggestion as to their origin. *Ibis*, **11** (4), 702–9.

Ashby, E. (1929). Notes on the unique methods of nidification of the Australian Mallee-fowl (*Leipoa ocellata*) with original data supplied by Bruce W. Leake, R.A.O.U. *Auk*, **46**, 294–305.

Assink, H. (1989). Grootpoothoenders; onbekende vogels uit Nieuw Guinea. *World Pheasant Association Nieuwsbrief Jubileumnummer*, **10**, 102–8.

Atyeo, W. T. (1990). Ornate feather mites (Acari, Pterolichidae) from Megapodiidae (Aves, Galliformes). *Entomologische Mitteilungen aus dem Zoologischen Museum Hamburg*, **10**, 67–74.

Atyeo, W. T. (1992). The pterolichoid feather mites (Acarina, Astigmata) of the Megapodiidae (Aves, Galliformes). *Zoologica Scripta*, **21**, 265–305.

Atyeo, W. T. and Perez, T. M. (1991*a*). *Phycoferus*, a new genus of pterolichid feather mites (Acarina, Pterolichidae) from the Megapodiidae (Aves). *Journal of Parasitology*, **77**, 32–7.

Atyeo, W. T. and Perez, T. M. (1991*b*). *Echinozonus*, a new genus of feather mites (Pterolichidae) from the Megapodiidae (Aves). *Entomologische Mitteilungen aus dem Zoologischen Museum Hamburg*, **10**, 113–26.

Austin, C. N. (1950). Further notes on the birds of Dunk Island, Queensland. *Emu*, **49**, 225–31.

Baird, R. F. (1991). Avian fossils from the Quaternary of Australia. In *Vertebrate palaeontology of Australasia*, (ed. P. Vickers-Rich, J. M. Monaghan, R. F. Baird, and T. H. Rich), pp. 810–70. Pioneer Design Studio, Lilydale.

Baker, E. C. S. (1928). *The fauna of British India*, Vol. 5, pp. 436–9. Taylor & Francis, London.

Baker, E. C. S. (1935). *The nidification of birds of the Indian empire*, Vol. 4, pp. 277–80. Taylor & Francis, London.

Baker, R. H. (1948). Report on collections of birds made by United States Naval Medical Research Unit No. 2 in the Pacific war area. *Smithsonian Miscellaneous Collections*, **107** (15), 1–74.

Baker, R. H. (1951). The avifauna of Micronesia, its origin, evolution, and distribution. *University of Kansas Publications, Museum of Natural History*, **3**(1), 1–359.

Balouet, J. C. (1987). Extinctions des vertébrés terrestres de Nouvelle-Caledonie. *Mémoires de la Société Géologique de France*, **150**, 177–83.

Balouet, J. C. and Olson, S. L. (1989). Fossil birds from late Quarternary deposits in New Caledonia. *Smithsonian Contributions to Zoology*, **469**, 1–38.

Baltin, S. (1969). Zur Biologie und Ethologie des Talegalla-Huhns (*Alectura lathami* Gray) unter besonderer Berücksichtigung des Verhaltens während der Brutperiode. *Zeitschrift für Tierpsychologie*, **26**, 524–72.

Baltin, S., Faust, I., and Faust, R. (1965). Beobachtungen über die Entwicklung von *Alectura lathami* (Talegalla-Huhn) bei natürlich und künstlich bebrüteten Eiern (Megapodiden, Grossfusshühner). *Die Naturwissenschaften*, **52** (9), 218–19.

Baltzer, M. C. (1990). A report on the wetland avifauna of South Sulawesi. *Kukila*, **5**, 27–55.

Banfield, E. J. (1913). Megapode mounds and pits. *Emu*, **12**, 281–3.

Bangs, O. and Peters, J. L. (1927). Birds from Maratua Island, off the east coast of Borneo. *Occasional Papers of the Boston Society of Natural History*, **5**, 235–42.

Barker, G. H. (1949). Branch reports: Queensland. *Emu*, **48**, 208–10.

Barker, R. D. and Vestjens, W. J. M. (1989). *The food of Australian birds*, Vol. 1. *Non-passerines*. CSIRO, Melbourne.

Barrett, C. (1943). *An Australian animal book*. Oxford University Press, Melbourne.

Barrett, C. and Crandall, L. S. (1931). Avian mound builders and their mounds. *New York Zoological Society Bulletin*, **34**, 106–27.

Baud, F. J. (1978). Oiseaux des Philippines de la collection W. Parsons II. Luzon, Mindoro et Palawan. *Revue Suisse de Zoologie*, **85**, 55–97.

Bayliss-Smith, T. P. (1972). The birds of Ontong Java and Sikaiana, Solomon Islands. *Bulletin of the British Ornithologists' Club*, **92**, 1–10.

Becker, R. (1959). Die Strukturanalyse der Gefiederfolgen von *Megapodius freyc. reinw.* und ihre Beziehung zu der Nestlingsdune der Hühnervögel. *Revue Suisse de Zoologie*, **66** (23), 411–527.

Beehler, B. M. (1978). *Upland birds of northeastern New Guinea*. Wau Ecology Institute, Wau, Papua New Guinea.

Beehler, B. M., Pratt, T. K., and Zimmerman, D. A. (1986). *Birds of New Guinea*. Princeton University Press.

Bell, H. L. (1969). Recent Papuan breeding records. *Emu*, **69**, 235–7.

Bell, H. L. (1982). A bird community of lowland rainforest in New Guinea. I. Composition and density of the avifauna. *Emu*, **82**, 24–41.

Bellamy, L. (1986). A man of the little desert offers Mallee fowl hope for survival. *The Age (Australia)*, 7 June.

Bellchambers, T. P. (1916). Notes on the Mallee Fowl *Leipoa ocellata rosinae*. *South Australian Ornithologist*, **2**, 134–40.

Bellchambers, T. P. (1917). Notes on the Malleefowl *Leipoa ocellata rosinae*. *South Australian Ornithologist*, **3**, 78–81.

Bellchambers, T. P. (1921). The Mallee Fowl of Australia. *Avicultural Magazine*, **13**, 12 (2), 19–24.

Belterman, R. H. R. and de Boer, L. E. M. (1984). A karyological study of 55 species of birds, including karyotypes of 39 species new to cytology. *Genetica*, **65**, 39–82.

Bennett, G. (1862). Letter addressed to the Secretary. *Proceedings of the Zoological Society of London*, **1862**, 246–8.

Bennett, K. H. (1884). On the habits of the Mallee Hen, *Leipoa ocellata*. *Proceedings of the Linnean Society of New South Wales*, **8**, 193–7.

Benshemesh, J. (1988). Report on a study of Malleefowl ecology. Unpublished report. Department of Conservation, Forests and Lands, Melbourne.

Benshemesh, J. (1989). Report on the establishment of sites for the monitoring of Malleefowl populations: Operation Raleigh 1989. Unpublished report. Department of Conservation, Forests and Lands, Melbourne.

Benshemesh, J. (1990a). Methods for monitoring Malleefowl populations. Unpublished report. Department of Conservation and Environment, Melbourne.

Benshemesh, J. (1990b). Management of Malleefowl with regard to fire. In *The Mallee lands: a conservation perspective*, (ed. J. C. Noble, P. J. Joss, and G. K. Jones). CSIRO, Melbourne.

Benshemesh, J. (1991). Evaluation of thermal sensing for locating Malleefowl nests. Unpublished report. Department of Conservation and Environment, Melbourne.

Benshemesh, J. (1992a). The conservation ecology of Malleefowl, with particular regard to fire. Thesis, Monash University, Clayton, Australia.

Benshemesh, J. (1992b). Further investigations into the feasibility of surveying Malleefowl populations using airborne thermal scanners. Report, Department of Conservation and Environment, Melbourne.

Bergman, S. (1963). Observations on the early life of *Talegalla* and *Megapodius* in New Guinea. *Nova Guinea, Zoology*, **17**, 347–57.

Bergman, S. (1964). Iakttagelser över storfothönsens levnadssätt på Nya Guinea. *Fauna och Flora*, **1964**, 81–93.

Beruldsen, G. (1991). Wattle colour on Australian Brush-turkey at Iron Range, Queensland. *Australian Bird Watcher*, **14** (4), 151.

Beste, H. and Beste, J. (1980). The egg … the bird … and the 3-ton nest. *International Wildlife*, **10** (6), 33–5.

Birks, S. (1990). Paternity in the Australian Brush-turkey: all that work and no cigar? *Acta XX Congressus Internationalis Ornithologici, Supplement*, 469.

Birks, S. M. (1991). Female mate choice in Australian Brush-turkeys. *World Pheasant Association Journal*, **33**, 21–6.

Birks, S. (1992). Mate choice in Australian Brush-turkeys *Alectura lathami*: a preliminary report. In *Proceedings of the first international megapode symposium, Christchurch, New Zealand, December 1990*, (ed. R. W. R. J. Dekker and D. N. Jones). *Zoologische Verhandelingen*, **278**, 43–52.

Bishop, D. (1978). A review of the information relating to the occurrence of *Megapodius freycinet* in the islands of Papua New Guinea. *World Pheasant Association Journal*, **3**, 22–30.

Bishop, K. D. (1980). Birds of the volcanoes—the Scrubfowl of West New Britain. *World Pheasant Association Journal*, **5**, 80–90.

Bishop, D. and Broome, L. (1979). The Scrubfowl *Megapodius freycinet eremita* in West New Britain. *Wildlife in Papua New Guinea*, **79/2**.

Blaber, S. J. M. (1990). A checklist and notes on the current status of the birds of New Georgia, Western Province, Solomon Islands. *Emu*, **90**, 205–14.

Blakers, M., Davies, S. J. J. F., and Reilly, P. N. (1984). *The atlas of Australian birds*. Melbourne University Press.

Blasius, W. (1882). On a collection of birds from the isle of Ceram made by Dr. Platen in November and December 1881. *Proceedings of the Zoological Society of London*, **1882**, 697–711.

Blasius, W. (1888). Die Vögel von Gross-Sanghir. *Ornis*, **4**, 527–651.

Blasius, W. (1896). Vögel von Celebes. *Mitteilungen der Geographischen Gesellschaft und des Naturhistorischen Museums in Lübeck*, **2** (10–11), 124–5.

Blasius, W. (1897). *Neuer Beitrag zur Kenntniss der Vogelfauna von Celebes*. Vieweg, Braunschweig.

Blasius, W. and Nehrkorn, A. (1883). Dr. Platen's ornithologische Sammlungen aus Amboina. *Verhandlungen der kaiserlich-königlichen zoologisch-botanischen Gesellschaft in Wien*, **32**, 411–34.

Board, R. G., Perrott, H. R., Love, G., and Seymour, R. S. (1982). A novel pore system in the eggshells of the Mallee Fowl, *Leipoa ocellata*. *Journal of Experimental Zoology*, **220**, 131–4.

Böhner, J. and Immelmann, K. (1987). Aufbau, Variabilität und mögliche Funktionen des Rufduetts beim Thermometerhuhn *Leipoa ocellata*. *Journal für Ornithologie*, **128**, 91–100.

Bolau, H. (1898). Die Typen der Vogelsammlung des Naturhistorischen Museums zu Hamburg. *Mitteilungen aus dem Naturhistorischen Museum in Hamburg*, **15**, 45–71.

Booth, D. T. (1984). Thermoregulation in neonate Mallee Fowl *Leipoa ocellata*. *Physiological Zoology*, **57**, 251–60.

Booth, D. T. (1985a). Thermoregulation in neonate Brush Turkeys (*Alectura lathami*). *Physiological Zoology*, **58**, 374–9.

Booth, D. T. (1985b). Ecological physiology of Malleefowl (*Leipoa ocellata*). Unpublished D. Phil. thesis, University of Adelaide.

Booth, D. T. (1986). Crop and gizzard contents of two Mallee fowl. *Emu*, **86**, 51–3.

Booth, D. T. (1987a). Effect of temperature on development of Mallee fowl *Leipoa ocellata* eggs. *Physiological Zoology*, **60**, 437–45.

Booth, D. T. (1987b). Home range and hatching success of Malleefowl, *Leipoa ocellata* Gould (Megapodiidae), in Murray mallee near Renmark, S.A. *Australian Wildlife Research*, **14**, 95–104.

Booth, D. T. (1987c). Metabolic response of Mallee fowl *Leipoa ocellata* embryos to cooling and heating. *Physiological Zoology*, **60**, 446–53.

Booth, D. T. (1987d). Water flux in Malleefowl, *Leipoa ocellata* Gould (Megapodiidae). *Australian Journal of Zoology*, **35**, 147–59.

Booth, D. T. (1988a). Respiratory quotient of Malleefowl (*Leipoa ocellata*) eggs late in incubation. *Comparative Biochemistry and Physiology*, **90A**, 445–7.

Booth, D. (1988b). Shell thickness in megapode eggs. *Megapode Newsletter*, **2**(3), 13.

Booth, D. T. (1989a). Growth rates of Malleefowl and an Australian Brush-turkey in captivity. *Megapode Newsletter*, **3**(2), 9–10.

Booth, D. T. (1989b). Metabolism in Malleefowl (*Leipoa ocellata*). *Comparative Biochemistry and Physiology*, **92A**, 207–9.

Booth, D. T. and Seymour, R. S. (1984). Effect of adding water to Malleefowl mounds during a drought. *Emu*, **84**, 116–18.

Booth, D. T. and Seymour, R. S. (1987). Effect of eggshell thinning on water vapor conductance of Malleefowl eggs. *Condor*, **89**, 453–9.

Booth, D. T. and Thompson, M. B. (1991). A comparison of reptilian eggs with those of megapode birds. In *Egg incubation: its effects on embryonic development in birds and reptiles*, (ed. M. W. J. Ferguson and D. C. Deeming), pp. 325–44. Cambridge University Press, Cambridge.

Bowler, J. and Taylor, J. (1989). An annotated checklist of the birds of Manusela National Park, Seram. (Birds recorded on the Operation Raleigh Expedition.) *Kukila*, **4**, 3–29.

Brandle, R. (1991). *Malleefowl mound distribution and status in an area of the Murray Mallee of South Australia*. Nature Conservation Society of South Australia, Adelaide.

Bravery, J. A. (1970). The birds of Atherton Shire, Queensland. *Emu*, **70**, 49–63.

Brazier, J. A. (1869). Notes on an egg of a species of *Megapodius*. *Proceedings of the Zoological Society of London*, **1869**, 528–9.

Bregulla, H. L. (1992). *Birds of Vanuatu*. Anthony Nelson Ltd, Oswestry, Shropshire.

Brickhill, J. (1980). Malleefowl. *Parks and Wildlife Journal, NSW*, 49–55.

Brickhill, J. (1982). Distribution and abundance of Malleefowl in New South Wales. Abstract, *Royal Australasian Ornithologists Union Congress*, Armidale 27–28 November 1982.

Brickhill, J. (1984). Malleefowl. A remarkable bird with an uncertain future. *Australian Natural History*, **21**, 147–51.

Brickhill, J. (1985a). An aerial survey of nests of Malleefowl *Leipoa ocellata* Gould (Megapodiidae) in Central New South Wales. *Australian Wildlife Research*, **12**, 257–61.

Brickhill, J. (1985b). Breeding success of Malleefowl. Poster abstract, *Royal Australasian Ornithologists Union Congress*, 27–29 September 1985, Toowoomba, Australia.

Brickhill, J. (1985c). Aerial survey of Malleefowl nests. Poster abstract, *Royal Australasian Ornithologists Union Congress*, 27–29 September 1985, Toowoomba, Australia.

Brickhill, J. (1987a). Breeding success of Malleefowl *Leipoa ocellata* in Central New South Wales. *Emu*, **87**, 42–5.

Brickhill, J. (1987b). The conservation status of Malleefowl in New South Wales. Master of Natural Resources thesis, University of New England, Armidale.

Brickhill, J. (1990). Malleefowl. In *Threatened birds of Australia*, (ed. J. Brouwer and S. Garnett). RAOU and ANPWS, Melbourne.

Brodkorb, P. (1964). Catalogue of fossil birds: part 2 (Anseriformes through Galliformes).

Bulletin of the Florida State Museum, **8**(3), 195–335.

Brom, T. G. (1991*a*). Variability and phylogenetic significance of detachable nodes in feathers of tinamous, galliforms and turacos. *Journal of Zoology, London*, **225**, 589–604.

Brom, T. G. (1991*b*). The diagnostic and phylogenetic significance of feather structures. Unpublished D. Phil. thesis, University of Amsterdam.

Brom, T. G. and Dekker, R. W. R. J. (1992). Current studies on megapode phylogeny. In *Proceedings of the first international megapode symposium, Christchurch, New Zealand, December 1990*, (ed. R. W. R. J. Dekker and D. N. Jones). *Zoologische Verhandelingen*, **278**, 7–17.

Brookes, G. B. (1919*a*). Report on investigations in regard to the spread of prickly-pear by the Scrub Turkey. *Queensland Agricultural Journal*, **11**, 26–8.

Brookes, G. B. (1919*b*). Report on investigations in regard to the spread of prickly-pear by the Scrub Turkey. *Emu*, **18**, 288–92.

Broome, L. S., Bishop, K. D., and Anderson, D. R. (1984). Population density and habitat use by *Megapodius freycinet eremita* in West New Britain. *Australian Wildlife Research*, **11**, 161–71.

Brown, P. (1977). World Pheasant Association census of cracids and megapodes 1977. *World Pheasant Association Journal*, **2**, 76–9.

Bruce, M. D. and McAllen, I. A. W. (1990). Some problems in vertebrate nomenclature. II. Birds. Part. I. *Bollettino Museo Regionale di Scienza Naturali Torino*, **8**, 453–85.

Brüggemann, F. (1876). Beiträge zur Ornithologie von Celebes und Sangir. *Abhandlungen des naturwissenschaftlichen Vereins zu Bremen*, **5**, 35–106.

Bryan, E. H. Jr (1936). Birds of Guam. *Guam Recorder*, **13**(2), 15.

Bull, J. J. (1980). Sex determination in reptiles. *Quarterly Review of Biology*, **55**, 3–21.

Buller, W. (1870). Notice of a species of megapode, in the Auckland Museum. *Transactions of the New Zealand Institute*, **3**, 14–15.

Burrows, I. (1987). Some notes on the birds of Lihir. *Muruk*, **2**, 40–2.

Cabanis, J. and Reichenow, A. (1876). Uebersicht der auf der Expedition Sr. Maj. Schiff 'Gazelle' gesammelten Vögel. *Journal für Ornithologie*, **24** (4), 319–32.

Cain, A. J. and Galbraith, I. C. J. (1956). Field notes on birds of the eastern Solomon Islands. *Ibis*, **98**, 100–34.

Calder, W. A., Parr, C. R., and Karl, D.P. (1978). Energy content of eggs of the Brown Kiwi *Apteryx australis*: an extreme in avian evolution. *Comparative Biochemistry and Physiology*, **60A**, 177–9.

Campbell, A. G. (1941). Lowans at Wyperfeld. *Emu*, **40**, 422–3.

Campbell, A. J. (1884). Malleehens and their egg mounds. *Victorian Naturalist*, **1**, 124–9.

Campbell, A. J. (1901). *Nests and eggs of Australian birds*. Pawson & Brailsford, Sheffield.

Campbell, A. J. (1903). The mound-building birds of Australia. *Bird-Lore*, **5**, 3–8.

Campbell, A. J. and Barnard, H. C. (1917). Birds of the Rockingham Bay District, North Queensland. *Emu*, **17**, 2–38.

Campbell, B. and Lack, E. (ed.) (1985). *A dictionary of birds*. Poyser, Calton.

Carpentier, J. (1968). Het hamerhoen. *Zoo, Antwerp*, **33**, 152–3.

Carter, T. (1921). Remarks and notes on some western Australian birds. *Emu*, **21**, 54–8.

Carter, T. (1923). Birds of the Broome Hill district. *Emu*, **23**, 125–42.

Chandler, L. G. (1913). Bird-life of Kow Plains (Victoria). *Emu*, **13**, 33–45.

Chandler, L. G. (1934). Notes on the Mallee Fowl. *Victorian Naturalist*, **50**, 199–201.

Chisholm, A. H. (1934). *Bird wonders of Australia*. Angus & Robertson, Sydney.

Chisholm, A. H. (1946). Observations and reflections on the birds of the Victorian mallee. *Emu*, **46**, 168–86.

Christian, F. (1926). The megapode bird and the story it tells. *Polynesian Society Journal*, **35**, 260.

Christidis, L. (1990). B. Aves. *Animal Cytogenetics*, **4**, (Chordata 3), 1–116.

Clark, G. A. (1960). Notes on the embryology and evolution of the megapodes (Aves: Galliformes). *Postilla*, **45**, 1–7.

Clark, G. A. (1964*a*). Ontogeny and evolution in the megapodes (Aves: Galliformes), *Postilla*, **78**, 1–37.

Clark, G. A. (1964*b*). Life histories and the evolution of megapodes. *The Living Bird*, **3**, 149–67.

Clarke, M. M. (1982). Notes on a visit to Bensbach. *Papua New Guinea Bird Society Newsletter*, **195–6**, 18–21.

Clay, T. (1938). A revision of the genera and species of Mallophaga occurring on Gallinaceous hosts. I. *Lipeurus* and related genera. *Proceedings of the Zoological Society of London*, **108** (B), 109–204.

Clay, T. (1940). Genera and species of Mallophaga occurring on gallinaceous hosts. II. *Goniodes*. *Proceedings of the Zoological Society of London*, **110** (B), 1–120.

Clay, T. (1947). Mallophaga miscellany IV. I. Notes on the Goniodidae. *Annals and Magazine of Natural History*, **14** (11), 540–52.

Cleland, J. B. (1912). Examination of contents of stomachs and crops of Australian birds. *Emu*, **12**, 8–18.

Coates, B. J. (1985). *The Birds of Papua New Guinea*, Vol. 1. Dove Publications, Alderley, Queensland.

Coates, B. J. and Swainson, G. W. (1978). Notes on the birds of Wuvulu Islands. *New Guinea Bird Society Newsletter*, **145**, 8–10.

Cohen, G. (1960). Les Mégapodes. Des oiseaux qui pratiquent l'incubation artificielle. *Nature (Paris)*, **3300**, 158–62.

Coles, C. (1937). Some observations on the habits of the Brush Turkey (*Alectura lathami*). *Proceedings of the Zoological Society of London*, **107**(A), 261–73.

Collar, N. J. and Andrew, P. (1988). *Birds to watch. The ICBP world check-list of threatened birds*. ICBP Technical Publication, **8**.

Collias, N. E. and Collias, E. C. (1984). *Nest building and bird behaviour*. Princeton University Press.

Condon, H. T. (1975). *Checklist of the birds of Australia I, Non-passerines*. RAOU, Melbourne.

Coomans de Ruiter, L. (1930). De Maleo [*Megacephalon maleo* (Hartl.)]. *Ardea*, **19**, 16–19.

Coomans de Ruiter, L. (1962). Grootpoothoenders (Megapodiidae). *Nederlands Nieuw Guinea*, **10** (6), 26–31.

Coombs, W. P., Jr (1989). Modern analogs for dinosaur nesting and parental behavior. In *Paleobiology of the dinosaurs*, (ed. J. O. Farlow). *Geological Society of America Special Paper*, **238**, 21–53.

Cooper, R. P. (1966a). The call-notes of the Mallee-Fowl. *Emu*, **66**, 29–31.

Cooper, R. P. (1966b). The call-notes of the Mallee-Fowl. *Australian Bird Watcher*, **2**, 29–30.

Cooper, R. P. (1974). The avifauna of Wilson's Promontory. *Australian Bird Watcher*, **5**, 137–74.

Cracraft, J. (1972a). The relationships of the higher taxa of birds: problems in phylogenetic reasoning. *Condor*, **74**, 379–92.

Cracraft, J. (1972b). Continental drift and Australian avian biogeography. *Emu*, **72**, 171–4.

Cracraft, J. (1973). Continental drift, paleoclimatology, and the evolution and biogeography of birds. *Journal of Zoology, London*, **169**, 455–545.

Cracraft, J. (1976). Avian evolution on southern continents: influences of palaeogeography and palaeoclimatology. *Acta XVI Congressus Internationalis Ornithologici*, 40–52.

Cracraft, J. (1980). Avian phylogeny and intercontinental biogeographic patterns. *Acta XVII Congressus Internationalis Ornithologici*, **2**, 1302–8.

Cracraft, J. and Mindell, D. P. (1989). The early history of modern birds: a comparison of molecular and morphological evidence. In *The hierarchy of life: molecules and morphology in phylogenetic analysis*, (ed. B. Fernholm, K. Bremer, and H. Jörnvall), pp. 389–403. Excerpta Medica, Amsterdam.

Cramp, S. (1988). *Handbook of the birds of Europe, the Middle East, and North Africa: the birds of the Western Palearctic*. Vol. V: *Tyrant Flycatchers to Thrushes*. Oxford University Press, Oxford.

Cramp, S. and Simmons, K. E. L. (1977). *Handbook of the birds of Europe, the Middle East, and North Africa: the birds of the Western Palearctic*. Vol. 1: *Ostrich to Ducks*. Oxford University Press, Oxford.

Croizat, L. (1958). *Panbiogeography*, 3 Vols. Salland, Deventer.

Crome, F. H. J. and Brown, H. E. (1979). Notes on social organization and breeding of the Orange-footed Scrubfowl *Megapodius reinwardt*. *Emu*, **79**, 111–19.

Crowe, T. M. and Short, L. L. (1992). A new gallinaceous bird from the Oligocene of Nebraska with comments on the phylogenetic position of the Gallinuloididae. In *Papers in avian paleontology. Honoring Pierce Brodkorb*, (ed. K. E. Campbell). Sciences Series, **36**, 179–85. Los Angeles.

Crowe, T. M. and Withers, P. C. (1979). Brain temperature regulation in Helmeted Guineafowl. *South African Journal of Science*, **75**, 362–5.

Curio, E. (1992a). Bericht über eine Expedition zum Erhalt des Tonga-Grossfusshuhns (*Megapodius pritchardii*). *World Pheasant Association Rundbrief (German Section)*, **55**, 14–17.

Curio, E. (1992b). Report on an expedition to save the Polynesian Megapode. *World Pheasant Association News*, **38**, 13–16.

Dahl, F. (1899). Das Leben der Vögel auf den Bismarckinseln. *Mitteilungen aus der Zoologischen Sammlung des Museums für Naturkunde in Berlin*, **I** (3), 107–222.

Darlington, P. J., Jr (1957). *Zoogeography: the geographical distribution of animals*. John Wiley & Sons, New York.

Davidson, P. J., Lucking, R. S., Stones, A. J., Bean, N. J., Raharjaningtrah, W., and Banjaransari, H. (1994). Report on an ornithological survey of Taliabu, Indonesia, with notes on the Babirusa Pig. Report, University of East Anglia.

Davies, W. (1983). *Brisbane Wildlife Survey Report. April 1980–October 1981*. Wildlife Preservation Society of Queensland, Brisbane.

Davis, T. A. (1983). Observations on some megapodes of Indonesia and Australia. *Proceedings of the Jubilee Symposium of the Bombay Natural History Society*, p. 25.

Davis, T. A. (1984). The thermometer birds. *Sanctuary*, **4**, 226–33.

Davis, T. A. (1985). The thermometer birds. *Garuda Magazine*, **5**(2), 13–15.

Dawson, W. R. and Hudson, J. W. (1970). Birds. In *Comparative physiology of thermoregulation*, (ed. G. C. Whitton), pp. 223–310. Academic Press, New York.

De Iongh, H., van Helvoort, B., Atmosoedirdjo, S., and Sutanto, H. (1982). An ecological survey of the Kangean island archipelago in Indonesia. Unpublished report.

Dekker, D. (1968). Boskalkoenen. *Artis*, **14** (4), 136–43.

Dekker, R. W. R. J. (1984). Population ecology and management of the Maleo *Macrocephalon maleo*. ICBP Conservation programme project proposal.

Dekker, R. (1987a). De Maleo, een kip met gouden eieren. *Panda*, **23** (5), 76–7.

Dekker, R. W. R. J. (1987b). Managing the Maleo. In *Spirit of Enterprise. The 1987 Rolex awards*, (ed. D. W. Reed), pp. 378–80. Van Nostrand Reinhold, Wokingham, UK.

Dekker, R. W. R. J. (1987c). English names of megapodes. *Megapode Newsletter*, **1** (3), 13–16.

Dekker, R. W. R. J. (1988a). English names of megapodes. *Megapode Newsletter*, **2** (1), 2–3.

Dekker, R. (1988b). Megapodes—from fairy tales to reality. *Oriental Bird Club Bulletin*, **7**, 10–13.

Dekker, R. W. R. J. (1988c). Megapodes under threat—Nicobar Scrubfowl close to extinction. *Megapode Newsletter*, **2** (3), 17–18.

Dekker, R. W. R. J. (1988d). Notes on ground temperatures at nesting sites of the Maleo *Macrocephalon maleo* (Megapodiidae). *Emu*, **88**, 124–7.

Dekker, R. W. R. J. (1989a). Maleos hatched at New York Zoological Societies Wildlife Survival Center, St Catherine's Island. *Megapode Newsletter*, **3** (1), 6–7.

Dekker, R. W. R. J. (1989b). The unfortunate Maleo. *Voice of Nature*, **69**, 59–60.

Dekker, R. W. R. J. (1989c). Predation and the western limits of megapode distribution (Megapodiidae; Aves). *Journal of Biogeography*, **16**, 317–21.

Dekker, R. W. R. J. (1989d). Over grootpoothoenders en eieren. *World Pheasant Association Nieuwsbrief jubileumnummer*, **10**, 67–71.

Dekker, R. W. R. J. (1990a). The distribution and status of nesting grounds of the Maleo *Macrocephalon maleo* in Sulawesi, Indonesia. *Biological Conservation*, **51**, 139–50.

Dekker, R. W. R. J. (1990b). Evolution of megapode incubation strategies. In 'Conservation and biology of megapodes (Megapodiidae, Galliformes, Aves)', pp. 105–29. Unpublished D. Phil. thesis, University of Amsterdam.

Dekker, R. W. R. J. (1990c). Conservation and biology of megapodes (Megapodiidae, Galliformes, Aves). Unpublished D. Phil. thesis, University of Amsterdam.

Dekker, R. W. R. J. (1991). The Moluccan megapode *Eulipoa wallacei* 'rediscovered'. *Megapode Newsletter*, **5**, 9–10.

Dekker, R. W. R. J. (1992). Status and breeding biology of the Nicobar Megapode *Megapodius nicobariensis abbotti* on Great Nicobar, India. Report, National Museum of Natural History, Leiden.

Dekker, R. W. R. J. (in press). Conservation and management of megapodes (Megapodiidae; Galliformes). In *Management methods for populations of threatened birds*. ICBP Technical Publication.

Dekker, R. and Argeloo, M. (1992). New Maleo nesting-grounds. *Megapode Newsletter*, **6**, 6.

Dekker, R. W. R. J. and Brom, T. G. (1990). Maleo eggs and the amount of yolk in relation to different incubation strategies in megapodes. *Australian Journal of Zoology*, **38**, 19–24.

Dekker, R. W. R. J. and Brom, T. G. (1992). Megapode phylogeny and the interpretation of incubation strategies. In *Proceedings of the first international megapode symposium, Christchurch, New Zealand, December 1990*, (ed. R. W. R. J. Dekker and D. N. Jones). *Zoologische Verhandelingen*, **278**, 19–31.

Dekker, R. W. R. J. and Jones, D. N. (ed.) (1992). *Proceedings of the first international megapode symposium, Christchurch, New Zealand, December 1990. Zoologische Verhandelingen*, **278**, 1–78.

Dekker, R. W. R. J. and Wattel, J. (1987). Egg and image: new and traditional uses for the Maleo (*Macrocephalon maleo*). In *The value of birds*, (ed.

A. W. Diamond and F. L. Filion), pp. 83–7. ICBP Technical Publication, **6**.

Delacour, J. (1935). Le Talégalle de Latham ou d'Australie. *L'Oiseau et la Revue Française d'Ornithologie*, **1**, 8–33.

De Schauensee, R. M. (1940*a*). Rediscovery of the megapode, *Aepypodius bruynii*. *Auk*, **57**, 83–4.

De Schauensee, R. M. (1940*b*). On a collection of birds from Waigeu. *Notulae Naturae*, **45**, 1–16.

De Schauensee, R. M. and DuPont, J. E. (1962). Birds from the Philippine Islands. *Proceedings of the Academy of Natural Sciences of Philadelphia*, **114**, 149–73.

Diamond, J. M. (1972). Avifauna of the eastern highlands of New Guinea. *Publications of the Nuttall Ornithological Club*, **12**, 1–438.

Diamond, J. (1983). The reproductive biology of mound-building birds. *Nature*, **301**, 288–9.

Diamond, J. M. (1985). New distributional records and taxa from the outlying mountain ranges of New Guinea. *Emu*, **85**, 65–91.

Diamond, J. M. and LeCroy, M. (1979). Birds of Karkar and Bagabag islands, New Guinea. *Bulletin of the American Museum of Natural History*, **164**, 467–531.

Diamond, J. M. and Marshall, A. G. (1976). Origin of the New Hebridean avifauna. *Emu*, **76**, 187–200.

Diamond, J. M. and Terborgh, J. W. (1968). Dual singing by New Guinea birds. *Auk*, **85**, 62–82.

Dickinson, E. C., Kennedy, R. S., and Parkes, K. C. (1991). *The birds of the Philippines*. BOU Check-list No. 12, British Ornithologists' Union, London.

Dillwyn, L. (1851). On an undescribed species of *Megapodius*. *Proceedings of the Zoological Society of London*, **1851** (1853), 118–19.

Dobbyn, A. (1989). Don't let the turkeys get you down. *The Road Ahead (RACQ)*, June, 13.

Dow, D. D. (1980). Primitive weaponry in birds: the Australian Brush-turkey's defence. *Emu*, **80**, 91–2.

Dow, D. D. (1988*a*). Dusting and sunning by Australian Brush-turkeys. *Emu*, **88**, 47–8.

Dow, D. D. (1988*b*). Sexual interactions by Australian Brush-turkeys away from the incubation mound. *Emu*, **88**, 49–50.

Downes, M. C. (1972). The wildfowl egg-grounds of West New Britain. *Harvest*, **2**(1), 1–4.

Drent, R. H. (1975). Incubation. In *Avian biology*, Vol. V, (ed. D. S. Farner and J. R. King), pp. 333–420. Academic Press, New York.

Drent, R. H. and Daan, S. (1980). The prudent parent: energetic adjustments in avian breeding. *Ardea*, **68**, 225–52.

Dubois, A. (1902–1904). *Synopsis Avium. Nouveau manuel d'ornithologie*, Vols 1 and 2. Lamertin, Bruxelles.

Dumont, C. (1823). Megapode. In *Dictionnaire des Sciences Naturelles*, **29**, pp. 414–18. Levrault, Paris.

DuPont, J. E. (1976). *South Pacific birds*. Monograph series **3**. Delaware Museum of Natural History.

DuPont, J. E. and Rabor, D. S. (1973). Birds of Dinagat and Siargao, Philippines. An expedition report. *Nemouria*, **10**, 1–111.

Dwyer, P. D. (1981). Two species of megapode laying in the same mound. *Emu*, **81**, 173–4.

Dyck, J. (1985). The evolution of feathers. *Zoologica Scripta*, **14**, 137–54.

Eck, S. (1976). Die Vögel der Banggai-Inseln, insbesondere Peleng (Aves). *Zoologische Abhandlungen Staatliches Museum für Tierkunde in Dresden*, **34**, 53–100.

Eckstein, F. W. (1990). Naturbrut vom Talegallahuhn oder Buschhuhn unter Zuhilfenahme eines kleines Tricks. *World Pheasant Association Rundbrief (German Section)*, **49**, 14–16.

Edkins, E. and Hansen, I. A. (1971). Diol esters from the uropygial glands of Mallee Fowl and Stubble Quail. *Comparative Biochemistry and Physiology*, **39B**, 1–4.

Elzanowski, A. (1985). The evolution of parental care in birds with reference to fossil embryos. *Acta XVIII Congressus Internationalis Ornithologici*, **1**, 178–83.

Elzanowski, A. (1988). Ontogeny and evolution of the ratites. *Acta XIX Congressus Internationalis Ornithologici*, **2**, 2037–46.

Emerson, K. C. and Price, R. D. (1972). A new genus and species of Mallophaga from a New Guinea Bush Fowl. *Pacific Insects*, **14**, 77–81.

Emerson, K. C. and Price, R. D. (1984). A new species of *Goniodes* (Mallophaga: Philopteridae) from the Mallee Fowl (Galliformes: Megapodiidae). *International Journal of Entomology*, **26**, 366–8.

Emerson, K. C. and Price, R. D. (1986). Two new species of Mallophaga (Philopteridae) from the Mallee Fowl (Galliformes: Megapodiidae) in Australia. *Journal of Medical Entomology*, **23**, 353–5.

Emerson, K. C. and Ward, R. A. (1958). Notes on Philippine Mallophaga. I. Species from Ciconiiformes, Anseriformes, Falconiformes, Galliformes, Gruiformes and Charadriiformes. *Fieldiana, Zoology*, **42** (4), 49–61.

Engbring, J. (1988). *Field guide to the birds of Palau*. Conservation Office, Koror, Palau.

Engbring, J. and Pratt, H. D. (1985). Endangered birds in Micronesia: their history, status and future prospects. In *Birds conservation*, Vol. 2, (ed. S. A. Temple), pp. 71–105. University of Wisconsin Press, Madison.

Engbring, J., Ramsey, F. L., and Wildman, V. J. (1986). Micronesian forest bird survey, 1982: Saipan, Tinian, Aguiguan, and Rota. Unpublished report. U.S. Fish and Wildlife Service, Honolulu.

Falanruw, M. V. C. (1975). Distribution of the Micronesian Megapode *Megapodius laperouse* in the northern Mariana Islands. *Micronesica*, **11**, 149–50.

Farabaugh, S. M. (1982). The ecological and social significance of duetting. In *Acoustic communication in birds*, Vol. 2, (ed. D. E. Kroodsma and E. H. Miller), pp. 85–124. Academic Press, New York.

Finsch, O. (1872). Zur Ornithologie der Samoa-Inseln. *Journal für Ornithologie*, **20**, 30–58.

Finsch, O. (1875). Zur Ornithologie der Sudsee-Inseln I. Die Vögel der Palau-Gruppe. *Journal des Museum Godeffroy*, **8**, 1–51.

Finsch, O. (1877). On a collection of birds from Niuafou Island, in the Pacific. *Proceedings of the Zoological Society of London*, **1877**, 782–7.

Finsch, O. (1901). Systematische Übersicht der Vögel der Südwest-Inseln. *Notes from the Leyden Museum*, **22**, 225–309.

Finsch, O. and Hartlaub, G. (1867). *Beitrag zur Fauna Centralpolynesiens. Ornithologie der Viti-, Samoa- und Tonga-Inseln.* Schmidt, Halle.

Fleay, D. H. (1937). Nesting habits of the Brush-Turkey. *Emu*, **36**, 153–63.

Fleay, D. (1960). *Living with animals*. Lansdowne Press, Melbourne.

Fleay, D. (1983). The busy builders. *Nature Notes, Courier-Mail*, 14 June.

Flieg, G. M. (1970). Breeding the Yellow-wattled Brush Turkey in North America. *Avicultural Magazine*, **76**, 161–3.

Flieg, G. M. (1971). Megapodes. *Game Bird Breeders', Pheasant Fanciers' and Aviculturists' Gazette*, **20** (7), 33–6.

Ford, H. (1989). *Ecology of birds—an Australian perspective.* Surrey-Beatty, Chipping Norton, Sydney.

Ford, J. (1988). Distributional notes on North Queensland birds. *Emu*, **88**, 50–3.

Ford, J. R. and Stone, P. S. (1957). Birds of the Kellerberrin/Kwolyin district, western Australia. *Emu*, **57**, 9–21.

Friedländer, B. (1899). Über die Nestlöcher des *Megapodius pritchardi* auf der Insel Niuafu. *Ornithologische Monatsberichte*, **7**, 37–40.

Friedmann, H. (1931). Observations on the growth rate of the foot in the mound birds of the genus *Megapodius*. *Proceedings of the United States National Museum*, **80**, 1–4.

Frith, H. J. (1955). Incubation in the Mallee Fowl (*Leipoa ocellata*, Megapodiidae). *Acta XI Congressus Internationalis Ornithologici*, 570–4.

Frith, H. J. (1956a). Temperature regulation in the nesting mounds of the Mallee-fowl, *Leipoa ocellata* Gould. *CSIRO Wildlife Research*, **1**, 79–95.

Frith, H. J. (1956b). Breeding habits in the family Megapodiidae. *Ibis*, **98**, 620–40.

Frith, H. J. (1956c). The mound builders. *Pacific Discovery*, **9**(5), 14–17.

Frith, H. J. (1956d). Wie regelt der Thermometervogel die Temperatur seines Nesthügels? *Die Umschau in Wissenschaft und Technik*, **8**, 238–9.

Frith, H. J. (1957). Experiments on the control of temperature in the mound of the Mallee-fowl, *Leipoa ocellata* Gould (Megapodiidae). *CSIRO Wildlife Research*, **2**, 101–10.

Frith, H. J. (1958). The Mallee Fowl. *Australian Museum Magazine*, **12**(9), 289–94.

Frith, H. J. (1959a). Incubator birds. *Scientific American*, **201**(2), 52–8.

Frith, H. J. (1959b). Breeding of the Mallee Fowl, *Leipoa ocellata* Gould (Megapodiidae). *CSIRO Wildlife Research*, **4**, 31–60.

Frith, H. J. (1962a). *The Mallee-fowl*. Angus & Robertson, Sydney.

Frith, H. J. (1962b). Conservation of the Mallee Fowl, *Leipoa ocellata* Gould (Megapodiidae). *CSIRO Wildlife Research*, **7**, 33–49.

Frith, H. J. (1968). Family Megapodiidae. In *Grzimek's animal life encyclopedia*, (ed. H. C. B. Grzimek). Van Nostrand Reinhold, New York.

Fürbringer, M. (1888). *Untersuchungen zur Morphologie und Systematik der Vögel, zugleich ein Beitrag zur Anatomie der Stütz- und Bewegungsorgane.* Van Holkema, Amsterdam.

Gaimard, P. J. (1823). Mémoire sur un nouveau genre de Gallinacés, etabli sous le nom de Mégapode. *Bulletin Général et Universel des Annonces et de Nouvelles Scientifiques*, **2**, 450–1.

Garnett, S. and Bredl, R. (1985). Birds in the vicinity of Edward river settlement. *Sunbird*, **15**, 6–23.

Gell, P. and Werren, G. (1982). Malleebirds under threat. *Habitat Australia*, **10**(4), 4–7.

Gibson, R. M. and Bradbury, J. W. (1985). Sexual selection in lekking Sage Grouse: phenotypic correlates of male mating success. *Behavioral Ecology and Sociobiology*, **18**, 117–23.

Giebel, C. G. (1872–1877). *Thesaurus ornithologiae*, 3 vols. Brockhaus, Leipzig.

Giffard, C. (1989). Raptor notes from Cape York. *Australian Raptor Association Newsletter*, **13**, 14.

Gill. H. B. (1970). Birds of Innisfail and hinterland. *Emu*, **70**, 105–16.

Gilliard, E. T. (1950). Notes on birds of southeastern Papua. *American Museum Novitates*, **1453**, 1–40.

Gilliard, E. T. and LeCroy, M. (1966). Birds of the Middle Sepik Region, New Guinea. Results of the American Museum of Natural History Expedition to New Guinea in 1953–1954. *Bulletin of the American Museum of Natural History*, **132**, 245–76.

Gilliard, E. T. and LeCroy, M. (1967*a*). Results of the 1958–59 Gilliard New Britain expedition. 4. Annotated list of birds of the Whiteman mountains, New Britain. *Bulletin of the American Museum of Natural History*, **135**, 173–216.

Gilliard, E. T. and LeCroy, M. (1967*b*). Annotated list of birds of the Adelbert mountains, New Guinea. Results of the 1959 Gilliard expedition. *Bulletin of the American Museum of Natural History*, **138**, 51–81.

Gilliard, E. T. and LeCroy, M. (1970). Notes on birds from the Tamrau Mountains, New Guinea. *American Museum Novitates*, **2420**, 1–28.

Glass, P. O. (1988). Micronesian Megapode surveys and research. In *Five year progress report 1983–1987*, (Pittman-Robertson federal aid in wildlife restoration program), pp. 131–53. Saipan, CNMI, Division of Fish and Wildlife.

Glass, P. and Villagomez, S. (1986). Field trip report. Guguan Island, 8–20 September 1986. Mimeographed report. Saipan, CNMI, Division of Fish and Wildlife.

Goodwin, D. (1974). Galliformes. In *Birds of the Harold Hall Australian expeditions*, (ed. B. P. Hall), pp. 60–2. Trustees of the British Museum (Natural History), London.

Gould, J. (1840). In minutes of the meeting of October 13, 1840. *Proceedings of the Zoological Society of London*, **1840**, 119–28.

Gould, J. (1842). Megapodius tumulus. *Proceedings of the Zoological Society of London*, **1842**, 20–1.

Gould, J. (1848). *The birds of Australia*. Vols. 1 and 5. Taylor, London.

Gould, J. (1865). *Handbook to the birds of Australia*, pp. 168–74. The author, London.

Gould, J. (1886). *The birds of New Guinea and the adjacent Papuan Islands, including many new species recently discovered in Australia*, Vol. **5**, Part 22. Sotheran, London.

Gräffe, E. (1870). Ornithologische Mitteilungen aus Central-Polynesien. I. Die Vogelwelt der Tonga-Inseln. *Journal für Ornithologie*, **18**, 401–20.

Grahame, D. (1980). WPA census of cracids and megapodes 1980. *World Pheasant Association Journal*, **5**, 60–3.

Gray, G. R. (1840). *A list of the genera of birds*. R. & J. Taylor, London.

Gray, G. R. (1849). *The genera of birds*, Vol. 3. Longman, Brown, Green, & Longmans, London.

Gray, G. R. (1859). *Catalogue of the birds of the tropical islands of the Pacific Ocean, in the collection of the British Museum*. London.

Gray, G. R. (1860). List of birds collected by Mr. Wallace at the Molucca Islands, with descriptions of new species, etc. *Proceedings of the Zoological Society of London*, **1860**, 341–66.

Gray, G. R. (1861). List of species composing the family Megapodiidae, with descriptions of new species, and some account of the habits of the species. *Proceedings of the Zoological Society of London*, **1861**, 288–96.

Gray, G. R. (1864*a*). On a new species of megapode. *Proceedings of the Zoological Society of London*, **1864**, 41–4.

Gray, G. R. (1864*b*). On a species of megapode. *The Annals and Magazine of Natural History, including Zoology, Botany, and Geology*, **14**, 378–9.

Gray, G. R. (1870). Descriptions of new species of birds from the Solomon and Bank's groups of islands. *The Annals and Magazine of Natural History, including Zoology, Botany, and Geology*, Ser. 4, 327–31.

Greenway, J. C. (1966). Birds collected on Batanta, off western New Guinea, by E. Thomas Gilliard in 1964. *American Museum Novitates*, **2258**, 1–27.

Greenway, J. C. (1967). *Extinct and vanishing birds of the world*. Dover Publications, New York.

Griffiths, F. J. (1954). Survey of the Lowan or Mallee-fowl in New South Wales. *Emu*, **54**, 186–9.

Grzimek, B. (1962). Vögel bauen ihren eigenen Brutapparat: Australische Grossfusshühner. *Das Tier*, **1962** (8), 42–3.

Guillemard, F. H. H. (1885). Report on the collection of birds obtained during the voyage of the yacht 'Marchesa'. IV. Celebes. *Proceedings of the Zoological Society of London*, **1885**, 542–61.

Guillemard, F. H. H. (1886). *The cruise of the Marchesa to Kamschatka and New Guinea*, Vol. 2. Murray, London.

Gyldenstolpe, N. (1955). Birds collected by Dr. Sten Bergman during his expedition to Dutch New Guinea 1948–1949. *Arkiv för Zoologi*, Ser. 2, **8**, 183–397.

Hachisuka, M. (1930). Contributions to the birds of the Philippines. No. 2, Part 6. *The Ornithological Society of Japan, Supplementary Publication*, **14**, 141–222.

Hachisuka, M. (1931–1932). *The birds of the Philippine islands*, Vol. 1. Witherby, London.

Hamilton-Smith, E. (1965). Scrub Turkey at Chillagoe, Queensland. *Emu*, **65**, 118.

Harding, E. (1982). Birds of Nissan and Pinipel Island, north Solomons Province. *Papua New Guinea Bird Society Newsletter*, **195–6**, 4–12.

Harris, B. (1979). Watch on mystery turkey influx. *Telegraph, Brisbane*, 3 November.

Harrison, C. J. O. and Frith, C. B. (1970). Nests and eggs of some New Guinea birds. *Emu*, **70**, 173–8.

Harrison, L. (1916). The genera and species of Mallophaga. *Parasitology*, **9**, 1–156.

Harrisson, T. (1965). A future for Borneo's wildlife? *Oryx*, **8**, 99–104.

Hartert, E. (1896). List of a collection of birds made in Lombok by Mr. Alfred Everett. *Novitates Zoologicae*, **3**, 591–9.

Hartert, E. (1898a). On the birds collected by Mr. Everett in South Flores. II. *Novitates Zoologicae*, **5**, 42–50.

Hartert, E. (1898b). On the birds of the Marianne Islands. *Novitates Zoologicae*, **5**, 51–69.

Hartert, E. (1898c). List of a collection of birds made in the Sula islands by William Doherty. *Novitates Zoologicae*, **5**, 125–36.

Hartert, E. (1898d). On the birds of Lomblen, Pantar, and Alor. *Novitates Zoologicae*, **5**, 455–65.

Hartert, E. (1902). The birds of the Kangean Islands. *Novitates Zoologicae*, **9**, 419–43.

Hartert, E. (1903). On the birds collected on the Tukang-Besi islands and Buton, south-east of Celebes, by Mr. Heinrich Kühn. *Novitates Zoologicae*, **10**, 18–38.

Hartert, E. (1904). The birds of the South-west islands Wetter, Roma, Kisser, Letti and Moa. *Novitates Zoologicae*, **11**, 174–221.

Hartert, E. (1930). List of the birds collected by Ernst Mayr. *Novitates Zoologicae*, **36**, 27–128.

Hartert, E., Paludan, K., Rothschild, W., and Stresemann, E. (1936). Die Vögel des Weyland-Gebirges und seines Vorlandes. *Mitteilungen aus dem Zoologischen Museum in Berlin*, **21**, 165–240.

Hartin, M. H. (1961). Birds of Guam. *Elepaio*, **22** (3), 34–8.

Hartlaub, G. (1867). On a collection of birds from some less-known localities in the western Pacific. *Proceedings of the Zoological Society of London*, **1867**, 828–32.

Hartlaub, G. and Finsch, O. (1868a). On a collection of birds from the Pelew Islands. *Proceedings of the Zoological Society of London*, **1868**, 4–9.

Hartlaub, G. and Finsch, O. (1868b). Additional notes on the ornithology of the Pelew Islands. *Proceedings of the Zoological Society of London*, **1868**, 116–18.

Hartlaub, G. and Finsch, O. (1872). On a fourth collection of birds from the Pelew and Mackenzie Islands. *Proceedings of the Zoological Society of London*, **1872**, 87–114.

Haywood, J. L. (1970). Report of a Malleefowl survey. *Mid Murray Field Naturalist*, **3**, 14–16.

Hediger, H. (1934). Zur Biologie und Psychologie der Flucht bei Tieren. *Biologisches Zentralblatt*, **54**, 21–40.

Heilmann, G. (1926). *The origin of birds*. Witherby, London.

Heinrich, G. (1932). *Der Vogel Schnarch*. Reimer & Vohsen, Berlin.

Heinrich, G. (1956). Biologische Aufzeichnungen über Vögel von Halmahera und Batjan. *Journal für Ornithologie*, **97**, 31–40.

Heinroth, O. (1902). Ornithologische Ergebnisse der 'I. Deutschen Südsee Expedition von Br. Mencke'. *Journal für Ornithologie*, **50**, 390–457.

Heinroth, O. (1922). Die Beziehungen zwischen Vogelgewicht, Eigewicht, Gelegegewicht und Brutdauer. *Journal für Ornithologie*, **70**, 172–285.

Hellebrekers, W. Ph. J. and Hoogerwerf, A. (1967). A further contribution to our oological knowledge of the island of Java (Indonesia). *Zoologische Verhandelingen, Leiden*, **88**, 1–164.

Hellmayr, C. E. (1914). *Zoologie von Timor I, die avifauna von Timor*. Haniel, Stuttgart.

Holmes, D. A. (1989). Status report on Indonesian Galliformes. *Kukila*, **4**, 133–43.

Hoogerwerf, A. (1948). Contribution to the knowledge of the distribution of birds on the island of Java with remarks to some new birds. *Treubia*, **19**, 83–137.

Hoogerwerf, A. (1949). *Een bijdrage tot de oölogie van het eiland Java*. Ponsen & Looyen, Wageningen.

Hoogerwerf, A. (1955). Iets over de vogels van de eilanden Komodo, Padar en Rintja, het land van *Varanus komodoensis*. *Limosa*, **28**, 96–112.

Hoogerwerf, A. (1971). On a collection of birds from the Vogelkop, near Manokwari, north-western New Guinea. *Emu*, **71**, 1–12.

Hopkins, G. H. E. and Clay, T. (1952). *Check list of the genera and species of Mallophaga*. British Museum (Natural History), London.

Horton, H. (1989). A comparative look at megapodes. *Queensland Naturalist*, **29**, 49–56.

Howe, F. E. and Tregellas, T. H. (1914). Rarer birds of the Mallee. *Emu*, **14**, 71–84.

Howell, T. R. (1979). Breeding biology of the Egyptian Plover *Pluvianus aegyptius*. *University of California Publications in Zoology*, **113**.

Howes, C. A. (1969). A survey of extinct and nearly extinct birds in the Royal Albert Memorial Museum, Exeter. *Bulletin of the British Ornithologists' Club*, **89**, 89–92.

Hudson, G. E. and Lanzillotti, P. J. (1964). Muscles in the pectoral limb in galliform birds. *American Midland Naturalist*, **71**, 1–113.

Hume, A. (1874). The islands of the Bay of Bengal. *Stray Feathers*, **2**, 29–324.

Hutton, F. W. (1869). Letter from the Auckland Museum. *Ibis*, **5**, 352–3.

Huxley, T. H. (1867). On the classification of birds; and on the taxonomic value of modifications of certain of the cranial bones observable in that class. *Proceedings of the Zoological Society of London*, **1867**, 415–72.

Huxley, T. H. (1868). On the classification and distribution of Alecteromorphae and Heteromorphae. *Proceedings of the Zoological Society of London*, **1868**, 294–319.

Hyem, E. L. (1936). Notes on the birds of 'Mernot', Barrington, NSW. *Emu*, **36**, 109–27.

Immelmann, K. (1963). Review of: H. J. Frith 1962. The Mallee-Fowl. *Journal für Ornithologie*, **104**, 255–6.

Immelmann, K. and Böhner, J. (1984a). Beobachtungen am Thermometerhuhn (*Leipoa ocellata*) in Australien. *Journal für Ornithologie*, **125**, 141–55.

Immelmann, K. and Böhner, J. (1984b). Rufduet beim Thermometerhuhn (*Leipoa ocellata*). *Verhandlungen der Deutschen Zoologischen Gesellschaft*, **77**, 295.

Immelmann, K. and Sossinka, R. (1986). Parental behaviour in birds. In *Parental behaviour*, (ed. W. Sluckin and M. Herbert), pp. 8–43. Blackwell, Oxford.

Indrawan, M. (1992). News on the Maleo colony at Bakiriang, eastern Sulawesi, Indonesia. *Megapode Newsletter*, **6**(2), 13–14.

Indrawan, M., Fujita, M. S., Masala, Y., and Pesik, L. (1992). Status and conservation of the Sula Scrubfowl *Megapodius bernsteinii*: a report from Banggai islands, Indonesia. Draft report for EMDI/KLH and PHPA, Bogor.

Iredale, T. (1956). *Birds of New Guinea*, Vol. 1. Griffin Press, Adelaide.

Iredale, T. and Whitley, G. (1943). Plain dwellers and mound builders. The *Pedionomus* puzzle. *Emu*, **42**, 246–9.

Jacobi, E. F. (1970). Die Zucht von Talegallahühnern (*Alectura lathami* Gray) mit electrischer Bruthitze. *Der Zoologische Garten*, **39**, 129–32.

Jakoby, J. R. and Kosters, J. (1986). Die Ordnung Huhnervögel—Artenvielfalt und Haltungsprobleme aus tierarztlicher Sicht. *Praktische Tierarzt*, **67**, 205–8.

Jarman, H. (1965). The mound-builders. *Victoria's Resources*, **7**, 22–5.

Jerrard, C. H. H. (1933). Brush Turkeys and their young. *Emu*, **33**, 52–3.

Jobling, J. A. (1991). *A dictionary of scientific bird names*. Oxford University Press.

Johnston, B. (1971). Some observations on the Lowan at Wychitella. *Victorian Naturalist*, **88**, 116–17.

Jones, D. (1979). Notes of the breeding habits of the Brush Turkey. *Sunbird*, **10**, 8–10.

Jones, D. N. (1985a). Selection of incubation sites in the Australian brush-turkey *Alectura lathami*. Abstract, Ecological Society of Australia Open Forum, 13–15 May, Armidale, Australia.

Jones, D. (1985b). Mating system of the Australian Brush-turkey. Abstract, Royal Australasian Ornithologists Union Congress, 27–29 September, Toowoomba, Australia.

Jones, D. (1985c). Contrasting behavioural ecology of Australian megapodes. Poster abstract, Royal Australasian Ornithologists Union Congress, 27–29 September, Toowoomba, Australia.

Jones, D. (1986). Telling tales on turkeys. *Wildlife Australia*, **23**, 8–9.

Jones, D. (1987a). Animals using the incubation mounds of the Australian Brush-turkey. *Sunbird*, **17**, 32–5.

Jones, D. N. (1987b). Behavioural ecology of reproduction in the Australian Brush-turkey *Alectura lathami*. Unpublished D. Phil. thesis, Griffith University, Brisbane.

Jones, D. N. (1988a). Construction and maintenance of the incubation mounds of the Australian Brush-turkey *Alectura lathami*. *Emu*, **88**, 210–18.

Jones, D. (1988b). Selection of incubation mound sites by the Australian Brush-turkey *Alectura lathami*. *Ibis*, **130**, 251–60.

Jones, D. N. (1988c). Hatching success of the Australian Brush-turkey *Alectura lathami* in south-east Queensland. *Emu*, **88**, 260–3.

Jones, D. N. (1988d). Megapode mounds and mating systems: parental care or resource defence? Abstract, International Comparative Psychology/ Australasian Society for the Study of Animal Behaviour Joint Conference, 26–28 August, Sydney, Australia.

Jones, D. N. (1988e). "Waifs, without care or guidance!" Questions about learning in extremely precocial megapode hatchlings. Poster abstract, International Comparative Psychology/ Australasian Society for the Study of Animal Behaviour Joint Conference, 26–28 August, Sydney, Australia.

Jones, D. (1989a). Mounds and mates: the breeding strategies of the Australian Brush-Turkey. *Australian Science Magazine*, **31**(2), 13–16.

Jones, D. N. (1989b). Modern megapode research: a post-Frith review. Australasian Bird Reviews No. 1. *Corella*, **13**, 145–54.

Jones, D. N. (1990a). Social organization and sexual interactions in Australian Brush-turkeys (*Alectura lathami*): implications of promiscuity in a mound-building megapode. *Ethology*, **84**, 89–104.

Jones, D. N. (1990b). Sexual conflict in a promiscuous megapode. *Acta XX Congressus Internationalis Ornithologici, Suppl.* 407.

Jones, D. N. (1990c). Megapode research and conservation: recent advances and future priorities. *Acta XX Congressus Internationalis Ornithologici, Suppl.* 441.

Jones, D. N. (1990d). Male mating tactics in a promiscuous megapode: patterns of incubation mound ownership. *Behavioral Ecology*, **1**(2), 107 15.

Jones, D. N. (1992). An evolutionary approach to megapode mating systems. In *Proceedings of the first international megapode symposium, Christchurch, New Zealand, December 1990*, (ed. R. W. R. J. Dekker and D. N. Jones). *Zoologische Verhandelingen*, **278**, 33–42.

Jones, D. and Birks, S. (1992). Megapodes: recent ideas on origins, adaptations and reproduction. *Trends in Ecology and Evolution*, **7**(3), 88–91.

Jones, D. N. and Everding, S. E. (1991). Australian Brush-turkeys in a suburban environment: implications for conflict and conservation. *Australian Wildlife Research*, **18**, 285–97.

Jones, D. N., Everding, S. E., and Nattrass, R. (1993). Suburban brush-turkeys: managing conflict with recalcitrant mound-builder. *Wingspan*, **11**, 20–1.

Jones, J. (1963a). Malleefowl mount count. *Bird Observer*, **374**, 3–4.

Jones, J. (1963b). Malleefowl mount count. *Bird Observer*, **375**, 5–7.

Jones, M. and Banjaransari, H. (1990). The ecology and conservation of the birds of Sumba and Buru. Unpublished report: 25pp.

Junge, G. C. A. (1937). *The birds of South New Guinea*, Vol. 1. *Non passeres*. Brill, Leiden.

Kaveney, M. (1958). Notes on the Brush Turkey. *Emu*, **58**, 152–3.

Kazacos, K. R., Kazacos, E. A., Rinder, J. A., and Thacker, H. L. (1982). Cerebrospinal nematodiasis and visceral larva migrans in an Australian (Latham's) bush turkey. *American Veterinary Medical Association Journal*, **181**, 1295–8.

Keartland, G. A. (1901). Talegallus hybrids. *The Victorian Naturalist*, **17**, 172.

Keys, M. G. (1990). Relocation of Australian Brush-turkeys. *Sunbird*, **20**, 33–6.

Kimber, R. G. (1985). The history of the Malleefowl in Central Australia. *RAOU Newsletter*, **64**, 6–8.

King, W. B. (1979). *Red data book*, Vol. 2. Aves. IUCN, Morges.

Kirch, P. V. and Yen, D. E. (1982). Tikopia: the prehistory and ecology of a Polynesian outlier. *Bernice Pauahi Bishop Museum Bulletin*, **238**, 1–396 (282).

Kisokau, K. (1976). A study in the biology of the megapodes of West New Britain. *Papua New Guinea Bird Society Newsletter*, **121**, 18–20.

Kitchener, D. J., Boeadi, Charlton, L., and Maharadatunkamsi. (1990). Wild mammals of Lombok Island. *Records of the Western Australian Museum, Supplement*, **33**, 1–129.

Kloska, C. (1986). Untersuchungen zur Brutbiologie des Kamm-Talegalla (*Aepypodius arfakianus* Salvad.). Unpublished student report, University of Hamburg.

Kloska, C. and Nicolai, J. (1988). Fortpflanzungsverhalten des Kamm-Talegalla (*Aepypodius arfakianus* Salvad). *Journal für Ornithologie*, **129**, 185–204.

Korn, T. (1986). Malleefowl *Leipoa ocellata* attacked by Brown Goshawk *Accipiter fasciatus*. *Australian Bird Watcher*, **11**, 274–5.

Korn, T. (1989). The Malleefowl of the Goonoo Forest, Dubbo. *National Parks Journal*, 22–4.

Krebs, C. J. (1963). To fugle fra New Britain. *Naturens Verden*, April, 102–6.

Krohn, J. (1982). Unexpected sighting of a Malleefowl. *Australian Bird Watcher*, **9**, 175–6.

Kukila (1990). Sulawesi bird report. *Kukila*, **5**, 4–26.

Lack, D. (1968). *Ecological adaptations for breeding in birds*. Methuen, London.

Lakshminarayana, K. V., Vijayalakshmi, S., and Talukdar, B. (1980). The chewing-lice (Phthiraptera: Insecta) from Andaman and Nicobar Islands with remarks on some host relationships. *Records of the Zoological Survey of India*, **77**, 31–7.

Laskowski, M. and Fitch, W. M. (1989). Evolution of avian ovomucoids and of birds. In *The hierarchy of life: molecules and morphology in phylogenetic analysis*, (ed. B. Fernholm, K. Bremer, and H. Jörnvall), pp. 371–87. Excerpta Medica, Amsterdam.

Layard, E. L. C. (1876). Notes on the birds of the Navigators' and Friendly islands, with some additions to the ornithology of Fiji. *Proceedings of the Zoological Society of London*, **1876**, 490–506.

Layard, E. L. C. (1880). Notes of a collecting-trip in the New Hebrides, the Solomon Islands, New Britain, and the Duke-of-York Islands. *Ibis*, Ser. 4, vol. 4, 290–309.

Layard, E. L. and Layard, E. L. C. (1878). Notes on some birds collected or observed by Mr. E. Leopold C. Layard in the New Hebrides. *Ibis*, Ser. 4, vol. 4, **2**, 267–80.

Lea, A. M. and Gray, J. T. (1935). The food of Australian birds, Part 1. *Emu*, **34**, 275–92.

Lemke, T. O. (1983*a*). Recent observations on the avifauna of the Northern Mariana Islands north of Saipan. Unpublished report. Saipan, CNMI, Division of Fish and Wildlife: 40pp.

Lemke, T. O. (1983*b*). Micronesian Megapode surveys and inventories. Pittman-Robertson job progress report FY 1983. Unpublished report. Saipan, CNMI, Division of Fish and Wildlife.

Lemke, T. O. (1984). Micronesia Megapode surveys and inventories. Pittman-Robertson job progress report FY 1984. Unpublished report. Saipan, CNMI, Division of Fish and Wildlife: 106–10.

Le Souëf, D. (1898). On some birds and eggs lately collected at Cape York, Queensland, by Mr. H. G. Barnard. *Ibis*, Ser. 7, vol. 4, 51–9.

Le Souëf, D. (1899). On the habits of the mound-building birds of Australia. *Ibis*, Ser. 7, vol. 5, 9–19.

Lesson, R. P. (1828). *Manuel d'ornithologie, ou description des genres et des principales espèces d'oiseaux*, Vol. 2, pp. 1–448. Roret, Paris.

Lesson, R. P. (1831). *Traité d'ornithologie, ou tableau méthodique des ordres, sous-ordres, familles, tribus, genres, sous-genres et races d'oiseaux*. Levrault, Paris, pp. 1–659.

Lesson, R. P. and Garnot, P. (1826). *Voyage autour du monde, exécuté par ordre du roi, sur la corvette de sa majesté, La Coquille, pendant les années 1822, 1823, 1824 et 1825. (Zoologie)*, Vol. 1. Bertrand, Paris.

Lewis, F. (1939). The breeding habits of the Lowan in Victoria. *Emu*, **39**, 56–62.

Lewis, F. (1940). Notes on the breeding habits of the Mallee-Fowl. *Emu*, **40**, 97–110.

Lewis, F. (1950). Some factors relating to the survival, or otherwise, of the Lowan or Mallee Fowl (*Leipoa ocellata*). *Victorian Naturalist*, **67**, 142–3.

Lilljeborg, W. (1866). Outline of a systematic review of the class of birds. *Proceedings of the Zoological Society of London*, **1866**, 5–20.

Lincoln, G. A. (1974). Predation of incubator birds (*Megapodius freycinet*) by Komodo dragons (*Varanus komodoensis*). *Journal of Zoology, London*, **174**, 419–28.

Lindsey, A. I. G. (1979). A feeding association between Australian Fernwren and Orange-footed Scrubfowl. *Sunbird*, **10**(2), 47.

Linsley, L. N. (1935). Curious things about Guam: the mountain chicken. *Guam Recorder*, **12**, 249–50.

Lint, K. C. (1967). The Maleo ... A mound builder from the Celebes. *Zoonooz*, **40**, 4–8.

Lint, K. C. (1975). The Maleo ... A mound builder from the Celebes. *Kukila*, **1**, 29–35.

Lister, J. J. (1911*a*). The distribution of the avian genus *Megapodius* in the Pacific Islands. *Proceedings of the Zoological Society of London*, **52**, 749–59.

Lister, J. J. (1911*b*). On the distribution of the Megapodiidae in the Pacific. *Proceedings of the Cambridge Philosophical Society*, **16**, 148–9.

Lister, J. J. (1911*c*). The distribution of the Megapodiidae in the Pacific. *Nature, London*, **86**, 33.

Low, H. (1851). Extract from a letter from Mr. Hugh Low, dated Labuan, 4th of July, 1850. *Proceedings of the Zoological Society of London*, **1851**, (1853), 119–20.

Lucas, A. H. S. and le Souëf, W. H. D. (1911). *The birds of Australia*. Whitcombe & Tombs, Melbourne.

Lucas, A. M. and Stettenheim, P. R. (1972). *Avian anatomy integument.* II. Agriculture Handbook 362, US Department of Agriculture, Washington, DC.

Lucking, R. S., Davidson, P. J. A., and Stones, A. J. (1992). The status and ecology of the Sula Scrubfowl *Megapodius bernsteinii* on Taliabu, Maluku, Indonesia. *Megapode Newsletter*, **6** (2), 15–22.

MacDonald, J. D. (1973). *Birds of Australia.* Reed, Sydney.

Mace, G. M. and Lande, R. (1991). Assessing extinction threats: toward a reevaluation of IUCN threatened species categories. *Conservation Biology*, **5**, 148–57.

Macgillivray, W. (1914). Notes on some North Queensland birds. *Emu*, **13**, 132–86.

McGilp, J. N. (1935). Birds of the Musgrave Ranges. *Emu*, **34**, 163–76.

McGregor, R. C. (1903). On birds from Luzon, Mindoro, Masbate, Ticao, Cuyo, Culion, Cagayan Sulu, and Palawan. *Bulletins of the Philippine Museum*, **1**, 3–12.

McGregor, R. C. (1905*a*). Birds from the islands of Romblon, Sibuyan, and Cresta de Gallo. *Publication of the Bureau of Government Laboratories, Manila*, **25**, 5–34.

McGregor, R. C. (1905*b*). Birds from Mindoro and small adjacent islands. *Publication of the Bureau of Government Laboratories, Manila*, **34**, 1–27.

Mackay, R. D. (1977). Birds recorded in Manus and New Ireland provinces between 21/2/77 and 11/3/77. *New Guinea Bird Society Newsletter*, **137**, 4–6.

MacKinnon, J. (1978). Sulawesi megapodes. *World Pheasant Association Journal*, **3**, 96–103.

MacKinnon, J. (1979). A glimmer of hope for Sulawesi. *Oryx*, **15**, 55–9.

MacKinnon, J. (1980). *Cagar Alam Gunung Tangkoko-Dua Saudara, Sulawesi Utara. Management Plan 1981–1986.* WWF/IUCN, Bogor.

MacKinnon, J. (1981). Methods for the conservation of Maleo birds, *Macrocephalon maleo* on the island of Sulawesi, Indonesia. *Biological Conservation*, **20**, 183–93.

Mack, G. (1934). The Malleefowl (*Leipoa ocellata*, Gould). *Victorian Naturalist*, **50**, 202–3.

Maitland, R. T. (1893). Oiseaux qui habitent les Iles de Kei. *Tijdschrift van het Koninklijk Nederlandsch Aardrijkskundig Genootschap*, **1893**, 1–22.

Marshall, J. T. (1949). The endemic avifauna of Saipan, Tinian, Guam and Palau. *Condor*, **51**, 200–21.

Marshall, W. (1939). The Brush-Turkey and 'turning'. *Emu*, **38**, 489–91.

Martin, K. (1894). *Reisen in den Molukken*, in *Ambon, den Uliassern, Seran (Ceram) und Buru.* Brill, Leiden.

Maschlanka, H. (1972). Proportionsanalyse von Huhnervögeln. *Zeitschrift für Wissenschaftlichen Zoologie*, **183**, 206–52.

Masters, G. (1876). Zoology of the 'Chevert'. Ornithology. 1. *Proceedings of the Linnean Society of New South Wales*, **1**, 44–64.

Mathews, G. M. (1910–1911). *The birds of Australia*, Vol. 1. Witherby, London.

Mathews, G. M. (1912*a*). A reference-list to the birds of Australia. *Novitates Zoologicae*, **18** (3), 171–455.

Mathews, G. M. (1912*b*). Additions and corrections to my reference list to the birds of Australia. *The Austral Avian Record*, **1** (2), 24–52.

Mathews, G. M. (1927). *Systema Avium Australasianarum*, Vol. 1, pp. 12–17. British Ornithologists' Union, London.

Mathews, G. M. (1929). In minutes of the annual general meeting. *Bulletin of the British Ornithologists' Club*, **50**, 2–11.

Mathews, G. M. and Iredale, T. (1921). *A manual of the birds of Australia*, Vol. **1**. Witherby, London.

Mattingley, A. H. E. (1909). Thermometer bird or Mallee Fowl. *Emu*, **8**, 53–61.

Mayr, E. (1930). Beobachtungen über die Brutbiologie der Grossfusshühner von Neuguinea (*Megapodius*, *Tallegallus* und *Aepypodius*). *Ornithologische Monatsberichte*, **38**, 101–6.

Mayr, E. (1931). Die Vögel des Saruwaged- und Herzoggebirges (NO-Neuguinea). *Mitteilungen aus dem Zoologischen Museum in Berlin*, **17**, 639–723.

Mayr, E. (1933). Die Vögelwelt Polynesiens. *Mitteilungen aus dem Zoologischen Museum in Berlin*, **19**, 306–23.

Mayr, E. (1938). Birds collected during the Whitney South Sea Expedition. 39. Notes on New Guinea birds. 4. *American Museum Novitates*, **1006**, 1–16.

Mayr, E. (1941). *List of New Guinea birds. A systematic and faunal list of the birds of New Guinea and adjacent islands.* The American Museum of Natural History, New York.

Mayr, E. (1944). The birds of Timor and Sumba. *Bulletin of the American Museum of Natural History*, **83** (2), 123–94.

Mayr, E. (1945). *Birds of the southwest Pacific.* Macmillan, New York.

Mayr, E. (1949). Birds collected during the Whitney South Sea Expedition. 57. Notes on the birds of northern Melanesia. 2. *American Museum Novitates*, **1417**, 1–38.

Mayr, E. (1963). *Animal species and evolution.* Harvard University Press, Cambridge, Mass.

Mayr, E. (1972). Continental drift and the history of the Australian bird fauna. *Emu,* **72,** 26–8.

Mayr, E. and Rand, A. L. (1937). Results of the Archbold Expeditions. 14. The birds of the 1933–1934 Papuan expedition. *Bulletin of the American Museum of Natural History,* **73,** 1–248.

Mayr, E. and De Schauensee, R. M. (1939). Zoological results of the Denison–Crockett South Pacific Expedition for the Academy of Natural Sciences of Philadelphia, 1937–38. V. Birds from the western Papuan Islands. *Proceedings of the Academy of Natural Sciences of Philadelphia,* **91,** 145–63.

Mearns, E. A. (1909). A list of birds collected by Dr. Paul Bartsch in the Philippine Islands, Borneo, Guam, and Midway Island, with descriptions of three new forms. *Proceedings of the United States National Museum,* **36,** 463–78.

Mees, G. F. (1965). The avifauna of Misool. *Nova Guinea, Zoology,* **31,** 139–203.

Mees, G. F. (1982). Birds from the lowlands of southern New Guinea (Merauke and Koembe). *Zoologische Verhandelingen, Leiden,* **191,** 1–188.

Meise, W. (1930). Die Vögel von Djampea und benachbarten Inseln nach einer Sammlung Baron Plessens (II). *Journal für Ornithologie,* **78,** 180–214.

Meise, W. (1941). Über die Vogelwelt von Noesa Penida bei Bali nach einer Sammlung von Baron Viktor von Plessen. *Journal für Ornithologie,* **89,** 345–76.

Mellor, J. W. (1911*a*). Mallee-Fowl on Kangaroo Island. *Emu,* **11,** 35–7.

Mellor, J. W. (1911*b*). Mallee-Fowl for a Sanctuary. *Emu,* **11,** 110–14.

Mey, E. (1982). Zur Taxonomie und Biologie der Mallophagen von *Talegalla jobiensis longicaudus* A. B. Meyer, 1891 (Aves, Megapodiidae). *Reichenbachia, Staatliches Museum für Tierkunde in Dresden,* **20** (29), 223–46.

Mey, E. (1986). Ischnozere Mallophagen (Insecta: Phthiraptera) von *Leipoa ocellata* Gould, 1840 (Aves: Galliformes: Megapodiidae). *Zoologische Jahrbücher (Syst.),* **113,** 525–39.

Mey, E. (1990). Zur Taxonomie der auf Grossfusshühnern (Megapodiidae) schmarotzenden *Oxylipeurus*—Arten (Insecta, Phthiraptera, Ischnocera: Lipeuridae). *Zoologische Abhandlungen, Staatliches Museum für Tierkunde Dresden,* **46** (6), 103–16.

Meyer, A. B. (1874*a*). Über neue und ungenügend bekannte Vögel von Neu-Guinea und den Inseln der Geelvinksbai. *Sitzungsberichte der Mathematisch-Naturwissenschaftlichen Classe der Kaiserlichen Akademie der Wissenschaften,* **69** (1–5), 74–91.

Meyer, A. B. (1874*b*). Über neue und ungenügend bekannte Vögel von Neu-Guinea und den Inseln der Geelvinksbai. *Sitzungsberichte der Mathematisch-Naturwissenschaftlichen Classe der Kaiserlichen Akademie der Wissenschaften,* **69** (1–5), 202–18.

Meyer, A. B. (1879). Field-notes on the birds of Celebes. *Ibis,* Ser. 4, vol. 3, 125–47.

Meyer, A. B. (1884). Notizen über Vögel, Nester und Eier aus dem Ostindischen Archipel, Speziel über die durch Hernn C. Ribbe von den Aru-Inseln jüngst erhaltenen. *Zeitschrift für die gesammte Ornithologie,* **1,** 269–96.

Meyer, A. B. (1890*a*). Brush-turkeys on the smaller islands north of Celebes. *Nature,* **41,** 514–15.

Meyer, A. B. (1890*b*). Notes on birds from the Papuan region, with descriptions of some new species. *Ibis,* Ser. 6, vol. 2, 412–24.

Meyer, A. B. (1891). Ueber Vögel von Neu Guinea und Neu Britannien. *Abhandlungen und Berichte des Königlichen Zoologischen und Anthropologisch-Ethnographischen Museums zu Dresden,* **4,** 1–17.

Meyer, A. B. (1892). Beitrag zur Kenntnis der Vogelfauna von Kaiser Wilhelms-Land. *Journal für Ornithologie,* **1892,** 255–66.

Meyer, A. B. and Wiglesworth, L. W. (1898). *The birds of Celebes and the neighbouring islands,* Vol. 2. Friedländer, Berlin.

Meyer, O. and Stresemann, E. (1928). Zur Kenntnis der Entwicklung von *Megapodius* und *Oxyura* im Ei. *Ornithologische Monatsberichte,* **36,** 65–71.

Meyer, P. O. (1930*a*). Untersuchungen an den Eiern von *Megapodius eremita. Ornithologische Monatsberichte,* **38,** 1–5.

Meyer, P. O. (1930*b*). Über die Dauer der Embryonalentwicklung von *Megapodius eremita. Ornithologische Monatsberichte,* **38,** 6–7.

Meyer, P. O. (1933). Vogeleier und Nester aus Neubritannien, Südsee. *Beiträge zur Fortpflanzungsbiologie der Vögel mit Berücksichtigung der Oologie,* **9,** 122–35.

Milligan, A. W. (1904). Notes on a trip to the Wongan Hills, Western Australia, with a description of a new Ptilotis. *Emu,* **3,** 217–26.

Mjöberg, E. (1910). Studien über Mallophagen und Anopluren. *Arkiv för Zoologi,* **6** (3–4), 1–296.

Mourer-Chauviré, C. (1982). Les oiseaux fossiles des phosphorites du Quercy (Éocène Supérieur a Oligocène Supérieur): implications paléobiogéographiques. *Géobios, mémoire spécial*, **6**, 413–26.

Mourer-Chauviré, C. (1992). The Galliformes (Aves) from the Phosphorites du Quercy (France): systematics and biostratigraphy. In *Papers in avian paleontology. Honoring Pierce Brodkorb*, (ed. K. E. Campbell), Sciences Series, Vol. 36, pp. 67–95. Los Angeles.

Mourer-Chauviré, C. and Poplin, F. (1985). Le mystère des tumulus de Nouvelle-Calédonie. *La Recherche*, **16** (169), 1094.

Müller, S. (1846). Ueber den Charakter der Thierwelt auf den Inseln des indischen Archipels, ein Beitrag zur zoologischen Geographie. *Archiv für Naturgeschichte*, **12** (1), 109–28.

Murphy, E. C. and Haukioja, E. (1986). Clutch size in nidicolous birds. In *Current ornithology*, Vol. 4, (ed. R. F. Johnston), pp 141–80. Plenum Press, New York.

Nandi, N. C. and Mandal, A. K. (1980). *Haemoproteus megapodius* sp. nov. in *Megapodius freycinet abbotti* Oberholser (Megapodiidae) from the South Nicobar. *Records of the Zoological Survey of India*, **77**, 51–4.

Neill, W. T. (1971). *The last of the ruling reptiles*. Columbia University Press, New York.

Neumann, O. (1939). Six new races from Peling. *Bulletin of the British Ornithologists' Club*, **59**, 104–8.

Nice, M. M. (1962). Development of behaviour in precocial birds. *Transactions of the Linnaean Society of New York*, **8**, 1–212.

North, A. J. (1901–1914). Nests and eggs of birds found breeding in Australia and Tasmania. *Australian Museum Special Catalogue*, **1**.

Oates, E. W. (1901). *Catalogue of the collection of birds' eggs in the British Museum (Natural History)*, Vol. 1, pp. 15–19. Taylor & Francis, London.

Oberholser, H. C. (1917). Birds collected by Dr. W. L. Abbott on various islands in the Java Sea. *Proceedings of the United States National Museum*, **54**, 177–200.

Oberholser, H. C. (1919). The races of the Nicobar Megapode, *Megapodius nicobariensis* Blyth. *Proceedings of the United States National Museum*, **55**, 399–402.

Oberholser, H. C. (1924). Descriptions of new Treronidae and other non-passerine birds from the East Indies. *Journal of the Washington Academy of Sciences* **14** (13), 294–303.

Ogilvie-Grant, W. R. (1893). *Catalogue of the birds in the British Museum*, Vol. 22, pp. 445–72. Taylor & Francis, London.

Ogilvie-Grant, W. R. (1897). *A hand-book to the game-birds*, Vol. 2. Lloyd, London.

Ogilvie-Grant, W. R. (1915). Report on the birds collected by the British Ornithologists' Union expedition and the Wollaston Expedition in Dutch New Guinea. *Ibis*, Ser. 10, Jubilee Supplement, **2**, 319–25.

Olson, S. L. (1980). The significance of the distribution of the Megapodiidae. *Emu*, **80**, 21–4.

Olson, S. L. (1985). The fossil record of birds. In *Avian biology*, Vol. 8, (ed. D. S. Farner., J. R. King, and K. C. Parkes), pp. 80–238. Academic Press, New York.

Oring, L. W. (1982). Avian mating systems. In *Avian biology*, Vol. 6, (ed. D. S. Farner, J. R. King, and K. C. Parkes), pp. 1–92. Academic Press, New York.

Oustalet, E. (1878). Sur quelques oiseaux de la Papouasie. *Bulletin hebdomadaire de l'Association Scientifique de France*, **21**, 247–8.

Oustalet, E. (1879–1880). Monographie des oiseaux de la famille des Mégapodiidés. *Annales des Sciences Naturelles (Zoologie et Paléontologie)*, **6** (10), 1–60.

Oustalet, E. (1880). Monographie des oiseaux de la famille des Mégapodiidés. *Bibliothèque de l'Ecole des Hautes Etudes*, **22**(5).

Oustalet, E. (1881). Monographie des oiseaux de la famille des Mégapodiidés. *Annales des Sciences Naturelles (Zoologie et Paléontologie)*, **6** (11), 1–182.

Oustalet, E. (1891). Note sur la Mégapode de la Pérouse. *Annales des Sciences Naturelles*, **17** (11), 196.

Oustalet, E. (1896). Les mammifères et les oiseaux des iles Mariannes. *Nouvelles Archives du Muséum d'Histoire Naturelle*, **3** (8), 25–74.

Parker, S. (1967a). Some eggs from the New Hebrides, south-west Pacific. *Bulletin of the British Ornithologists' Club*, **87**, 90–1.

Parker, S. (1967b). The eggs of the Wattled Brush Turkey *Aepypodius arfakianus* (Salvadori) (Megapodiidae). *Bulletin of the British Ornithologists' Club*, **87**, 92.

Pearson, L. (1943). The Brush Turkey. *North Queensland Naturalist, Cairns*, **11** (69), 4.

Pedersen, H. C. and Steen, J. B. (1979). Behavioural thermoregulation in Willow Ptarmigan chicks *Lagopus lagopus*. *Ornis Scandinavica*, **10**, 17–21.

Pérez, T. L. and Atyeo, W. T. (1990). New taxa of feather mites (Acarina, Pterolichidae) from megapodes (Aves, Megapodiidae). *Tijdschrift voor Entomologie*, **133**, 245–9.

Peters, J. L. (1934). *Check-list of birds of the world*, Vol. 2. Harvard University Press, Cambridge.

Piaget, E. (1880). *Les Pédiculines. Essai monographique*. Brill, Leiden.

Piaget, E. (1890). Quelques Pédiculines nouvelles. *Tijdschrift voor Entomologie*, **33**, 223–59.

Pockley, E. (1937). Notes on nesting holes of a megapode. *Emu*, **37**, 63–5.

Poplin, F. (1980). *Sylviornis neocaledoniae* n. g., n. sp. (Aves), Ratite éteint de la Nouvelle-Calédonie. *Comptes Rendus hebdomadaires des Séances de l'Academie des Sciences, Paris*, **290**, 691–4.

Poplin, F. and Mourer-Chauviré, C. (1985). *Sylviornis neocaledoniae* (Aves, Galliformes, Megapodiidae), oiseau géant éteint de l'Ille des Pins (Nouvelle-Calédonie). *Geobios, Lyon*, **18**, 73–97.

Poplin, F., Mourer-Chauviré, C., and Evin, J. (1983). Position systématique et datation de *Sylviornis neocaledoniae*, Mégapode géant (Aves, Galliformes, Megapodiidae) éteint de la Nouvelle-Calédonie. *Comptes Rendus hebdomadaires des Séances de l'Academie des Sciences, Paris*, **297**, II, 301–4.

Portmann, A. (1938). Beiträge zur Kenntnis der postembryonalen Entwicklung der Vögel. *Revue Suisse de Zoologie et Annales du Musee d'Histoire Naturelle de Genève*, **45**, 273–348.

Portmann, A. (1955). Die postembryonale Entwicklung der Vögel als Evolutionsproblem. *Acta XI Congressus Internationalis Ornithologici* (Basel 1954), 138–51.

Portmann, A. (1963a). Die Vogelfeder als morphologisches Problem. *Verhandlungen der Naturforschenden Gesellschaft in Basel*, **74**, 106–32.

Portmann, A. (1963b). Die Vogelfeder als morphologisches Problem. *Journal für Ornithologie*, **104**, 285–7

Prager, E. M. and Wilson, A. C. (1980). Phylogenetic relationships and rates of evolution in birds. *Acta XVII Congressus Internationalis Ornithologici* (Berlin 1978), 1209–14.

Pramono, A. H. (1991). Maleo on Buton. *Kukila*, **5**, 150.

Pratt, H. D. and Bruner, P. L. (1978). Micronesian Megapode rediscovered on Saipan. *Elepaio*, **39**, 57–9.

Pratt, H. D., Bruner, P. L., and Berrett, D. G. (1979). America's unknown avifauna: the birds of the Mariana Islands. *American Birds*, **33**, 227–35.

Pratt, H. D., Engbring, J., Bruner, P. L., and Berrett, D. G. (1980). Notes on the taxonomy, natural history, and status of the resident birds of Palau. *Condor*, **82**, 117–31.

Pratt, H. D., Bruner, P. L., and Berrett, D. G. (1987). *The birds of Hawaii and the tropical Pacific*. Princeton University Press.

Preston, F. W. (1969). Shape of birds' eggs: extant North American families. *Auk*, **86**, 246–64.

Price, R. D. and Beer, J. R. (1964). Species of *Colpocephalum* (Mallophaga: Menoponidae) parasitic upon the Galliformes. *Annals of the Entomological Society of America*, **57**, 391–402.

Price, R. D. and Emerson, K. C. (1966). The genus *Kelerimenopon* Conci with the description of a new subgenus and six new species (Mallophaga: Menoponidae). *Pacific Insects*, **8**, 349–62.

Price, R. D. and Emerson, K. C. (1984). A new species of *Megapodiella* (Mallophaga: Philopteridae) from the Mallee Fowl of Australia. *Florida Entomologist*, **67**, 160–3.

Priddel, D. (1989). Conservation of rare fauna: the Regent Parrot and the Malleefowl. In *Mediterranean landscapes in Australia: Mallee ecosystems and their management*, (ed. J. C. Noble and R. A. Bradstock), pp. 243–9. CSIRO, Melbourne.

Priddel, D. (1990). Conservation of the Malleefowl in New South Wales: an experimental management strategy. In *The Mallee lands: a conservation perspective*, (ed. J. C. Noble, P. J. Joss, and G. K. Jones), pp. 71–4. CSIRO, Melbourne.

Priddel, D. and Wheeler, R. (1988). Use of reflective glass balls to deter predatory birds. *Corella*, **12**, 61–2.

Priddel, D. and Wheeler, R. J. (1990a). Conservation of the endangered Malleefowl *Leipoa ocellata*. *Acta XX Congressus Internationalis Ornithologici, Supplement*, 483.

Priddel, D. and Wheeler, R. (1990b). Survival of Malleefowl *Leipoa ocellata* chicks in the absence of ground-dwelling predators. *Emu*, **90**, 81–7.

Pycraft, W. P. (1900). A contribution towards our knowledge of the pterylography of the Megapodii. In *Zoological results based on material from New Britain, New Guinea, collected during the years 1895, 1896 and 1897*, Part IV, (ed. A. Willey). Cambridge University Press, Cambridge.

Pycraft, W. P. (1902). The bird's wing, and the problem of diastataxy. *Transactions of the Norfolk and Norwich Naturalists' Society*, **7**, 312–27.

Quinnell, S. (1985). Living with Scrub-Turkeys. *Habitat*, **13** (6), 23.

Quoy, J. R. C. and Gaimard, P. J. (1824–1826). Voyage autour du monde. Entepres par l'ordre du Roi. Exécuté sur les corvettes de S. M. l'Uranie et la Physicienne, pendant les années 1817, 1818, 1819, et 1820. Par M. Louis de Freycinet, Capitaine de Vaisseau. *Paris Zoologie*, 1–712 (127).

Quoy, J. R. C. and Gaimard, P. J. (1825). Notice su.. es mammiferes et les oiseaux des iles Timor, Rawalk, Bone, Vaigiou, Guam, Rota, et Tinian. *Annales des Sciences Naturelles*, **6**, 138–50.

Quoy, J. R. C. and Gaimard, P. J. (1830). *Voyage de découvertes de l'Astrolabe exécuté par ordre du roi, pendant les années 1826–1827–1828–1829, sous le commandement de M. J. Dumont d'Urville. Zoologie* Vol. 1. Tastu, Paris.

Raethel, H. S. (1988). *Hühnervögel der Welt*. Neumann-Neudamm, Melsungen.

Rand, A. L. (1942a). Results of the Archbold Expeditions. No. 42. Birds of the 1936–1937 New Guinea Expedition. *Bulletin of the American Museum of Natural History*, **79**, 289–366.

Rand, A. L. (1942b). Results of the Archbold Expeditions. No. 43. Birds of the 1938–1939 New Guinea Expedition. *Bulletin of the American Museum of Natural History*, **79**, 425–515.

Rand, A. L. and Gilliard, E. T. (1967). *The handbook of the New Guinea birds*. Weidenfeld & Nicolson, London.

Rappart, F. W. and Karstel, H. R. (1960). Eiwitbehoefte, jacht en inzameling. *Nederlands Nieuw-Guinea*, **8** (5), 14–17.

Reichel, J. D. and Glass, P. O. (1991). Checklist of the birds of the Mariana islands. *Elepaio*, **51** (1), 3–10.

Reichel, J. D., Taisacan, S., Villagomez, S. C., Glass, P. O., and Aldan, D. T. (1987). Field trip report. Northern islands, 27 May–5 June 1987. Mimeographed report. Saipan, CNMI, Division of Fish and Wildlife.

Reichenow, A. (1882). *Die Vögel der Zoologischen Gärten*, Vol. 1. Kittler, Leipzig.

Reichenow, A. (1899). Die Vögel der Bismarckinseln. *Mitteilungen aus der Zoologischen Sammlung des Museums für Naturkunde in Berlin*, 1 (3), 1–106.

Reichenow, A. (1913). *Die Vögel. Handbuch der systematischen Ornithologie*, Vol. 1, pp. 271–4. Enke, Stuttgart.

Rensch, B. (1931a). Die Vogelwelt von Lombok, Sumbawa und Flores. *Mitteilungen aus dem Zoologischen Museum in Berlin*, **17**, 451–637.

Rensch, B. (1931b). Ueber einige Vogelsammlungen des Buitenzorger Museums von den kleinen Sunda-Inseln. *Treubia*, **13**, 371–400.

Renshaw, G. (1915). Rare birds in continental zoos. *Avicultural Magazine*, Ser. 3 (6), 160–3.

Renshaw, G. (1917). The Celebean Maleo. *Avicultural Magazine*, Ser. 3 (8), 168–70.

Rich, P. V. (1975). Antarctic dispersal routes, wandering continents, and the origin of Australia's non-passeriform avifauna. *Memoirs of the National Museum of Victoria*, **36**, 63–126.

Rich, P. V. and Van Tets, G. F. (1985). *Kadimakara. Extinct vertebrates of Australia*. Pioneer Design Studio, Lilydale.

Rich, P. V., Van Tets, G. F., and Balouet, C. (1985). The birds of western Australasia. *Acta XVIII Congressus Internationalis Ornithologici*, **1**, 200–26.

Rifai, A. and Soehjar, M. B. (1976). Metoda perbaikan habitat burung maleo (*Macrocephalon maleo* Müller) di Tanjung Batikolo, Sulawesi Tenggara. Unpublished report. Directorat Perlindungan dan Pengawetan Alam, Bogor.

Riley, J. H. (1924). A collection of birds from north and north-central Celebes. *Proceedings of the United States National Museum*, **64**, 1–118.

Rinke, D. (1986a). The status of wildlife in Tonga. *Oryx*, **20**, 146–51.

Rinke, D. (1986b). Notes on the avifauna of Niuafo'ou Island, Kingdom of Tonga. *Emu*, **86**, 82–6.

Rinke, D. R. (1991). Birds of 'Ata and Late, and additional notes on the avifauna of Niuafo'ou, Kingdom of Tonga. *Notornis*, **38**, 131–51.

Rinke, D. (1994). The Malau on Fonualei in northern Tonga. *World Pheasant Association News*, **44**, 7–8.

Ripley, S. D. (1941). Notes on a collection of birds from Northern Celebes. *Occasional Papers of the Boston Society of Natural History*, **8**, 343–58.

Ripley, S. D. (1957). New birds from the western Papuan Islands. *Postilla*, **31**, 1–4.

Ripley, S. D. (1960). Distribution and niche differentiation in species of megapodes in the Moluccas and Western Papuan area. *Acta XII Congressus Internationalis Ornithologici*, **1**, 631–40.

Ripley, S. D. (1964). A systematic and ecological study of birds of New Guinea. *Peabody Museum of Natural History, Yale University Bulletin*, **19**, 1–85.

Ripley, S. D. and Beehler, B. M. (1989). Ornithogeographic affinities of the Andaman and Nicobar Islands. *Journal of Biogeography*, **16**, 323–32.

Robiller, F., Gerstner, R., and Trogisch, K. (1985). Naturbrut von Halsbandtalegalla oder Jobi-Maleo *Talegalla jobiensis* Meyer, 1874. *Gefiederte Welt*, **109**, 214–16.

Robinson, A. C., Casperson, K. D., and Copley, P. B. (1990). Breeding records of Malleefowl (*Leipoa ocellata*) and Scarlet-chested Parrots (*Neophema splendida*) within the Yellabinna Wilderness Area, South Australia. *South Australian Ornithologist*, **31**, 8–12.

Rogers, K., Rogers, A., and Rogers, D. (1990). *Bander's aid*, Supplement 1. Royal Australasian Ornithologists Union, Moonee Ponds.

Roper, D. S. (1983). Egg incubation and laying behaviour of the incubator bird *Megapodius freycinet* on Savo. *Ibis*, **125**, 384–9

Roper, D. and Roper, J. (1984). The egg game. *Geo*, **5**, 86–95.

Ross, J. A. (1919). Six months' record of a pair of Mallee-Fowl. *Emu*, **18**, 285–8.

Roselaar, C. S. (1994). Systematic notes on Megapodiidae (Aves, Galliformes), including the description of five new subspecies. *Bulletin Zoölogisch Museum, Universiteit van Amsterdam*, **14**, 9–36.

Rothschild, W. and Hartert, E. (1901). Notes on Papuan birds. *Novitates Zoologicae*, **8**, 55–88, 102–62.

Rothschild, W. and Hartert, E. (1913). List of the collections of birds made by Albert S. Meek in the lower ranges of the Snow Mountains, on the Eilanden River, and on Mount Goliath during the years 1910 and 1911. *Novitates Zoologicae*, **20**, 473–527.

Rothschild, W. and Hartert, E. (1914a). On a collection of birds from Goodenough island. *Novitates Zoologicae*, **21**, 1–9.

Rothschild, W. and Hartert, E. (1914b). On the birds of Rook island, in the Bismarck archipelago. *Novitates Zoologicae*, **21**, 207–18.

Rothschild, W. and Hartert, E. (1914c). The birds of the Admiralty islands, north of German New Guinea. *Novitates Zoologicae*, **21**, 281–98.

Rothschild, W., Stresemann, E., and Paludan, K. (1932). Ornithologische Ergebnisse der Expedition Stein 1931–1932. *Novitates Zoologicae*, **28**, 127–247.

Ruempler, G. (1988). Erster Nachwuchs bei den Buschhühnern *Alectura lathami* im Allwetterzoo Münster. *Gefiederte Welt*, **112**, 254–5.

Ryves, V. W. (1955). The nesting of the megapode in North Borneo. *Sarawak Museum Journal*, **6**, 316–17.

Safford, W. E. (1902). The birds of the Marianne Islands and their vernacular names. 1. *Osprey*, **6**, 39–42, 65–70.

Safford, W. E. (1904). Extracts from the notebook of a naturalist on the island of Guam. *The Plant World*, 7, 265.

Salvadori, T. (1877). Intorno alle specie del genere *Talegallus*, Less. *Annali del Museo Civico di Storia Naturale di Genova*, **9**, 327–34.

Salvadori, T. (1882). *Ornitologia della Papuasia e delle Molucche*, Vol. 3. Paravia, Torino.

Salvadori, T. (1891). *Aggiunte alla ornitologia della Papuasia e delle Molucche*, Vol. 3. Clausen, Torino.

Sarasin, P. and Sarasin, F. (1894). Reisebericht aus Celebes I. Überlandreise von Menado nach Gorontalo. *Zeitschrift der Gesellschaft für Erdkunde zu Berlin*, **29**, 351–401.

Sarasin, P. and Sarasin, F. (1905). *Reisen in Celebes ausgeführt in den Jahren 1893–1896 und 1902–1903*, Vol. 1. Kreidel's Verlag, Wiesbaden.

Sasaki, M., Nishida, C., and Hori, H. (1982). Banded karyotypes of the green-backed guan, *Penelope jacquacu granti* (Cracidae), with notes on the karyotypic relationship to the maleo fowl (Megapodiidae) and domestic fowl (Phasianidae) (Galliformes: Aves). *Chromosome Information Service*, **32**, 26–8.

Schenkel, R. (1958). Zur Deutung der Balzleistungen einiger Phasianiden und Tetraoniden. Zweiter Teil. *Der Ornithologische Beobachter*, **55**, 65–95.

Schlegel, H. (1862). De Maleo. *Megacephalon maleo*. *Artis Jaarboekje*, **1862**, 185–92.

Schlegel, H. (1866a). Observations zoologiques. II. *Nederlandsch Tijdschrift voor de Dierkunde*, **3**, 249–58.

Schlegel, H. (1866b). Notice sur les espèces du genre *Megapodius* habitant l'archipel Indien. *Nederlandsch Tijdschrift voor de Dierkunde*, **3**, 259–64.

Schlegel, H. (1879). Note XXXIX. On *Talegallus pyrrhopigius*. *Notes from the Royal Zoological Museum of the Netherlands at Leyden*, **1**, 159–61.

Schlegel, H. (1880a). Revue méthodique et critique des collections déposées dans cet établissement. *Muséum d'Histoire Naturelle des Pays-Bas*, **8**, 52–86.

Schlegel, H. (1880b). On an undescribed species of black-legged megapode, *Megapodius sanghirensis*. *Notes from the Royal Zoological Museum of the Netherlands at Leyden*, **2**, 91–2.

Schmitt, B. L. (1985). Micronesian Megapode survey and research, pp 44–6. In Annual Report.

Pittman-Robertson federal aid in wildlife restoration program. Saipan, CMNI, Division of Fish and Wildlife, 65pp.

Schodde, R. (1977). Contributions to Papuasian ornithology. VI. Survey of the birds of southern Bougainville Island, Papua New Guinea. *CSIRO Australian Division of Wildlife Research Technical Paper*, **34**, 1–103.

Schodde, R. and Tidemann, S. C. (ed.). (1986). *Complete book of Australian birds*. Readers' Digest Services, Sydney.

Schönwetter, M. (1961). *Handbuch der Oologie*, Vol. 1, No. 4, pp. 196–201. Akademie-Verlag, Berlin.

Schönwetter, M. (1985). *Handbuch der Oologie*, Vol. 41, p. 49. B. Mathematischer Teil. Berechnungen für Zwecke der Oologie. Akademie-Verlag, Berlin.

Schultze-Westrum, T. (1976). Auf dem brodelnden Vulkan schlüpfen Grossfusshühner von selbst aus dem Ei. *Das Tier*, **16**(2), 4–7.

Sclater, P. L. (1869). Notes on an egg of a species of *Megapodius* sent by Mr. J. Brazier. *Proceedings of the Zoological Society of London*, **1869**, 528–9.

Sclater, P. L. (1877). Report on the collection of birds made during the voyage of H. M. S. 'Challenger'. III. On the birds of the Admiralty islands. *Proceedings of the Zoological Society of London*, **1877**, 551–7.

Sclater, P. L. (1880). Remarks on the present state of the *Systema Avium. Ibis*, Ser. 4, vol. 4, 399–411.

Sclater, P. L. (1883). On birds collected in the Timor-Laut or Tenimber group of islands by Mr. Henry O. Forbes. *Proceedings of the Zoological Society of London*, **1883**, 48–58.

Seale, A. (1901). Report of a mission to Guam. *Bernice Pauahi Bishop Museum Occasional Papers*, **1**, 17–128.

Seebohm, H. (1888). An attempt to diagnose the suborders of the great Gallino-Gralline group of birds by the aid of osteological characters alone. *Ibis*, Ser. 5, vol. 6, 415–35.

Serventy, V. (1966). *A continent in danger*. André Deutsch, London.

Serventy, D. L. and Whittell, H. M. (1967). *Birds of Western Australia*, (4th edn). Lamb Publications, Perth.

Seth-Smith, D. (1930). The megapodes or mound builders. *Avicultural Magazine*, **8**, 319–22.

Seth-Smith, D. (1934). Brush-turkeys. *Avicultural Magazine*, **12**, 15–16.

Seymour, R. S. (1984). Patterns of lung aeration in the perinatal period of Domestic Fowl and Brush Turkey. In *Respiration and metabolism of embryonic vertebrates*, (ed. R. S. Seymour), pp. 319–32. Junk, Dordrecht.

Seymour, R. S. (1985). Physiology of megapode eggs and incubation mounds. *Acta XVIII Congressus Internationalis Ornithologici*, **2**, 854–63.

Seymour, R. S. (1991). The Brush Turkey. *Scientific American*, December, 68–74.

Seymour, R. S. and Ackerman, R. A. (1980). Adaptations to underground nesting in birds and reptiles. *American Zoologist*, **20**, 437–47.

Seymour, R. S. and Bradford, D. F. (1992). Temperature regulation in the incubation mounds of the Australian Brush-turkey. *Condor*, **94**, 134–50.

Seymour, R. S. and Rahn, H. (1978). Gas conductance in the eggshell of the mound-building Brush Turkey. In *Respiratory function in birds, adult and embryonic*, (ed. J. Piiper), pp. 243–6. Springer-Verlag, Berlin.

Seymour, R. S., Vleck, D., and Vleck, C. M. (1986). Gas exchange in the incubation mounds of megapode birds. *Journal of Comparative Physiology*, **156 B**, 773–82.

Seymour, R. S., Vleck, D., Vleck, C. M., and Booth, D. T. (1987). Water relations of buried eggs of mound building birds. *Journal of Comparative Physiology*, **157 B**, 413–22.

Seymour Sewell, R. B. (1922). A survey season in the Nicobar Islands on the R. I. M. S. 'Investigator', October, 1921, to March, 1922. *Journal of the Bombay Natural History Society*, **28**, 970–89.

Sharpe, R. B. (1874). On a new species of megapode. *The Annals and Magazine of Natural History, including Zoology, Botany, and Geology*, Ser. 4, 13, 448.

Sharpe, R. B. (1875). On a collection of birds from Labuan. *Proceedings of the Zoological Society of London*, **1875**, 99–111.

Sharpe, R. B. (1877). On the birds collected by Professor J. B. Steere in the Philippine Archipelago. *Transactions of the Linnean Society, Zoology*, Ser. 2, 1 (6), 307–55.

Sharpe, R. B. (1899). *A hand-list of the genera and species of birds. (Nomenclator avium tum fossilium tum viventium.)*, Vol. 1. British Museum, London.

Sharpe, R. B. (1900). On a collection of birds made by Captain A. M. Farquhar, R. N., in the New Hebrides. *Ibis*, Ser. 7, vol. 6, 337–51.

Shufeldt, R. W. (1919). Material for a study of the Megapodiidae. *Emu*, **19**, 10–28, 107–27, 179–92.

Sibley, C. G. (1946). Breeding habits of megapodes on Simbo, Central Solomon Islands. *Condor*, **48**, 92–3.

Sibley, C. G. (1976). Protein evidence of the origin of certain Australian birds. *Acta XVI Congressus Internationalis Ornithologici*, 66–70.

Sibley, C. G. and Ahlquist, J. E. (1972). A comparative study of the egg white proteins of non-passerine birds. *Bulletin of the Peabody Museum of Natural History*, **39**, 1–276.

Sibley, C. G. and Ahlquist, J. E. (1985). The relationships of some groups of African birds, based on comparisons of the genetic material. In *Proceedings of an international symposium of African vertebrates*, (ed. K. L. Schuchmann), pp. 115–61. Zoologisches Forschungsinstitut und Museum Alexander Koenig, Bonn.

Sibley, C. G. and Ahlquist, J. E. (1990). *Phylogeny and classification of birds: a study in molecular evolution*. Yale University Press, New Haven and London.

Sibley, C. G. and Frelin, C. (1972). The egg white protein evidence for ratite affinities. *Ibis*, **114**, 377–87.

Sibley, C. G. and Monroe, B. L., Jr (1990). *Distribution and taxonomy of birds of the world*. Yale University Press, New Haven and London.

Sibley, C. G., Ahlquist, J. E., and Monroe, B. L. (1988). A classification of the living birds of the world based on DNA–DNA hybridization studies. *Auk*, **105**, 409–23.

Siebers, H. C. (1930). Fauna Buruana; Aves. *Treubia*, **7** (5), 165–303.

Sielmann, H. (1979). Wie Vögel den Brutschrank erfanden. *Tierwelt*, **4**, 22–7.

Simonson, D. (1987). Morowali rainforest expedition 1985. Final report. Unpublished report. 23 pp.

Sims, R. W. (1956). Birds collected by Mr. F. Shaw-Mayer in the central highlands of New Guinea 1950-1951. *Bulletin of the British Museum (Natural History), (Zoology)*, **3** (10), 387–438.

Smithe, F. B. (1975). *Naturalist's color guide*. The American Museum of Natural History, New York.

Smithe, F. B. (1981). *Naturalist's color guide* (3rd edn). The American Museum of Natural History, New York.

Smythies, B. E. (1981). *The birds of Borneo* (3rd edn). The Sabah Society, Kuala Lumpur.

Sody, H. J. V. (1929). Uitbroeden van eieren zonder aanvoer van warmte van buitenaf. *Ardea*, **18**, 182–4.

Sorenson, E. S. (1920). Aboriginal names of birds. *Emu*, **20**, 32–3.

Sotherland, P. R. and Rahn, H. (1987). On the composition of bird eggs. *Condor*, **89**, 48–65.

Sperling, E. (1966). The Mallee Fowl. *Birds Illustrated*, **11**, 327.

Starck, J. M. (1988). Note on the skull morphology of *Macrocephalon maleo*. *Megapode Newsletter*, **2**(1), 5–7.

Steadman, D. W. (1989a). New species and records of birds (Aves: Megapodiidae, Columbidae) from an archeological site on Lifuka, Tonga. *Proceedings of the Biological Society of Washington*, **102**, 537–52.

Steadman, D. W. (1989b). Extinction of birds in eastern Polynesia: a review of the record, and comparisons with other Pacific island groups. *Journal of Archaeological Science*, **16**, 177–205.

Steadman, D. W. (1991). The identity and taxonomic status of *Megapodius stairi* and *M. burnabyi* (Aves: Megapodiidae). *Proceedings of the Biological Society of Washington*, **104**, 870–7.

Steadman, D. W. (1992). Extinct and extirpated birds from Rota, Mariana islands. *Micronesica*, **25**, 71–84.

Steadman, D. W. (1993). Biogeography of Tongan birds before and after human impact. *Proceedings of the National Academy of Sciences USA*, **90**, 818–22.

Steadman, D. W. (in press). Birds from the To'aga site, Ofu, American Samoa: prehistoric loss of seabirds and megapodes. *University of California, Berkeley, Archaeological Research Facility Contributions*.

Steadman, D. W., Pahlavan, D. S., and Kirch, P. V. (1990). Extinction, biogeography, and human exploitation of birds on Tikopia and Anuta, Polynesian outliers in the Solomon Islands. *Occasional Papers of the Bishop Museum*, **30**, 118–53.

Steiner, H. (1918). Das Problem der Diastataxie des Vogelflügels. *Jenaische Zeitschrift für Naturwissenschaft*, **55**, 221–496.

Stephan, B. (1970). Eutaxie, Diastataxie und andere Probleme der Befiederung des Vogelflügels. *Mitteilungen aus dem Zoologischen Museum in Berlin*, **46**, 339–437.

Stinson, D. W. (1989). *Megapodius laperouse* in the Mariana Islands: current research. *Megapode Newsletter*, **3**(3), 18–21.

Stinson, D. W. and Glass, P. O. (1992). The Micronesian megapode *Megapodius laperouse*: conservation and research needs. *Zoologische Verhandelingen*, **278**, 53–5.

Stocker, G. C. (1971). The age of charcoal from old jungle fowl nests and vegetation change on Melville Island. *Search*, **2**, 28–30.

Stone, T. (1989). Origins and environmental significance of shell and earth mounds in northern Australia. *Archaeology in Oceania*, **24**, 59–64.

Stone, T. (1991). Megapode mounds and archaeology in northern Australia. *Emu*, **91**, 255–6.

Stresemann, E. (1913). Über eine Vogelsammlung aus Misol. *Journal für Ornithologie*, **61**, 597–611.

Stresemann, E. (1914a). Die Vögel von Seran (Ceram). *Novitates Zoologicae*, **21**, 25–153.

Stresemann, E. (1914b). Beiträge zur Kenntnis der Avifauna von Buru. *Novitates Zoologicae*, **21**, 358–400.

Stresemann, E. (1922). Neue Formen aus dem papuanischen Gebiet. *Journal für Ornithologie*, **70**, 405–8.

Stresemann, E. (1927–1934). Sauropsida: Aves. In *Handbuch der Zoologie*, 7(2), (ed. W. Kükenthal). De Gruyter, Berlin.

Stresemann, E. (1941). Die Vögel von Celebes. *Journal für Ornithologie*, **89**, 1–102.

Stresemann, E. (1965). Die Mauser der Hühnervögel. *Journal für Ornithologie*, **106**, 58–64.

Stresemann, E. and Paludan, K. (1935). Ueber eine kleine Vogelsammlung aus dem Bezirk Merauke (Süd-Neuguinea), angelegt von Dr. H. Nevermann. *Mitteilungen aus dem Zoologischen Museum in Berlin*, **20**, 447–63.

Stresemann, E. and Stresemann, V. (1966). Die Mauser der Vögel. *Journal für Ornithologie*, **107**, 1–448.

Stroud, P. (1992). A viable Malleefowl mound at the Adelaide Zoo. *Thylacinus*, **17**, 20–3.

Studer, T. (1877a). Über die Bildung der Federn bei dem Goldhaarpinguin und *Megapodius*. *Verhandlungen der Schweizerischen Naturforschenden Gesellschaft*, **60**, 240–6.

Studer, T. (1877b). *Die Forschungsreise S. M. S. 'Gazelle' in den Jahren 1874 bis 1876*. **3**. Zoologie und Geologie. Berlin.

Studer, T. (1878). Beiträge zur Entwicklungsgeschichte der Feder. *Zeitschrift für Wissenschaftliche Zoologie*, **30**, 421–36.

Stuebing, R. and Zazuli, J. (1986). The megapodes of Pulau Tiga. *Sabah Museum Journal*, **I**(1), 16–49.

Sturkie, P. D. (1976). *Avian physiology*. Springer-Verlag, New York.

Sundevall, C. J. (1872). *Methodi naturalis avium disponendarum tentamen. Försök till fogelklassens naturenliga uppställning*. Samson & Wallin, Stockholm.

Sutter, E. (1963). Nachwuchs bei den Talegalla hühnern. *Zolli, Bulletin des Vereins der Freunde des Zoologischen Gartens Basel*, **11**, 10–12.

Sutter, E. (1965). Zum Wachstum der Grossfusshühner (*Alectura* und *Megapodius*). *Der Ornithologische Beobachter*, **62**, 43–60.

Sutter, E. (1966). Zur Jugendmauser des Grossgefieders bei Pfau und Tallegallahuhn. *Journal für Ornithologie*, **107**, 408–9.

Taka-Tsukasa, N. (1932). *The birds of Nippon*, Vol. 1, part 1. Witherby, London.

Taka-Tsukasa, N. and Kuroda, N. (1915). A list of birds collected in the North Pacific islands by the Ornithological Society of Japan. *Tori*, **1**, 49–55.

Tarr, H. E. (1965). The Mallee-fowl in Wyperfeld National Park. *Bird Watcher*, **2**, 140–4.

Taschenberg, O. (1882). Die Mallophagen, mit besonderer Berücksichtigung der von Dr. Mayer gesammelten Arten systematisch bearbeitet. *Nova Acta Academiae Caesarea Leopoldino-Carolinae Germanicum Naturae Curiosorum*, **44**, 1–244.

Temminck, C. J. (1826). *Nouveau recueil de planches coloriées d'oiseaux*, Vol. 5, livre 69. Levrault, Paris.

Tendeiro, J. (1980). Études sur les Goniodidés (Mallophaga, Ischnocera) des Galliformes. I—Genre *Homocerus* Kéler, 1939. *Garcia de Orta (Zoology)*, **9**, 71–80.

Tendeiro, J. (1981–1982). Études sur les Goniodidés (Mallophaga, Ischnocera) des Galliformes. II—Un nouveau genre, *Aurinirmus* nov., pour cinq espèces parasites des Mégapodiidés. *Garcia de Orta (Zoology)*, **10**, 115–24.

Thienemann, F. A. L. (1856). Einhundert Tafeln colorirter Abbildungen von Vogeleiern. *Zur Fortpflanzungsgeschichte der gesammten Vögel*, Vols. 1 and 2.

Thompson, G. B. (1947). A list of type-hosts of the Mallophaga and the Lice described from them (cont.). *Annals and Magazine of Natural History*, **11** (14), 737–67.

Thomson, A. L. (ed.) (1964). *A new dictionary of birds*. Nelson, London.

Tinbergen, N. (1951). *The study of instinct*. Oxford University Press, London.

Todd, D. M. (1978). Preliminary study of Pritchard's Megapode, *Megapodius pritchardii*. Unpublished report.

Todd, D. (1983). Pritchard's Megapode on Niuafo'ou Island, Kingdom of Tonga. *World Pheasant Association Journal*, **8**, 69–88.

Tollan, A. (1989). Goshawks hunting megapodes. *Australian Raptor Association News*, **13**, 13.

Toxopeus, L. J. (1922). Eenige ornithologische mededeelingen over Boeroe. *Tweede Nederlandsch-Indisch Natuurwetenschappelijk Congres, Bandoeng*, **1922**, 1–6.

Tristram, H. B. (1879). Notes on collections of birds sent from New Caledonia, from Lifu (one of the Loyalty Islands), and from the New Hebrides by E. L. Layard, C. M. G. and c. *Ibis*, Ser. 4, vol. 3, 180–95.

Tristram, H. B. (1889). *Catalogue of a collection of birds belonging to H. B. Tristram*. Durham, pp. 1–278.

Troy, S. and Elgar, M. A. (1991). Brush-turkey incubation mounds: mate attraction in a promiscuous mating system. *Trends in Ecology and Evolution*, **6**, 202–3.

Tweeddale, A. (1877). Contributions to the ornithology of the Philippines. II. On the collection made by Mr. A. H. Everett in the island of Zebu. *Proceedings of the Zoological Society of London*, **1877**, 755–69.

Uchida, S. (1918). Mallophaga from birds of the Ponapé I. (Carolines) and the Palau Is. (Micronesia). *Annotationes Zoologicae Japonenses*, **9**, 481–93.

Uno, A. (1949). Het natuurmonument Panoea (N. Celebes) en het maleohoen (*Macrocephalon maleo* Sal. Müller) in het bijzonder. *Tectona*, **39**, 151–65.

US Fish and Wildlife Service (1970). Conservation of endangered species and other fish or wildlife. *Federal Register*, **35** (106), 8491–8.

Van Balen, J. H. (1926). *De dierenwereld van Insulinde in woord en beeld*, Vol. 2, pp. 103–15. Thieme, Zuphen.

Van Bemmel, A. C. V. (1947). Two small collections of New Guinea birds. *Treubia*, **19**, 1–45.

Van Bemmel, A. C. V. (1948). A faunal list of the birds of the Moluccan Islands. *Treubia*, **19**, 323–402.

Van Bemmel, A. C. V. and Hoogerwerf, A. (1940). The birds of Goenoeng Api. *Treubia*, **17**, 421–72.

Van den Berg, A. B. and Bosman, C. A. W. (1986). Supplementary notes on some birds of Lore Lindu Reserve, Central Sulawesi. *Forktail*, **1**, 7–13.

Van Dyck, S. (1987). A pile of compost is a turkey's pride and joy. *Courier Mail*, 25 September.

Van de Giessen, R. (1984). Zand erover. *Panda*, **20**, 20–2.

Van Ewijk, T. (1988). Een meter diep geboren. *Grasduinen*, **1988** (12), 48–51.

Van Tets, G. F. (1974). A revision of the fossil Megapodiidae (Aves), including a description of a new species of *Progura* De Vis. *Transactions of the Royal Society of South Australia*, **98**, 213–24.

Verheyen, R. (1956). Contribution à l'anatomie et á la systematique des Galliformes. *Bulletin Institut Royal des Sciences Naturelles de Belgique*, **32**, (42), 1–24.

Vernon, D. P. (1968). *Birds of Brisbane and environs*. Queensland Museum, Brisbane.

Vleck, D. (1986). Energetics of embryonic development in the megapode birds, Mallee Fowl *Leipoa ocellata* and Brush Turkey *Alectura lathami*. Abstract, *Mechanisms of very early behaviour development in animals and man*, 24–26 November, Bielefeld University, Germany.

Vleck, D., Vleck, C. M., and Seymour, R. S. (1980). Megapode eggs: energy and water loss during incubation. *American Zoologist*, **20**, 906.

Vleck, D., Vleck, C. M., and Seymour, R. S. (1984). Energetics of embryonic development in the megapode birds, Mallee Fowl *Leipoa ocellata* and Brush-turkey *Alectura lathami*. *Physiological Zoology*, **57**, 444–56.

Von Kéler, S. (1939). Baustoffe zu einer Monographie der Mallophagen. II. Überfamilie der Nirmoidea (1). *Nova Acta Leopoldina*, **8** (51), 1–254.

Von Kéler, S. (1958). The genera *Oxylipeurus* Mjöberg and *Splendoroffula* Clay and Meinertzhagen (Mallophaga). *Deutsche Entomologische Zeitschrift (NF)*, **5**, 299–362.

Von Rosenberg, H. (1878). *Der Malayische Archipel*. Wiegel, Leipzig.

Vorderman, A. G. (1889). Over het voorkomen van eene loophoendersoort in den Kangean-Archipel. *Natuurkundig Tijdschrift voor Nederlandsch-Indië*, **49**, 71–3.

Vorderman, A. G. (1890). Nog iets over het loophoen van den Kangean Archipel. *Megapodius duperrei* (Less.). *Natuurkundig Tijdschrift voor Nederlandsch-Indië*, **50**, 520–4.

Vorderman, A. G. (1898a). Celebes-vogels. *Natuurkundig Tijdschrift voor Nederlandsch-Indië*, **58**, 26–121.

Vorderman, A. G. (1898b). Molukken-vogels. *Natuurkundig Tijdschrift voor Nederlandsch-Indië*, **58**, 169–252.

Vuilleumier, F. (1965). Relationships and evolution within the Cracidae (Aves: Galliformes). *Bulletin of the Museum of Comparative Zoology at Harvard College, Cambridge*, **134**, 1–27.

Wakefield, N. (1960). Australian wonder-birds—mound builders. *Victorian Naturalist*, **77**, 132–5.

Walker, T. A. (1987). Birds of Bushy Island (with a summary of nesting status of bird species on southern Great Barrier Reef cays). *Sunbird*, **17**, 52–8.

Wallace, A. R. (1860). The ornithology of northern Celebes. *Ibis*, **2** 140–7.

Wallace, A. R. (1863). List of birds collected on the island of Bouru (one of the Moluccas), with descriptions of the new species. *Proceedings of the Zoological Society of London*, **1863**, 18–36.

Wallace, A. R. (1869). *The Malay Archipelago*. Macmillan, London.

Warham, J. (1962). Incubator birds. *Animal Kingdom*, **65**, 104–10.

Watling, D. (1981). Maleo egg omelettes... conservation or exploitation? *Conservation Indonesia*, **5** (2), 2–3.

Watling, D. (1982). *Birds of Fiji, Tonga and Samoa*. Millwood Press, Wellington.

Watling, D. (1983a). Ornithological notes from Sulawesi. *Emu*, **83**, 247–61.

Watling, D. (1983b). Sandbox incubator. *Animal Kingdom*, **86** (3), 28–35.

Watling, D. and Mulyana, Y. (1981). *Lore Lindu National Park Management Plan 1981–1986*. WWF/IUCN, Bogor.

Weathers, W. W. and Sullivan, K. A. (1989). Juvenile foraging proficiency, parental effort, and avian reproductive success. *Ecological Monographs*, **59** (3), 223–46.

Weathers, W. W., Weathers, D. L., and Seymour, R. S. (1990). Polygyny and reproductive effort in the Malleefowl *Leipoa ocellata*. *Emu*, **90**, 1–6.

Weathers, W. W., Seymour, R. S., and Baudinette, R. V. (1993). Energetics of mound-tending behaviour in the malleefowl, *Leipoa ocellata* (Megapodiidae). *Animal Behaviour*, **45**, 333–41.

Webb, H. P. (1992). Field observations of the birds of Santa Isabel, Solomon Islands. *Emu*, **92**, 52–7.

Weir, D. G. (1973). Status and habits of *Megapodius pritchardii*. *Wilson Bulletin*, **85**, 79–82.

West, J., Koentjoro, Hariyanto, S., Soekarno, N., Soetarsono, Sutopo, H., and Ariep, H. (1980). Studi Macan Tutul dan Burung Gosong di Kepulauan Kangean, Jawa Timur. Unpublished report. Directorat Perlindungan dan Pengawetan Alam, Bogor.

West, J., Madinah, and Hasan, M. (1981). Artificial incubation of the Moluccan Scrub Hen *Eulipoa wallacei*. *International Zoo Yearbook*, **21**, 115–18.

Wetmore, A. (1930). A systematic classification for the birds of the world. *Proceedings of the United States National Museum*, **76**, 24, 1–8.

Wetmore, A. (1960). A classification for the birds of the world. *Smithsonian Miscellaneous Collections*, **139** (11), 1–37.

White, C. M. N. (1938). Notes on Australian birds. *Ibis*, **14** (2), 761–4.

White, C. M. N. and Bruce, M. D. (1986). *The birds of Wallacea*. BOU Check-list No. 7, British Ornithologists' Union, London.

Whitehead, J. (1888). Notes on some oriental birds. *Ibis*, Ser. 5, vol. 6, 409–13.

Whitmee, S. J. (1875). List of Samoan birds, with notes on their habits and c. *Ibis*, Ser. 3, vol. 5, 436–47.

Wiglesworth, L. W. (1891). Aves Polynesiae. A catalogue of the birds of the Polynesian Subregion. *Abhandlungen und Berichte aus dem Staatlichen Museum für Tierkunde in Dresden*, **5**, 1–92 (58).

Wiles, G. J. and Conry, P. J. (1990). Terrestrial vertebrates of the Ngerukewid Islands Wildlife Preserve, Palau Islands. *Micronesica*, **23**, 41–66.

Wiles, G. J., Beck, R. E., and Amerson, A. B. (1987). The Micronesian Megapode on Tinian, Mariana Islands. *Elepaio*, **47**, 1–3.

Wiljes-Hissink, E. A. de (1953). A Meleo (*Eulipoa*) breeding-place in South Ceram. *Madjalah Ilmu Alam untuk Indonesia*, **109**, 217–24.

Williams, K. A. W. (1967). Scrub turkey versus black snake. *Queensland Naturalist*, **18** (3–4), 73.

Wilson, E. O. (1975). *Sociobiology—the new synthesis*. Harvard University Press, Cambridge.

Wilson, F. E. (1912). Oologists in the Mallee. *Emu*, **12**, 30–9.

Wind, J. (1984). *Management plan 1984–1989 Dumoga-Bone National Park, North Sulawesi province, district Bolaang-Mongondow, district Gorontalo*. WWF/IUCN, Bogor.

Winn, B. (1992). Captive breeding of the Maleo *Macrocephalon maleo*. *Megapode Newsletter*, **6**, 3–5.

Wiriosoepartho, A. S. (1979). *Pengamatan habitat dan tingkah laku bertelur maleo* (Macrocephalon maleo *Sal. Müller*) *di komplek hutan Dumoga, Sulawesi Utara*. Project No. 315. Lembaga Penelitian Hutan, Bogor.

Wiriosoepartho, A. S. (1980). *Penggunaan habitat dalam berbagai macam activitas oleh* Macrocephalon maleo *Sal. Müller, di Cagar Alam Panua, Sulawesi Utara. Project No. 356.* Lembaga Penelitian Hutan, Bogor.

Wittenberger, J. F. (1981). *Animal social behaviour.* Duxbury Press, Boston.

Wolff, T. (1965). Volcanic heat incubation in *Megapodius freycinet eremita* Hartl. *Dansk Ornitologisk Forenings Tidsskrift*, **59**, 74–84.

Wolters, H. E. (1976). *Die Vogelarten der Erde*, Vol. 2, pp. 81–160. Paul Parey, Hamburg and Berlin.

Woodford, C. M. (1888). General remarks on the zoology of the Solomon Islands, and notes on Brenchley's Megapode. *Proceedings of the Zoological Society of London*, **1888**, 248–50.

Wood-Gush, D. G. M. (1956). The agonistic and courtship behaviour of the brown leghorn cock. *British Journal of Animal Behaviour*, **4**, 133–42.

World Pheasant Association (1988). *A conservation strategy for the Galliformes 1988–1990.* Report World Pheasant Association.

Yamashina, Y. (1932). On a collection of birds' eggs from Micronesia. *Tori*, 7, 393–413.

Yamashina, Y. (1940). Some additions to the 'list of the birds of Micronesia'. *Tori*, **10**, 673–9.

Yong, D. (1990). Recent information on the Sula Scrubfowl *Megapodius bernsteinii*. *Megapode Newsletter*, 4 (1), 3–5.

Zieren, M. (1985). Maleo birds *(Macrocephalon maleo)* in the Dumoga-Bone National Park. An analysis of the habitat suitability of nesting grounds. Unpublished report. Wageningen Agricultural University.

Ziswiler, V. (1962). Die Afterfeder der Vögel. Untersuchungen zur Morphogenese und Phylogenese des sogenannten Afterschaftes. *Zoologische Jahrbücher (Anatomie)*, **80**, 245–308.

Index

Species accounts can be found on the page numbers set in bold type.

abbreviations xiv
Acacia 54, 126–7
adaptations 52–64
  of eggshells 8, 59–62
  of embryos 8, 60–3
  of hatchlings 64
  to underground incubation 7–8, 60–3
Adelaide Zoological Garden 10
Adelbert Range, New Guinea 213
Admiralty Islands 27, 198–201, 204
Aepypodius 17, 20, 22, 74–6, 96, 109, 114, 120, 122, 129, 130
  arfaki, see Wattled Brush-turkey
  arfakianus, see Wattled Brush-turkey
  bruijnii, see Bruijn's Brush-turkey
  pyrrhopygius, see Wattled Brush-turkey
aggression, see behaviour
Alecthelia urvillii, see Dusky Megapode
Alectura 17, 19, 22, 74–6, 89, 101–2, 122, 129, 130; see also Australian Brush-turkey
  lathami, see Australian Brush-turkey
altitudinal segregation
  in *Megapodius* 28–30
  in *Talegalla* 28
Amahai 144
Ambon 142–3, 145, 194–5
American Museum of Natural History 10
anatomy xviii–xix, 3–4
  neck sac 41
  temperature receptors 40
Andaman Islands 22, 162–3
annual cycles 33–8
antisocial behaviour 41
Arfak Mountains, New Guinea 97, 100, 102
Aroa River, New Guinea 119
Aru Islands 215–18, 221–2
Astrolabe Bay, New Guinea 119, 211–12, 215–16, 222
Australian Brush-turkey (*Alectura lathami*) 9, 19, 33–6, 38–43, 45–6, 51, 54–61, 63–4, 67, 70, 72, 74, 79, 84, **89–96**, 98, 100–2; Plate 1
  Alectura lathami lathami 17, 89–93, 96
  Alectura lathami purpureicollis 17, 89–93, 96

Babelthuap 157
Bacan 144–5, 189
Bagabag Island 31, 200–1, 204, 211–12

Bakiriang 137
Bali 22, 81
Banda Island (Sea) 22, 30–1, 218, 220–2
Banggai Islands 83, 175, 176–7
Bangkan, Kangean Archipelago 26, 133
Banks Islands, Vanuatu 205–8
bare parts, see species accounts
Bare-faced Scrubfowl, see Melanesian Megapode
Barn Owl (*Tyto alba*) 150
Bar-winged Rail (*Nesclopeus poecilopterus*) 148
basal metabolic rate 58
Batang Kecil 181, 185, 187
Batanta 30, 105, 181, 186–87
behaviour 33–43, 65, 67, 69–71, 73; see also species accounts
  agonistic 40–1
  antisocial 41
  cockfights 40–1, 71
  courtship 65–6
  communication 41–3, 69
  copulation 66–7
  defensive 40, 65; see also mound defence
  displays 67
  dust bathing 40
  feeding 40
  greeting ceremony 65–6
  at incubation site 40–1, 69, 73
  parental care 63
  perching 40
  postures 40
  raking 40
  reproductive 65–72
  roosting 40
  submissive 40–1
  temperature testing 40, 60
Biak Megapode (*Megapodius geelvinkianus*) 18, 19, 30, 45, 83, 146, 152, **189–93**, 194, 196; Plate 7
biochemical analysis 13
biogeography 21–32
  competition theory 25–8
  predation theory 25–8
  separation of New Zealand and Australia 24
Bismarck Archipelago 27, 31, 48, 212–13
Black-billed Talegalla (*Talegalla fuscirostris*) 18, 19–29, 45, 79, 105, 109–10, **110–16**, 111, 114, 116–17; Plate 2
  Talegalla fuscirostris aruensis 15, 18, 110–14

Black-billed Talegalla (*Talegalla fuscirostris*) (*cont.*)
  Talegalla fuscirostris fuscirostris 18, 110–14
  Talegalla fuscirostris meyeri 15, 18, 110–14
  Talegalla fuscirostris occidentis 18, 109–14
Bloomfield River 215, 222
body weight 34
Bombari Peninsula, New Guinea 100
Booby Island 215, 219, 221–2
booming, *see* vocalizations
Borneo 22, 25
Bougainville 200–4
'brazieri', *see* Vanuatu Megapode
breeding grounds, *see* burrow nesting
breeding seasons 34–7, 48, 68; *see also* incubation periods; *species accounts*
Brown-collared Talegalla (*Talegalla jobiensis*) 18, 19–20, 36, 45, 50, 79, 105–7, 110, 114, **116–21**; Plate 2
  Talegalla jobiensis jobiensis 18, 114, 116–20
  Talegalla jobiensis longicauda 18, 114, 116–20
Brown Kiwi (*Apteryx australis*) 59
Bruijn's Brush-turkey (*Aepypodius bruijnii*) 17, 19, 45, 80, 70, 96, **103–5**, 187; Plate 1
burrow nesters, *see* burrow nesting
burrow nesting 6–7, 34–7, 39, 44, 48–50, 73, 79
  distribution of burrow nesting species 48
  distribution of sites 48–9
  evolution of 74–7
  geothermal heat sources 48–9
  incubation processes 53–4
  less susceptible to predation 26, 77
  substrate 49, 62
burrows 6, 35, 37, 48–9, 81
  gas diffusion 62
  heat sources 6, 48–9
  moisture content 62
  substrate 49, 62
  water-logging 62
Buru 27, 31, 144–5, 195, 197
bustards (Otididae) 38
buttonquails 12
Butung (Buton) 133

Cairncross Island 215, 219, 221–2
Cairns 219, 225
Cape Tribulation 215, 220
Cape York 89, 215, 219–20, 226
captive breeding, *see* conservation
Catheturinae 12
Catheturus australis, *see* Australian Brush-turkey
Catheturus novaehollandiae, *see* Australian Brush-turkey
cats 25, 77
Cedar Bay, Australia 215, 222
central islands, Vanuatu 206–7
Central Range, New Guinea 28, 30, 114, 119, 212–13, 222
Ceram 27, 31–2, 142–4, 186, 194, 197,
Charadriiformes 11
chick, *see* hatchlings
Chimbu Province, New Guinea 28, 120
Chionididea 12
Chosornis praeteritus 23
chromosomes 12
cinder fields 49

civet-cats 25, 77
classification, *see* taxonomy
clutch size, *see* eggs
Cobourg peninsula 222
colonization, *see* dispersal
Columbiformes 11, 24
Common Moorhen (*Gallina chloropus*) 224
communal egg grounds, *see* burrows
communal nesting 54; *see also* burrows
communication 41–3, 69; *see also* vocalizations
  visual signals 41
  vocalizations 41
competition theory, *see* biogeography
competitive exclusion 25, 28–30
compost heaps 47
conservation 9, 78–85
  captive breeding 34–5, 41, 80–1, 83–5
  endangered and threatened species 80–3
  recovery plans 9
  reintroduction 82
  translocation 82
construction, *see* mounds
construction phases, *see* mounds
contact zones 30–2
Cooktown 89, 215, 219, 222
copulations 35, 66–7
  extra-pair 66–7
  patterns 66–7
Coturnix 25
courtship behaviour 65–6
Cracidae 11, 12, 26, 77
crocodiles 52–3; *see also* reptiles
cuckoldry 72
Cuming's Megapode, *see* Philippine Megapode
curassows, *see* Cracidae
Cyclops Mountains, New Guinea 100, 102

Dampier Island 17, 27
Daribi 102
decaying roots as incubation sites, *see* mounds,
D'Entrecasteaux Island 27, 31, 220, 215–21, 222
diastataxis, *see* wings
diet 38, 127–8
Dinornithiformes 24
dispersal 27
  colonization 28–30
  over water 25
  routes 27
distribution 4, 21–32; *see also* species accounts
  altitudinal segregation 28–30
  centre of origin 26–8
  contact zones 30–2
  fossil records 21–4
  of megapodes 21–3
  in the Pacific 21–5
  present 22
DNA–DNA hybridization 12
domestication 24
Dorei, Vogelkop peninsula 30
drainage 45
Duchateau Island 31
Dumoga-Bone National Park 81–2, 133–5
Dunk Island 222

Dusky Megapode *(Megapodius freycinet)* 18, 19, 27, 29, 45, 79, 105, 139, 141, 143, 164, 172, 177, 180, **181–9**, 189–92, 196, 199, 202, 206, 212, 222–4; Plate 7
   Megapodius freycinet freycinet 18, 181–2, 184, 186–7
   Megapodius freycinet oustaleti 18, 181–2, 185–7, 192
   Megapodius freycinet quoyi 18, 30, 181–2, 184, 186–8
dust bathing 40

eagles (Accipitridae) 38
ecophysiology 52–64
ectoparasites 12
egg burying in other species 52–3
egg grounds, *see* burrows
egg tooth 62
eggs 8, 14, 34, 59, 62–3, 68, 78; *see also species accounts*
   airspace 60, 62
   calcium 60
   chorioallantois 62–3
   clutch size 8, 33–4, 59, 68
   colour 14, 33
   comparison with reptile eggs 53
   development, *see* embryos
   egg whites 13
   energy content 59
   energy to water ratio 59
   fragility 59
   gas exchange 60–2
   hatching success (hatchability) 50, 67–8
   incubation temperatures 50–1
   laying behaviour 39–40, 48–9
   laying depth 49
   laying intervals 8, 34–6
   mass 59
   partial pressure 60
   pores 60–1
   production 39, 68, 71
   shape 33
   shell 13, 33, 59–63
   shell conductance 60–1
   shell thinning 60, 63
   size 8, 23, 33–4
   tissue formation 59
   variation 33
   water loss 59–60
   yolk content 14, 34, 53, 59, 68, 76, 78
eggshell 13, 59–63; *see also* eggs
   adaptations 8, 59–62
Egyptian Plover *(Pluvianus aegyptius)* 52
embryonic development, *see* embryos
embryos 51, 67
   adaptations 8, 60–3
   dehydration 61
   development 51, 59, 61–2, 73, 84
   gas exchange 60–2
   metabolic demands 61
   skeletal formation 60
Enaena, New Guinea 102
endangered and threatened species 80–3
English names 18
environmental variables 35–7, 39
Eocene deposits 24; *see also* distribution
Etna Bay, New Guinea 110
etymology 18–20
'Eua, Kingdom of Tonga 24, 148
Eucalyptus 45, 54, 57–8, 74–6, 94, 126
Eulipoa 15–17, 18, 19, 45, 96, 121–2, 137, 139–40, 146, 151, 192, 196–7, 204, 208, 213; *see also* Moluccan Megapode
   wallacei, *see* Moluccan Megapode
eutaxic, *see* wings
evolution 52–3, 73–7
   of family 52–3
   of incubation method 59, 67–8, 73–7
experimental manipulation of mounds 56
exploitation, *see* human use
extermination 22–5, 49
external pipping 62

Falconiformes 11
feather mites 14–15; *see also* ectoparasites
feather structure 13
feathers, *see* plumage
feeding ecology 38–9, 127; *see also species accounts*
   diet 38
   of hatchlings 39
   insect foods 127
   nutritional requirements 38–9, 68
   plant foods 127–8
Felidae, *see* cats
Felis 25
female defence monogamy 68–70
fermentation, *see* heat sources
field characters, *see species accounts*
Fiji 48
Fitzroy Island 222
flight 22
Flores, Indonesia 216, 218, 220–4, 226
Fly River, New Guinea 110, 112–14, 116, 226
Fonualei 148
food, *see* feeding ecology
Forsten's Megapode *(Megapodius forstenii)* 18, 19, 27, 31–2, 45, 79, 139, 141, 143–4, 146, 166–7, 177, 186, 188, 192, **193–7**, 202, 209, 216, 223; Plate 8
   Megapodius forstenii buruensis 18, 193, 195–7
   Megapodius forstenii forstenii 18, 139, 186, 193, 195–7
fossilized mounds 46
fossils 23–4, 26; *see also* distribution
fox *(Vulpes vulpes)* 80
Frankfurt Zoological Gardens 10
fungi 45

Gag Island 181, 184–7
Galela 144
Galliformes 11, 12, 37, 40, 74–6
   evolutionary relationships 73–7
Gallinae 12
Gallinuloididae 12, 24, 26
Gallomorphae 12
Gamkonora 144
Garu, New Britain 9, 48, 202, 204
gas exchange, *see* eggs
Gauttier Mountains, New Guinea 100
Geelvink Bay, New Guinea 28, 30–1, 100, 112, 114, 119, 192, 211–13, 215, 218

Index

general biology   33–43
general habits,   *see species accounts*
geothermal activity   44; *see also* heat sources
geothermal incubation sites   6, 39–40, 62, 68, 81
giant megapodes   23–4
Gondwana   26
Goodenough Islands   31
Gorong   31
Great Sangi   133
Greater Sunda Islands   25
Green Jungle Fowl (*Gallus varius*)   25
greeting ceremony   65–6
Griffith, Australia   125
Guadalcanal   200, 202
Guam   157
guans,   *see* Cracidae,
Guguan   152, 157–8
gular flutter   58, 64
Gulf of Bengal   22
Gunung Api   22, 30

Ha'apai group, Kingdom of Tonga   22, 23, 148
habitats,   *see species accounts*
Halmahera   32, 144–5, 181–8, 189
Hartlaub's Scrubfowl,   *see* Melanesian Megapode
Haruku   142–5, 196–7
harvesting   49, 78–9, 81–3; *see also* human use
hatchlings   8, 34, 37–9, 41, 62–4, 67, 84; *see also species accounts*
  adaptations   64
  behaviour   37–9, 41
  cost of emergence   63
  diet   39
  digging   63
  emergence time   63
  energy reserves   64
  foraging ability   63
  hatching   62–3
  lack of parental care   63
  lungs   62–3
  mass   64
  maturity   84
  metabolic rate   64
  plumage   37–8, 64
  precociality   59, 67, 75
  predation   63
  respiratory overlap   62
  survival   8, 63–4
  thermoneutral zone   64
  thermoregulation   63
heat production   54–8
heat sources   44–51; *see also species accounts*
  fermentation   45
  geothermal activity   44, 62
  microbial respiration   44–5, 61
  solar radiation   44
Helmeted Guineafowl (*Numida meleagris*)   138
historic distribution   22–4
home ranges   69, 71
Hoop Pine (*Araucaria*)   94
human use   9, 49, 78–83
  harvesting of eggs   78–9, 81–3
Humboldt, New Guinea   118–20, 210–11

Huon Peninsula, New Guinea   119–20, 209–13, 222
hybridization   29–32
  secondary   31
Hydrographer Mountains, New Guinea   102, 222

incubation duration,   *see* incubation period
incubation heat sources,   *see* heat sources
incubation methods   44–51; *see also species accounts*
  evolution   73–7
incubation mounds,   *see* mounds
incubation period   53; *see also species accounts*
  gas exchange   60–1
  in megapodes   53, 59
  in reptiles   53
incubation processes   34–6, 53–8, 73, 83–4
  in burrows   53–4
  in mounds   54–8
incubation sites   4, 34, 37, 39, 44–51, 54–8, 65, 68, 78–9; *see also species accounts*
  impact of rain, 55
  multiple use   50
  substrate   45–7
  temperatures   51, 53
internal pipping   62
introductions   24, 26
Iwaka River, New Guinea   28, 109

Japen Island   17, 31, 96–102, 107, 114, 118–21, 186, 209, 211–13
Java   22, 25, 81, 220, 215, 218, 222

Kai Islands   216, 218, 222
Kailolo   143–5
Kajoa Island   181, 184–5, 187
Kamaroa   137
Kamuai (Schildpad Island)   183, 185, 187
Kangaroo Island   55–8, 124
Kangean Islands (Archipelago)   25, 77, 225–6
Karimui area, New Guinea   28, 119–20
Karkar Island   27, 31, 200–1, 204
karyology   12
Kasiui   31
Kermadec Islands   22
Kingdom of Tonga,   *see* Tonga
Kofiau Island   181, 186–7
Komodo   22–6

Labobo Island   178
Labuan Island, Borneo   47
Lake Kutubu, New Guinea   28, 212–13
Lakemba, Fiji   24
Lantana camara   94
laying behaviour   48; *see also* behaviour
leaf litter   47–8
Leipoa   18, 19, 22, 96, 121–2, 130; *see also* Malleefowl
  ocellata,   *see* Malleefowl
  ocellata rosinae   122, 124
  penicillata   122
Leopard (*Panthera pardus*)   25, 26
Leopard Cat (*Felis bengalensis*)   26

Lesser Sunda Islands   26, 30, 213, 216–17, 220–2
Lifuka, Kingdom of Tonga   23
lithospheric plates   48
Little Civet (*Viverricula indica*)   26
Little Desert National Park   9
Lombok   26, 213, 215–18, 221–2, 224
Lord Howe Island   22
Lore Lindu National Park   137
Lorentz, New Guinea   110, 113
Louisiade Archipelago   27, 31, 215–16, 220, 222
Lucipara Islands   22, 30
lyrebirds   11

Macrocephalon   74–6, 96, 130, 138, 144, 151; *see also* Maleo
  maleo, see Maleo
Macrogalidia   25
Macronyches   12
Mafula, New Guinea   102
Magnificent Ground Pigeon (*Otidiphaps nobilis*)   105
maintenance phase   46
Malau,   *see* Polynesian Megapode
Maleo (*Macrocephalon maleo*)   9, 18–22, 25, 33–4, 36–7, 39–40, 42, 45, 48–50, 54, 59, 63, 66, 69, 74, 78–9, 81–4, **130–9**, 158, 161, 172, 174, 188, 192, 197, 204, 208, 213, 226; Plate 4
Mallee   22, 80
Malleefowl (*Leipoa ocellata*)   9, 18, 19, 33–6, 39–43, 45, 47, 50, 53–8, 63–5, 71, 79–80, 80–1, 82–4, **122–30**; Plate 3
Mamberamo, New Guinea   118–20, 210–13
Manam   27, 211–12
Maneao Range, New Guinea   121
Manokwari, Vogelkop peninsula   30
Manuk   22, 30
Mare Island   181, 184–5, 187
Mariana Islands   22, 24, 29–30, 49, 153–4, 158
mating systems   65–72; *see also* species accounts
  cuckoldry   72
  female defence monogamy   68–70
  monogamy   65, 68, 70–1
  polyandry   71
  resource defence monogamy   71–2
  resource defence polygyny   68–71
measurements   xvi, 28; *see also* species accounts
Megacephalon, see Macrocephalon
Megapode,   *see* species accounts
Megapodii   12
Megapodiidae   4, 12, 121–2
Megapodiinae   12
Megapodius   15–17, 19, 22, 26, 34, 36–8, 40, 42, 47–9, 66, 69–70, 74–7, 82–3, 96, 98, 101–2, 110, 116, 120–1, 129, 137, 139, 140–6, 149, 155, 158, 163, 169, 171–4, 178, 181, 188, 206–28, 213, 220, 223–7
  affinis, see New Guinea Megapode
  alimentum   23–4
  amboinensis, see Orange–footed Megapode
  andersoni   22–3, 207
  aruensis,   see Orange-footed Megapode
  assimilis,   see Orange-footed Megapode
  bernsteinii,   see Sula Megapode
  burnabyi   22, 46

brenchleyi,   see Melanesian Megapode
brunneiventris,   see New Guinea Megapode
cumingii,   see Philippine Megapode
decollatus,   see New Guinea Megapode
duperreyi,   see Orange-footed Megapode
eremita,   see Melanesian Megapode
forstenii,   see Forsten's Megapode
freycinet,   see Dusky Megapode
geelvinkianus,   see Biak Megapode
gouldi,   see Orange-footed Megapode
hueskeri,   see Melanesian Megapode
huonensis,   see New Guinea Megapode
jobiensis,   see New Guinea Megapode
laperouse,   see Micronesian Megapode
layardi,   see Vanuatu Megapode
melvillensis,   see Orange-footed Megapode
molistructor   23–4, 207
nicobariensis,   see Nicobar Megapode
pritchardii,   see Polynesian Megapode
reinwardt,   see Orange-footed Megapode
reinwardt tumulus,   see Orange-footed Megapode
rubrifrons,   see Melanesian Megapode
rubripes,   see Orange-footed Megapode
speciation   28–32
stairi   22, 146
tenimberensis,   see Tanimbar Megapode
variation among species   29–30
Melanesian Megapode (*Megapodius eremita*)   18, 19, 29, 31, 33, 37, 39, 45, 47–8, 50, 54, 78–9, 144, 146, 164, 171–2, 188–9, 196–7, **198–205**, 207, 209, 212, 221–3; Plate 7
Melanesian Scrubfowl,   *see* Melanesian Megapode
Meliagris lindesayii,   see Australian Brush-turkey
Melville Island   215, 220, 222
Menuridae   11
Merauke, New Guinea   110, 113, 114
micro-organisms   45, 54
microbial respiration   44, 54, 61; *see also* heat sources
Micronesian Megapode (*Megapodius laperouse*)   18, 19, 24, 29–31, 33, 40, 45, 48–9, 82, 140, 145, 149, **152–9**, 189, 192; Plate 5
  Megapodius laperouse laperouse   18, 152–3, 157–8
  Megapodius laperouse senex   18, 20, 152–3, 157–8
Micronesian Scrubfowl,   *see* Micronesian Megapode
middens   24
Milne Bay, New Guinea   119, 222
Mimika–Setakwa area, New Guinea   30, 113, 116, 212
Mindanao   167
Miocene   26
Mios Num, Geelvink Bay   31, 139
Misol Island   96, 100–2, 106, 108–10, 142, 181–9
moas   24
moisture conservation   47–8
Moluccan islands   22, 27, 30, 32, 78, 142, 186, 213
Moluccan Megapode (*Eulipoa wallacei*)   18, 19–20, 33–4, 45, 49–50, 69, 78, 82, **140–5**, 158, 174, 181, 183, 187–8, 226; Plate 5
Moluccan Scrubfowl,   *see* Moluccan Megapode
monogamy   65, 68, 70–1
Morotai Island   181–7
moult   xvi, 37–8; *see also* species accounts
mound builders,   *see* mound building
mound building   4–6, 34–5, 44–8, 73; *see also* mounds
  evolution of   50, 73–7

260 Index

mound building (cont.)
  susceptible to predation 25, 50, 77
mound homeothermy 56–8
mound/s 4, 34–6, 68–70; see also species accounts
  annual 6
  attendance 35, 42–3
  composition 45–8, 47, 58
  construction by males 47
  construction phases 35–6, 46, 69, 70–1
  decaying roots as incubation site 45
  defence 40, 65, 70, 71
  drainage 46
  energetics 58
  heat loss 48, 56–8
  heat sources 5, 35, 69
  homeothermy 56–8
  maintenance 35–6, 40, 47, 57, 60, 63, 70
  moisture conservation 47–8, 57, 61
  oxygen consumption 57
  perennial 6, 70
  shape 46–8, 55
  site 6, 46–8, 70
  size 6, 46, 48, 57–8
  substrate 45–8, 63
  temperatures 50–1, 55
  temperature regulation 54–6
Mount Sisa, New Guinea 28, 102, 119–21, 212–13
Mount Tamborine 57
Mount Elephant, New Guinea 114
Mount Enassa, New Guinea 120
Mount Giluwe, New Guinea 98, 102
Mount Simpson, New Guinea 224–6
museum collections 10

Naingani, Fiji 24
Nappan, New Guinea 114
National Museum of Natural History, Leiden 10
Natural History Museum, Tring 10, 23
Navigator's Islands 22
Negros, Philippines 25
Neiufu group, Samoa 24
Neofelis 25
nesting grounds, see burrow nesting
nesting sites, see incubation sites
New Britain 6, 26, 37, 39, 48, 78–9, 199, 201–4
New Caledonia 22, 23, 207
New Georgia 202
New Guinea 9, 17, 22, 27, 30, 36, 48–9, 78, 80, 96–7, 99–101, 105, 109, 113, 209, 211–13, 216–23, 225–6
New Guinea Megapode (*Megapodius decollatus*) 18, 19, 31, 45, 50, 79, 120, 171, 186, 191–2, 201–2, **208–13**, 222; Plate 8
New Holland Vulture 11; see also Australian Brush-turkey
New Hope Island 22
New Ireland 47, 201, 204
New Zealand 22
Ngerukewid Island Wildlife Reserve 157–8
Nicobar Islands 4, 22, 24, 25, 26, 78, 83, 145, 159, 164
Nicobar Megapode (*Megapodius nicobariensis*) 18, 19, 24, 29, 39, 45, 47, 79, 83, 136, 145, 149, **159–65**, 171, 176, 179–80, 202, 224–5; Plate 6

Megapodius nicobariensis abbotti 18–19, 160–2
Megapodius nicobariensis nicobariensis 18, 159–62
Nicobar Scrubfowl, see Nicobar Megapode
Niuafo'ou Scrubfowl, see Polynesian Megapode
Niuafo'ou, Kingdom of Tonga 22, 24, 26, 29, 49, 78, 82, 148, 150
nostrils 14
Numfoor Island 30–1
Numididae 12
Nusa Laut 143
nutritional requirements 38–9, 68

Obi Islands 32, 181–2, 184–7
observing megapodes 9–10
Odontophoridae 12
Ofu, Samoa 24
oil gland 13–14
Oligocene deposits 24; see also distribution
Orange-footed Megapode (*Megapodius reinwardt*) 9, 18, 19–20, 26, 29–32, 33–4, 36, 40, 42, 45–7, 51, 66, 79, 114, 121, 139, 145, 149, 170, 176, 180, 186–7, 191, 194–6, 207–9, 211–13, **213–27**; Plate 8
  Megapodius reinwardt castanonotus 18, 19, 214–15, 220–2, 224–5
  Megapodius reinwardt macgillivrayi 18, 19, 211, 214–15, 219–22, 224, 227
  Megapodius reinwardt reinwardt 18, 213–14, 217, 220–3, 226
  Megapodius reinwardt tumulus 18, 20, 214–15, 220–2, 224–6
  Megapodius reinwardt yorki 18, 20, 214–15, 221–2, 224–6
Orangerie Bay 219, 222
organic matter 34–6, 46
origins 24–8
  northern centre theory 26
  southern centre theory 26–7
ovomucoids 12; see also eggs: egg whites

Painted Scrubfowl, see Moluccan Megapode
pair-bonds 8–9, 65–6, 69, 71
palaeognaths 74
Palaeopelargus nobilis 23
Palau Islands 22, 24, 29, 154–5, 157
Palawan, Philippines 25, 167–74
Panay, Philippines 25
Papua New Guinea, see New Guinea
Papua, see New Guinea
parapatric forms 29
parasitism of mounds 45, 49–50
parental care 68
Passim, New Guinea 107
Pelelui 157–8
Peleng 31, 175, 177
Penicillum 54
penis 14
Penju Islands 30
Perth Zoological Garden 10
phasianids 25; see also pheasants
Phasianoidea 12
pheasants 12, 25

Philippine Megapode *(Megapodius cumingii)* 3, 18, 19, 25, 29, 45, 47, 50, 134, 145, 163, **165–74**, 177, 179–80, 187, 194, 225; Plate 6
   Megapodius cumingii cumingii 18, 19, 165, 167–8, 170–4
   Megapodius cumingii dillwyni 18, 19, 165, 167–71, 174
   Megapodius cumingii gilbertii 18, 19, 31, 134, 139, 165–8, 170–7, 194, 216
   Megapodius cumingii pusillus 18, 19, 165, 167–71, 174
   Megapodius cumingii sanghirensis 19, 20, 165, 167–8, 171
   Megapodius cumingii tabon 18, 20, 165, 167–71, 174
   Megapodius cumingii talautensis 18, 20, 165, 167–71, 174
Philippine Scrubfowl, *see* Philippine Megapode
Philippines 22, 25, 48, 78, 82–3, 169–74
phylogeny 14–15; *see also* relationships
Pigafetta, Antonio 3
pigeons 11
pits 49; *see also* burrow nesting
plant foods 127–8
plumage 13, 37–8; *see also* species accounts
   feather structure 13
   of hatchlings 37–8
   taxonomic importance 13
Pokili, New Britain 9, 48, 202, 204
polyandry 71
Polynesian Islands 24
Polynesian Megapode *(Megapodius pritchardii)* 18, 22, 24, 33, 39–40, 45, 49–50, 59, 66, 74, 78, 82, 116, 119, 136, 140, 143, 144–5, **146–51**, 158, 164, 181, 188, 192, 197, 204, 208–13, 226; Plate 5
   egg mass 59
   incubation temperatures 51
Polynesian Scrubfowl, *see* Polynesian Megapode
Port Moresby, New Guinea 28, 116, 119, 226
postures, *see* behaviour
precociality of hatchlings 59, 67, 75
predation 25, 80
predation theory, *see* biogeography
preening 40
primitive characters 12
Prince of Wales Island 215, 219, 221–2
Progura gallinacea 23
Progura naracoortensis 23
pterolichoid mites 14–15; *see also* ectoparasites
Pulau Pombo 142, 144–5
Pulau Tiga 173–4
Pulletop Nature Reserve, NSW, Australia 125

Quercymegapodiidae 24, 25
Quercymegapodius brodkorbi 24
Quercymegapodius depereti 24

rainfall 35–7
Rallidae 24
Rasores 12
ratites 23
Rau Island 181, 184–7
Red-billed Talegalla *(Talegalla cuvieri)* 18, 19–20, 27–8, 45, 79–80, 105, **106–10**, 109–11, 114, 116–17, 119, 187; Plate 2

Talegalla cuvieri cuvieri 18, 19, 106–9, 113
Talegalla cuvieri granti 15, 19, 106–9, 114
reed warbler *(Acrocephalus)* 147
relationships 11–20
   current 14–18
   earlier taxonomies 13–15
   interfamilial 11–13
   intrafamilial 13–15
   phylogenies 14–15
reproductive behaviour 65–72; *see also* species accounts
reptiles 52–3, 73–4
   comparison of eggs with megapodes 53
   contrast to megapodes 52–3, 73–4
   incubation sites 52–3
   incubation temperatures 52
resource defence monogamy 71–2
resource defence polygyny 68–71
respiratory overlap 62–3
Round Hill Nature Reserve, NSW, Australia 124
Rubi, New Guinea 114, 208

Sabah 167–8, 173–4
Sabang, New Guinea 110
Saebus, Kangean Archipelago 26
Saiko, New Guinea 102
Saipan 155, 157
Salahutu Mountains 143
Salawati, New Guinea 30, 106, 108, 109–10, 181, 183, 185–7
Samoa 22
San Cristobal 200–1, 203
sandgrouse 12
Santa Cruz group, Solomon Islands 24, 207
Saruwaged Mountains, New Guinea 100
Saubi, Kangean Archipelago 26
Savaii Island, Samoa 148
Savo 203–4
Scrubfowl, *see Megapodius*; species accounts
seed-snipes 12
Sepanjang, Kangean Archipelago 26
Sepik Basin, New Guinea 119–20, 210–13
Sepik Mountains, New Guinea 100
serially descendant moult 37–8
Seventy Islands 155
sexual dimorphism 70
sheathbills 12
silaceous sand 49; *see also* burrow nesting
Siriwo basin, New Guinea 110, 114
sister group 13; *see also* relationships
skeleton 12
Snow Mountains, New Guinea 28, 99–100, 102, 106, 108–9, 112–13, 118–19, 211, 220, 222
social behaviour 8–9; *see also* species accounts
   of hatchlings 8, 34
   pair bonding 8–9,
social organization 69–70; *see also* species accounts
   home ranges 69, 71
   pair bonds 66
solar-heated incubation sites 7, 39, 68; *see also* burrow nesting
solar radiation 4; *see also* heat sources
Solomon Islands 27, 31, 200–4, 207
Sorong, New Guinea 30, 183, 185–7, 212

South-East Islands   218, 221
Southern Highlands Province, New Guinea   28, 50
Southern Islands, Vanuatu   207
speciation   21–32; *see also species accounts*
Spotless Crake (*Porzana tabuensis*)   149
Staffelmauser   37–8
starlings (*Aplonis*)   147
storks (Ciconiidae)   37–8
subspecies   15–18; *see also species accounts*
  features used   15–16
substrate,   *see* incubation sites
suburban house yards   47
Sula   176–7
Sula Megapode *(Megapodius bernsteinii)*   18, 19, 31, 82, 145, 171–2, **175–8**, 188, 196, 202, 216; Plate 6
Sula Scrubfowl,   *see* Sula Megapode
Sulawesi   22, 31, 37, 39, 49, 78–9, 81, 130, 132–5, 167–8, 173–4
Sulu   171
Sumatra   22, 25
Sumba Islands   216, 218–23, 226
Sunday Island   22
Swan River, Western Australia   124
Sylviornis neocaledoniae   23
synonyms,   *see species accounts*

Tabon Scrubfowl,   *see* Philippine Megapode
Talegalla   18, 79–80; *see also species accounts*
  speciation   28
Talegalla   18–20, 22, 26, 42, 69–70, 74–6, 79, 96, 98, 101–2, 105–6, 110, 114, 119–20, 213
  cuvieri,   *see* Red-billed Talegalla
  fuscirostris,   *see* Black-billed Talegalla
  jobiensis,   *see* Brown-collared Talegalla
Tambun   134, 137
Tamrau Mountains, New Guinea   109–10
Tanami Desert, Australia   124
Tangkoko-Batuangus Nature Reserve   173
Tanimbar Islands   31, 180, 221
Tanimbar Megapode *(Megapodius tenimberensis)*   18, 31, 171, 176, **179–81**, 221; Plate 8
taxonomy   11–20; *see also* relationships
temperatures,   *see* incubation sites
Ternate Island   142–3, 181–2, 185–7
Tertiary deposits   26
Tetrao australis   22
thermal inertia of mounds   56
thermoregulation
  of hatchlings   63
  of mounds   54–8
Thinocoridae   12
threatened species   80–3
Tidore   181, 184–5, 187
Tikopia, Solomon Islands   24
tinamous, (Tinamidae)   11, 12
Tonga   4, 22, 23, 82
Tongan Megapode,   *see* Polynesian Megapode
Tor River, New Guinea   121

Torricelli Mountains, New Guinea   100
Trobriand Islands   31, 215, 222, 226
Truk Monarch (*Metabolus rugensis*)   147
turtles   52; *see also* reptiles

uropygial gland (secretion),   *see* oil gland
Utakwa River, New Guinea   109, 116, 212

Vanuatu   27, 205, 207–8
Vanuatu Megapode *(Megapodius layardi)*   18, 19, 27, 29, 45, 148, 149, 180, **205–8**; Plate 7
visual signals   41; *see also* communication
  neck sac   41
  skin colour   41
Viverridae,   *see* civet-cats
Viverricula   25
vocalizations   41–3, 69; *see also species accounts*
Vogelkop peninsula   30, 181, 185–7, 208, 211, 213, 216, 218, 220, 222–4
Vogelpark Walsrode, Germany   10
voice,   *see* vocalizations; *species accounts*
volcanic areas   48–9
Vulcan Island   27
vultures   11

waders   11
Wafawel   142
Waigeu Island   80, 96, 104, 181, 183–8, 189
Wallace's Line   22, 25
Wallace's Scrubfowl,   *see* Moluccan Megapode
Wandammen Mountains, New Guinea   100, 110
Wapona, New Guinea   121
Wattled Brush-turkey (*Aepypodius arfakianus*)   17, 19, 33, 36, 41, 43, 45, 50, 70, 79–80, 96, **97–103**, 105, 109, 213; Plate 1
  Aepypodius arfakianus arfakianus   17, 19, 97, 99–102
  Aepypodius arfakianus misoliensis   17, 19, 97, 99–101
wattles   14
Watubela Islands   31
weights,   *see species accounts*
Weyland, New Guinea   106, 109
Whitsunday Island   222
wing muscles   12
wings   14
Woodlark Island   215, 222
Wyperfield National Park, Vic., Australia   9, 125, 129

Yalgogrin, Australia   125
yearly egg mass   59; *see also* eggs
Yeppoon   215, 222
yolk content   14, 59, 76; *see also* eggs
yolk,   *see* eggs

zones of subduction   48